MACMILLAN WORK OUT SERIES

Dynamics

The titles in this series

- Dynamics
- Electric Circuits
- Electromagnetic Fields
- Electronics
- Elements of Banking
- Engineering Materials
- Engineering Thermodynamics
- Fluid Mechanics
- Mathematics for Economists
- Molecular Genetics
- Operational Research
- Organic Chemistry
- Physical Chemistry
- Structural Mechanics
- Waves and Optics

MACMILLAN WORK OUT SERIES

Dynamics

G. E. Drabble

First published 1987 by
MACMILLAN EDUCATION LTD
Houndmills, Basingstoke, Hampshire RG21 2XS
and London
Companies and representatives
throughout the world

ISBN 0–333–42131–0

A catalogue record for this book is available
from the British Library

Printed in Hong Kong

Reprinted 1990, 1992

Contents

Acknowledgements

The author and publishers are grateful to the authorities of the institutions listed below for permission to use questions from their examination papers. Questions used are identified in the text appropriately, as indicated in brackets.

University of London: University College (U. Lond. U.C.)
University of London: King's College (U. Lond. K.C.)
University of Manchester (U. Manchester)
University of Surrey (U. Surrey)
Polytechnic of Central London (PCL)
Thames Polytechnic (Thames Poly.)
Kingston Polytechnic (KP)

The authorities accept no responsibility for solutions to the questions used, or for the correctness of the answers.

The author and publishers also wish to thank Michael Barber for his careful checking of the manuscript.

Examinations

Some General Remarks

Although you have embarked on a course with the object of learning the theory and practice of engineering, you cannot afford to underestimate the importance of the annual examination. Notwithstanding various attempts by authorities to use other methods of assessing performance, such as continuous assessment, project work and the like, you are not going to qualify until you have shown that you can sit in a room for a few hours and show some reasonable competence in solving engineering problems. Mistrust those lecturers who tell you, 'Learn your subject thoroughly and the examination will take care of itself.' Such advice is likely to come either from brilliant young men and women who never had any trouble passing their own examinations or from old greybeards who can't even remember what taking an examination was like. Students given such advice might well be tempted to retort, 'Get me through this examination and I'll take care of the learning afterwards.'

Few people really like the examination system. Academics devise more or less abstract questions, often so contrived as to be removed almost completely from the real world of engineering, and throw them without warning at a class of students, incarcerated for three hours in a room for the purpose of trying to answer them. The experience bears practically no relation to the student's subsequent career. Studies have shown that there is little correlation between academic prowess and professional excellence. In spite of all this, there is a general consensus that there are certain things that would-be engineers should know, and that before they are allowed to work professionally they should demonstrate that they know these things by answering questions set by experts.

It is perhaps unfair to state that the best guarantee of passing an examination is to have done well in the examination previously. Nothing dogs a student's progress through a course more than weakness in certain areas at the outset. It is valuable to know that when a board of examiners meets for the purpose of deciding which students shall pass, and which fail, their prime consideration is not the marks scored in individual papers but the potential of the student (on the basis of those marks) for undertaking the work of the next year of the course. (For this reason, a final-year board is likely to be more lenient in its judgements than a first- or second-year board.) After passing a first-year examination a student will sometimes ask his tutor, 'What work do you advise me to do to prepare for next year?' My own answer to this question was always, 'Go through all the papers you have just taken, and make sure you can answer *every* question, not just five out of eight. Work on your weaknesses now, while you have a short respite before the next year begins.' Each year of the course builds on the work of the previous year.

However, we have to face the facts, and begin with how we find things at the moment, not how we should have them arranged if we could relive the past few years. Whatever your previous history and performances, you will have to face an examination at the end of the year. Set out below are some suggestions. Some are no doubt obvious; some perhaps not so obvious; some even might be of no use at all to you personally. The suggestions fall into two clear categories: what to do before the examination and what to do during the examination.

Before the Examination

1. If you are taking an internal examination, find out who the examiners are for your subjects. The subject teachers themselves are usually the examiners. For an external examination, you will not know who the examiners are and will not need to know. (An external examination is one which is set by a Board of Examiners who have nothing to do with the running of the course. The Ordinary and Advanced Level examinations of the General Certificate of Education are the most familiar examples.)
2. Obtain copies of previous examination papers in your subjects for the last two or three years. This will help to give you a 'feel' for the type of question you may expect to have to answer. But find out if the structure of the course has changed recently; if it has, then earlier papers might give you a misleading impression.
3. Check with your examiners that the conditions will be the same as for previous years, and, if they have changed, find out in what particulars.
4. Study the rubric of earlier papers carefully. The rubric is the name given to the general instructions at the head of the paper. This may vary from:

   > 'Answer five questions'

   to:

   > 'Answer ALL of Section A; not more than TWO questions from Section B, and not more than THREE questions from Section C. Marks appropriate to each question are shown alongside the question. Candidates are required to satisfy the examiners in ALL sections of the paper. Candidates are expected to spend not more than 40 minutes in answering Section A.'

   You can see that if you go into an examination and are unexpectedly faced with the almost legalistic verbosity of the second example, you may feel with some justification that you are entitled to some marks just for working out exactly what you are required to do!
5. Check that you will be allowed to use your own calculator, and whether there are any restrictions on, say, programmable calculators.
6. Find out what would be the consequences of your failing to attend an examination because of illness.
7. If you suffer from a chronic complaint, such as hay fever, or migraine, let your examiners or your tutors know, well before the time of the examination. Supplement your information with a note from your doctor. Excuses presented after an examination, even when genuine, are not likely to be accepted with sympathy; the examiners would have a tendency to suspect that the student kept the knowledge 'up his sleeve' in case he didn't do too well.
8. Shortly before the examination you might find it useful to decide on a strategy. You may tell yourself that it is quite unrealistic to revise every single topic effectively and that you will have to decide which topics to go for, and which to 'ditch'. A survey of earlier papers might help here. You may find that a question on, say, velocity diagrams has turned up every year for three years, and you may therefore, with some confidence, classify this topic as a 'banker'.

Make a list of all the topics on which questions could be set. If the paper normally contains eight questions, and you have covered only eight topics on the course, you are in the happy position of knowing with a fair degree of confidence that a question will turn up on any topic you choose. If on the other hand, twelve topics have been covered on the course, a simple calculation will tell you that in order to be sure of answering three questions, you must revise at least seven topics.

But a warning is necessary here. Some topics – a good example is work, energy and power – are sufficiently well-defined to justify the setting of a question. But such a topic is so fundamental that questions set in other topics require a knowledge of the principles (questions on momentum, for example). You should not forget that 'topics' are convenient subdivisions for lecturers, students and examiners (and writers of textbooks) for breaking down a subject, but that engineering problems very often cut across these rather artificial dividing lines. Many of the questions in this book will show this. So if you have to 'write off' some of your syllabus, then exercise considerable care in doing so.

9. It is not unknown nowadays for formula sheets to be issued in examinations, but it is certainly not universal practice. If you know that you will not receive such a sheet, part of your last-minute revision should be listing all the formulae you will need to know, and memorising them. This may not be easy for you. It will be easier, of course, if you have done a fair amount of work during the year on the subject. By the time you have reached this stage, certain basic formulae such as centripetal acceleration ($a = \omega^2 r = v^2/r$) and kinetic energy ($\frac{1}{2}mv^2$; $\frac{1}{2}I\omega^2$) should form part of your 'alphabet' of engineering. A working knowledge of dimensions (a topic that students tend to shun) also helps to avoid such mistakes as confusing the surface area of a sphere ($4\pi R^2$) with the volume of a sphere ($\frac{4}{3}\pi R^3$).

   But sheer memorising is difficult for some people. There are techniques which are really helpful in improving memory. But engineering students usually have so much work to do that they cannot afford the time to take up anything not directly related to the course. You should recognise that learning is not a passive business; you do not learn very effectively when sprawling in a comfortable chair, just reading your notes. To memorise a set of formulae – the contents of one of the Fact Sheets of this book, for example – first copy down on a sheet of paper all the items you have to recall, in words, leaving out the actual formula: for example:

   Moment of inertia of a uniform disc =
   Moment of inertia of a uniform rod about one end =
   Moment of inertia of a uniform rod about centre =

   When you have copied everything down, close the book, and try and fill in all the appropriate formulae, referring back to the book when you can't remember. Keep on doing this until you can write them all down correctly. (See item 3 in the section 'During the Examination'.)

10. Very shortly before the examination (say the night before), assemble all the necessary gear (pens, instruments, calculator, etc.) taking care to ensure that nothing contains any 'incriminating matter', such as a set of formulae stuck on the back of a ruler, and that your calculator is fully charged. If possible, include a watch or desk clock in your equipment.

During the Examination

Rules do vary from one institution to another. It is possible, though unlikely, that some of the hints that follow might be unacceptable in some places. If there is any possibility of doubt, check with your authority beforehand.

1. Take into the examination room only those items you are permitted to take. If you have used a formula sheet for last-minute revision, be sure to leave it outside. If by mistake you do take it in, declare the fact immediately to the invigilator.
2. Everything you write in the examination room should be written in the answer book, or on the paper provided. Anything you clearly cross out will not be read or marked by the examiner. You should therefore *not* require a sheet for rough work. Do all rough work and notes in the answer book, and cross it out when done with. DO NOT tear it out.
3. You may have particular trouble remembering a long and complex formula which you are sure you will need; you may have it written down on a sheet of paper outside the examination room, feverishly trying to memorise it. As soon as you are told to begin the examination, but not before, you may write the formula down into your answer book (if you can still remember it) even before you read the question paper. You then have it in front of you, whether you need it or not.
4. Now read the question paper carefully, first making sure that the rubric is of the expected form, and noting if it is not. Read every question, fairly quickly, to identify the topic covered, and also to make sure you have the full and correct complement of questions. (It is not unknown for some candidates to receive a question paper with one page blank.) If you find a question you were hoping to find, make a mark by it, as one that you are eventually going to attempt, but do not start to answer it until you have read through the paper. Some examiners allow ten minutes or so of 'reading time' added on to the statutory time for answering. If this is allowed, remember that reading-time is just that, and you may not write or make notes during that time.
5. Having read all the questions, you should have found some that you can attempt. Divide your allotted time by the number of questions you are asked to answer, thus calculating the approximate time allocation per question. This is usually about 30 minutes. Picking your question, read it through again, this time very carefully, making sure exactly what you are being asked. Candidates frequently come to grief on an answer by failing to answer the question asked, substituting instead an answer to a question they thought they should have been asked.
6. Try to keep your work neat. This is not so much for the benefit of the examiner (although he or she will be grateful) as for yourself. Mistakes in algebra and arithmetic tend to occur much more often with untidy work than with neat work. Needless to say, you should not leave it until this point to start being neat; it is a habit you should try to acquire right at the beginning. Remember that every slip will cost you marks.
7. Don't, however, be *too* neat. Orderly working, well spaced out, is a great help towards correctness. But under-ruling all subsidiary answers in green ink, and double-under-ruling the final answer in red ink will add nothing to the accuracy of the solution, and will earn you no more marks. If you have developed a habit of doing this, or something similar, try to break the habit. For the same reason if you write something you know is wrong, cross it out. Cross it out carefully and neatly by all means; a fierce scribble with the point of a ball-pen, penetrating two other pages, serves no purpose except to tell the examiner that the candidate is under stress. But *do not* carefully open a bottle of white

opaque ink and meticulously paint out the offending item. This again takes up valuable time, and earns no marks, quite apart from being somewhat expensive.

8. As a general, though not universal rule, solve your questions as much as possible in algebra, substituting arithmetical values at the end. This has three advantages. First, an algebraical error is more easily spotted than an arithmetical one. Second, you can reduce calculator work to a single operation at the end, instead of six or seven mini-calculations throughout the solution. Third, transferring long numbers from one line of working to the next can introduce error. Suppose you calculate the second moment of area of cross-section of a shaft as $0.000\,000\,076\,49\ m^4$ (a not unlikely value). Even if you call this 7.649×10^{-8}, it is still a long and tedious item to carry through a calculation; much more cumbersome than $(\pi/32)D^4$. The worked examples in this book will repeatedly illustrate this point.
9. Show all your working in your written solution. If you do all your calculations on a sheet of blotting-paper and only set down the final result your examiner might reasonably assume that you copied it from someone else.
10. Do not answer what you have not been asked. Read the question carefully. 'State the formula for . . . ' means just that; you don't have to prove it, and if you do prove it, you won't get marks for the proof. On the other hand, 'Derive a formula for . . .' means that you have to show all the working, i.e. prove it.
11. If the numerical answer to a question is clearly and obviously absurd – for example, if the deflection of a steel bar is several hundred metres, or the calculated speed of a car is several kilometres per second – leave it at this stage. It is probably a small slip, involving some carelessness with high powers of 10, and it may not cost you more than a mark or two. To go back over the whole calculation to try and retrieve the error might cost far more in precious time than the possible small gain of a mark or two justifies. If you have spare time at the end by all means come back and try to locate the mistake. But during the examination try to keep to the schedule of so many minutes per question.
12. If you find yourself in a tangle with a question that you were absolutely confident of being able to answer, and your allotted time is running out, then *leave it*, and try another question. Avoid getting angry and frustrated, if you can; this is likely to lead to further mistakes. Keep cool, and again, if you have time at the end, then come back and have another try, preferably starting from the beginning again.
13. You may reach a desperate stage. You may have limped through two questions, pulling in possibly 30 marks, and you know you need a minimum of 40 to pass, and you can't see another question that you can attempt with any confidence. You have come to the unenviable point where you have to pick up a few marks where you can. Many questions are set in two parts. The first part might consist of deriving a special formula and the second part, a substitution of numerical values in the formula. Well, you may feel incapable of deriving the formula, but you can substitute the numbers in it. If you do so, you won't get many marks for it; the examiner probably allots 16 marks to the first part, and only 4 to the second. But in your position, you can use those four marks, so you grab them where you can. They might mean the difference between a pass and a fail.
14. An excellent and almost universal rule is to adhere to the instructions in the rubric. If you are required to answer five questions, do not attempt six or seven. There was a time when an examiner would mark the statutory number of answers, in the order in which they were written in the answer-book, and then *not mark any further work*. Thus, an examinee might gain no marks for

a perfect answer to a question. It is most unlikely that any examiner nowadays would be so strict; a common procedure is to mark all the answers presented, and choose the ones bearing the best marks. But even so, answering more than you are told means that some of your work will not gain you marks. If you have answered the required number and have time to spare, use it in checking the answers you have done. There have been cases of examinees who have answered their questions perfectly, and have so much time to spare that they have answered another question, just to ensure that they score 100 per cent. But such people are rare, and do not require the sort of advice given here.

Examiners, Papers and Questions

Remember that most examiners are are reasonable people, and sometimes quite nice people. They are not always trying to catch you out. Boards of examiners do not like failing large numbers of students. Neither do they agree beforehand to fail a certain percentage of the candidates. This is a very common erroneous belief among students: that, say the fifteen poorest in a class of sixty are bound to fail, because this is the agreed policy. I can only state, firstly, that no board of which I have been a member has ever adopted this policy, and secondly, that so far as I know, no other board has. If fifteen out of sixty candidates fail, it is because they do not achieve the required pass standard, and not because they are the poorest fifteen of the bunch.

Of course, there have been, and there are, spiteful examiners. There is a story, which I am sure is untrue, of an examiner who, while setting a paper at his desk, leaned back with a satisfied smile and said to his colleague, 'Ah! They'll never do that one!' But even spiteful examiners have their papers scrutinised by moderators, to try to ensure that a paper is fair. Also, a degree examination has two functions. The first and obvious one is to find which students have learned enough to merit their progress to the next stage of the course. The second and less obvious one is that of grading. A degree examination is not a simple ability test such as you may find in training programmes for the armed forces, where you are taught a simple skill and have simply to show that you have learned it. The required pass mark in such a test may be from 70 per cent to as high as 90 per cent. I have known students with this 'ability test' mentality protest quite energetically because an examination question was not an almost exact copy of tutorial questions. But a high academic qualification is not earned by just learning to solve a range of routine problems. A good examination question is one which compels the examinee to think, and to work out how what has been learned on the course is to be applied to obtain a solution. In this way, the good candidate is distinguished from the mediocre or the poor one.

Finally, it must be confessed that luck may play an important part. The hard-working conscientious student who has worked steadily throughout the year may have an off-day, combined with a tough paper and score a mere 55 per cent, while a less industrious friend who began to work seriously only two weeks previously happened to 'mug up' two or three topics which turned up in the paper, and collects 60 per cent. There is no answer to this example of life's unfairness, except to wish you good luck!

How to Use the Book

First of all, this is a revision book, not a reference textbook. It is intended for use when you have completed, or nearly completed a first-year course in engineering mechanics and are about to prepare yourself for an examination. You should by now have a set of notes, either taken by yourself during the lectures, or in the form of hand-outs from lecturers, or received via a correspondence course. You may also have one or more textbooks. If you are taking a degree course, you have for some time been forcibly fed with concentrated doses of material in this and several other subjects, and in addition have probably been given sheets of examples which you have been advised to work through but just haven't found the time to do so. But the examination is now uncomfortably close; it is time to begin revising.

Here is a suggested procedure.

1. Choose a topic for revision.
2. Read up your notes on this topic and be sure you understand at least a good proportion of them reasonably well.
3. Look up the chapter in this book covering the topic. Each chapter begins with a Fact Sheet, which is a sheet of formulae and definitions covering the topic. Be sure that you can understand the contents of the Fact Sheet. You may encounter difficulties here. For example:
   (a) The Fact Sheet may include formulae or definitions which you do not have in your notes. For instance, the Fact Sheet may define moment of inertia, whereas your notes might assume, perhaps incorrectly, that you understood this before beginning the course.
   (b) Depending on your course, the course coverage of a topic may be wider and deeper than in this book.
   (c) Symbols and notations may be unfamiliar. If so you must try to be flexible, and adjust. If you can solve the equation

   $$x^2 - 9x - 30 = 0$$

   but cannot solve

   $$h^2 - 9h - 30 = 0$$

   without first writing 'Let $h = x$' you are up against a difficulty which only you can resolve. If you are one of those people who *do* find this a major hang-up, then, as a part of your revision, write out the Fact Sheet yourself completely, transposing everything to your preferred system of notation.
4. Now look at one of the worked examples. Do not just read it through. Begin each example by trying to solve it yourself. So have paper, pen and calculator to hand. Attempt a solution, and only when you come up against a difficulty refer to the book, to find out what you have forgotten, or where you have gone wrong. The book solutions include helpful comments and explanations, in contrasting type: these, of course, would not be required in an examination

solution. When a calculation uses material from the Fact Sheet this is usually indicated by quoting the appropriate key letter in brackets at the beginning of the line.

5. Having worked successfully through an example, look through your own example sheets, your textbooks, your old examination papers and the unworked problems at the end of the chapter, and select a problem you think you could solve. At this stage, select only those problems for which you have answers. If you fail to get the right answer then try to find out where the mistake is, but don't spend too much time doing this. It may be only a small slip which in an examination would cost you only a mark or two. But later, get someone to check your work for you, if you can. The mistake may be more serious, and in any case you need to know what it is.

Hopefully, by following this procedure, you should be able to achieve a fair degree of competence in answering questions in your chosen topic, and move to another. But do not make the mistake of supposing that possession of a book of worked examples releases you from the necessity of thoroughly learning your subject. On the contrary, one of the aims of this book is to help you to learn your subject. Used properly, it should reinforce and extend your knowledge of dynamics. But it is not intended, and should not be used, as a substitute for a textbook.

A word about the examples themselves. About a third of these are actual questions taken from the examination papers of various institutions. The rest are designed specifically for this book. Although an attempt has been made to provide questions of approximately examination standard, some of the questions are below examination standard and some above it. (In Chapter 1, the nature of the topic has dictated the use of examples which take too long to answer to form satisfactory examination questions.) For this reason, you should not necessarily be discouraged if you find that you require more time to obtain a solution than you would be allowed in an examination. On the other hand, of course, neither should you become complacent if you solve a question in much less than 'examination' time.

1 Kinematics of Mechanisms

Graphical and analytical treatment of mechanisms. Coriolis component of acceleration. Velocity and acceleration diagrams. Input and output of a machine. Efficiency.

1.1 The Fact Sheet

(a) General Vector Equation

$$\overline{V}_{ac} = \overline{V}_{ab} + \overline{V}_{bc} \qquad (\overline{V} \text{ denotes any vector}).$$

Displacement (or velocity, or acceleration) of C relative to A is the sum of displacement etc. of B relative to A and of C relative to B.

(b) Notation

Fig. 1.1

A vector ab, **read in that order**, represents displacement (or velocity or acceleration) of a point B relative to a point A (Fig. 1.1). The same vector, **read in the order** ba, represents displacement (or velocity or acceleration) of A relative to B.

Since vectors may be 'read' either way, they should not have arrows.

(c) Relative Velocity of Two Points on a Rigid Link

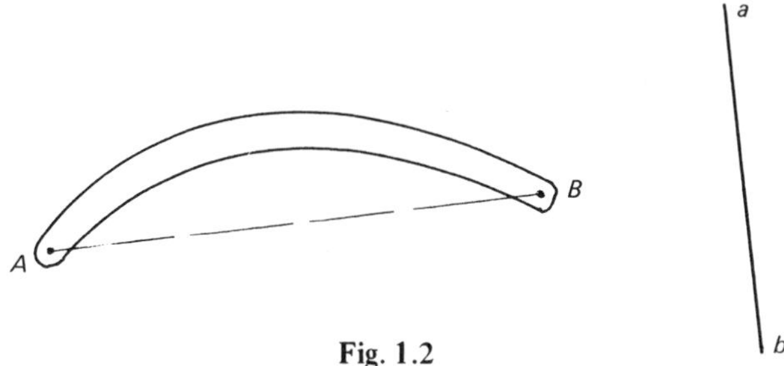

Fig. 1.2

When a rigid link AB moves in a plane, a point B on the link can move relative to a point A only in a circular path. The velocity of B relative to A is thus a vector ab which is perpendicular to the straight line joining A and B (Fig. 1.2).

(d) Velocity Image

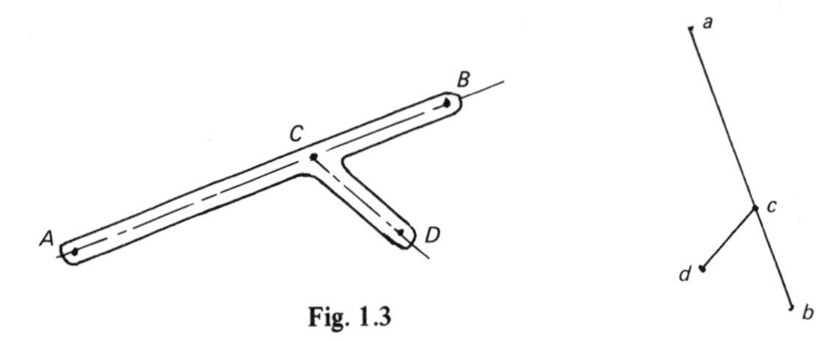

Fig. 1.3

When the relative velocity of two points A and B on a rigid link is known, and drawn as a vector ab, the relative velocity of other points on the link (e.g. C, D) may be determined by disposing points a, b, c, d on the vector geometrically similarly to the disposition of the points A, B, C, D on the link (Fig. 1.3). Vector $abcd$ is called the velocity image of link $ABCD$.

(e) Relative Acceleration of Two Points on a Rigid Link

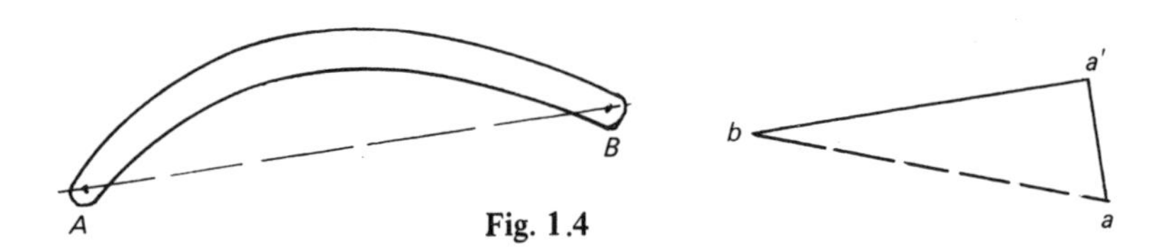

Fig. 1.4

When a rigid link AB moves in a plane, the acceleration of a point B relative to a point A (Fig. 1.4) comprises the following:

(i) a tangential component (aa') having direction perpendicular to the straight line joining AB, the magnitude given by

$$aa' = \alpha\,(AB),$$

where α is the angular acceleration of AB;

(ii) a centripetal component ($a'b$) having direction parallel to the straight line joining AB, the sense being from B to A, and the magnitude given by

$$a'b = \omega^2 R = \omega^2 (AB) = \frac{v^2}{R} = \frac{v^2}{AB}$$

where ω is the angular velocity of AB and v is the velocity of B relative to A.

Note: when the link turns with constant angular velocity, the tangential component will be zero.

(f) Acceleration Image

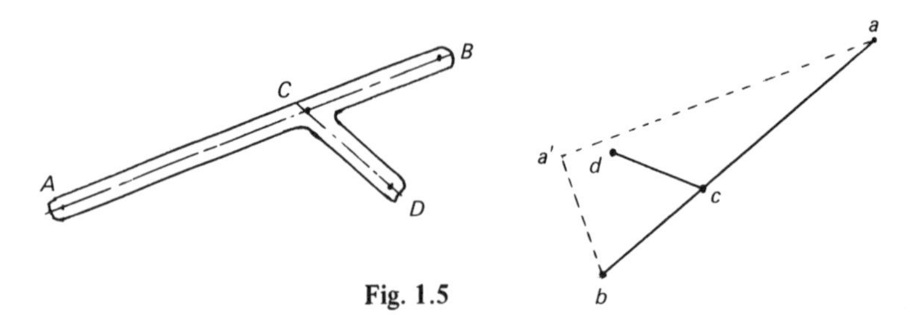

Fig. 1.5

When the relative acceleration of two points A and B on a rigid link is known, and drawn as a vector ab, then the relative acceleration of other points on the link (e.g. C, D) may be determined by disposing points a, b, c, d on the vector geometrically similarly to the disposition of the points A, B, C, D on the link (Fig. 1.5).

Vector $abcd$ is called the *acceleration image* of link $ABCD$.

(g) Coriolis Component

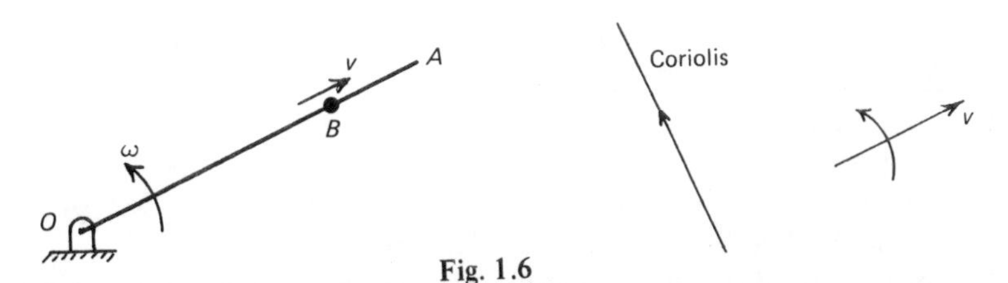

Fig. 1.6

When a point B moves with velocity v along a link CA which is turning with angular velocity ω then there is a component of acceleration of B relative to the link called the Coriolis component (Fig. 1.6).

(i) The magnitude of the component is $2v\omega$.
(ii) The line of action of the component is perpendicular to the direction of the velocity v.
(iii) The direction of the component is 90° in advance of the velocity vector, in the direction of rotation of the link.

(h) Velocity Diagram: Procedure

(i) Draw the link configuration diagram to a suitable scale.
(ii) Sketch the velocity diagram roughly to decide a suitable scale, and to arrange suitable disposition on the paper.
(iii) Preferably begin with a link rotating about a fixed point. Calculate the velocity of the free end and draw the vector in accordance with (c) above. If the velocity cannot be determined, draw the vector of an arbitrary length.
(iv) Proceed from one link to the next, in accordance with (c) above. Direction of velocity of points in slides will be that of the slide. Construct velocity images ((d) above) as required.

(i) Construction of Parallel and Perpendicular Lines

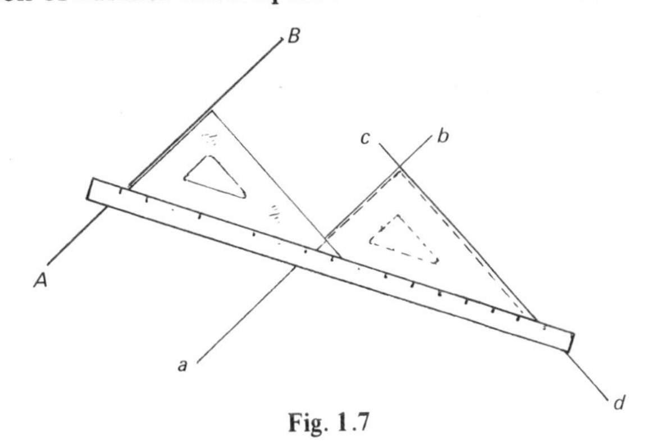

Fig. 1.7

Set one edge of the setsquare to line AB (Fig. 1.7). Set the rule up to the edge of the setsquare. Hold the rule firmly. Slide the setsquare along the rule. Draw lines ab, cd. Then ab will be parallel to AB and cd will be perpendicular to AB.

(j) Acceleration Diagram: Procedure

(i) Draw the velocity diagram as described in (h) above.
(ii) Calculate the relative centripetal accelerations for all links.
(iii) Calculate the Coriolis components where applicable.
(iv) Locate a starting-point (a fixed point on the mechanism). Begin by drawing centripetal vector of known length.
(iva) In general, centripetal components and Coriolis components will be known in magnitude, and tangential and sliding components will be unknown. The problem will include sufficient data to enable completion of the diagram. See the Worked Examples in this chapter.
(v) Proceed round the diagram from one link to an adjacent one, always drawing the vector of known length before that of unknown length. The latter will be a line of indefinite length.
(vi) When the diagram is complete, measure all the vector components required for the solution of the problem, using the scale chosen.
(vii) Calculate all the angular accelerations required:

$$\text{Angular acceleration} = \frac{\text{Tangential acceleration component}}{\text{Length of link}}.$$

(k) Analysis

$$v = \frac{dx}{dt}; \qquad a = \frac{dv}{dt} = \frac{d^2x}{dt^2}.$$

$$\omega = \frac{d\theta}{dt}; \qquad \alpha = \frac{d\omega}{dt} = \frac{d^2\theta}{dt^2}.$$

(l) The Mechanism as a Machine

$$\text{Power} = \text{Force} \times \text{Velocity} \ (Fv).$$

$$\text{Power} = \text{Torque} \times \text{Angular velocity} \ (T\omega).$$

$$\frac{\text{Total power out (i.e. Work done by machine)}}{\text{Total power in (i.e. Work done on machine)}} = \eta \ \text{(Efficiency)}.$$

1.2 Worked Examples

Example 1.1

The diagram shows a point P which has linear motion along a radial line which turns about the point O. The linear velocity and acceleration of P relative to the line are v and a respectively, and the angular velocity and acceleration of the radial line are ω and α respectively. The distance of P from O is 2 m.

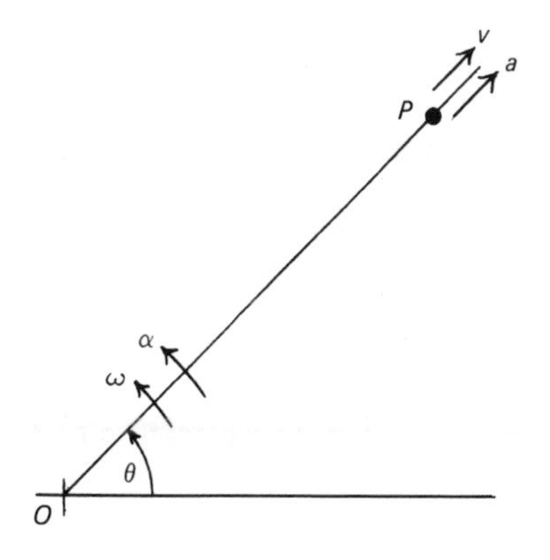

Draw the acceleration diagrams and hence determine the total acceleration of P with respect to O for the following two cases:
(a) $\omega = +4$ rad s^{-1}; $\alpha = +20$ rad s^{-2}; $v = +8$ m s^{-1}; $a = +40$ m s^{-2}; $\theta = 45°$.
(b) $\omega = -4$ rad s^{-1}; $\alpha = +16$ rad s^{-2}; $v = -4$ m s^{-1}; $a = +20$ m s^{-2}; $\theta = 90°$.
Take the directions shown in the diagram as positive.

Solution 1.1

Part (a)

All data are given, and we require only to calculate the various acceleration components before drawing the diagram.

Calling 'P_L' that point on the radial line which corresponds to P, then:
Centripetal acceleration (P_L relative to O) = $\omega^2 R = 4^2 \times 2 = 32$ m s^{-2}.
Tangential acceleration (P_L relative to O) = $\alpha R = 20 \times 2 = 40$ m s^{-2}.
Coriolis acceleration (P relative to P_L) = $2v\omega = 2 \times 8 \times 4 = 64$ m s^{-2}.
Sliding acceleration (P relative to P_L) (given) = 40 m s^{-2}.

Shown here is the acceleration diagram. Little explanation is needed; since all the

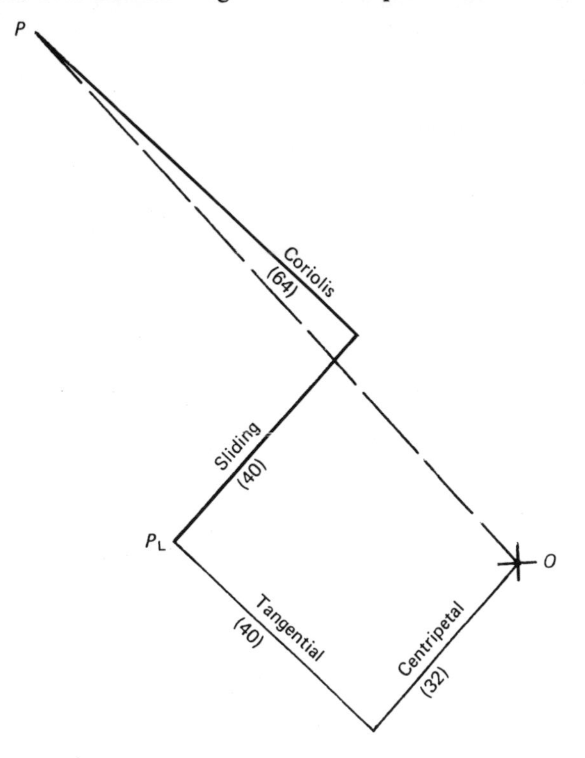

components are known, the vectors can be drawn in either order (for example, starting at O, the tangential component could be drawn before the centripetal). The final result would be the same. Observe that the direction of the Coriolis component follows the rule: the sliding velocity v (pointing to 2 o'clock) turns through 90° in the direction of rotation of the line, which is anticlockwise, to give the direction of the Coriolis component.

Scaling from the diagram, or using simple trigonometry to calculate:

Total acceleration of P relative to O = $\underline{104.3 \text{ m s}^{-2}}$.

Part (b)

Centripetal acceleration (P_L relative to O) = $\omega^2 R = 4^2 \times 2 = 32$ m s^{-2}.
Tangential acceleration (P_L relative to O) = $\alpha R = 16 \times 2 = 32$ m s^{-2}.
Coriolis acceleration (P relative to P_L) = $2v\omega = 2 \times 4 \times 4 = 32$ m s^{-2}.
Sliding acceleration (P relative to P_L) (given) = 20 m s^{-2}.

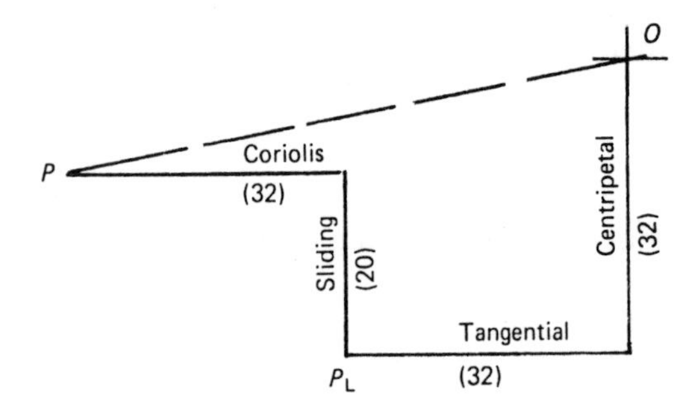

Note in the diagram that although v is now in the opposite direction, so also is ω with the result that the Coriolis component is still in the same direction relative to the radial line.

Total acceleration of P relative to O = $\underline{65.1 \text{ m s}^{-2}}$.

Example 1.2

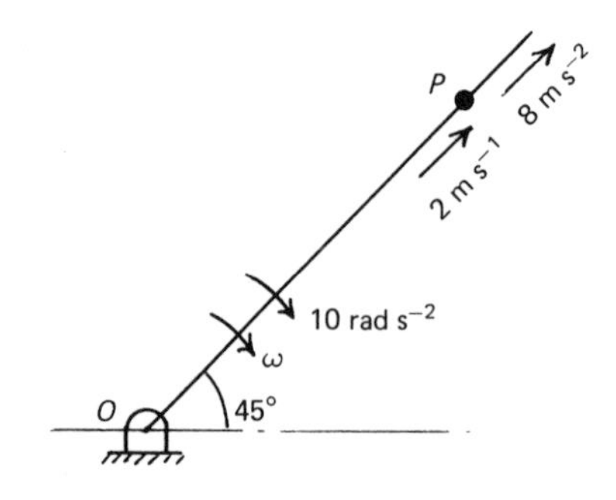

The diagram shows a light rigid link which turns in a vertical plane with an angular velocity ω and a clockwise angular acceleration of 10 rad s^{-2}. A point P on the link has a linear velocity relative to the link of 2 m s^{-1} and a linear acceleration relative to the link of 8 m s^{-2}, both radially outwards, as shown. At the instant shown the link is at 45° to the horizontal and the distance OP is 1.5 m. Determine two values of ω such that the total acceleration of P with respect to O is vertical. Calculate the value of the acceleration for one of these values of ω.

Solution 1.2

We first calculate the centripetal, tangential and Coriolis acceleration components, in terms of ω where necessary. Also define P_L coincident with P.

Centripetal acceleration (P_L relative to O) = $\omega^2 R = 1.5\omega^2$.
Tangential acceleration (P_L relative to O) = $\alpha R = 10 \times 1.5 = 15$ m s^{-2}.
Coriolis acceleration (P relative to P_L) = $2v\omega = 2 \times 2\omega = 4\omega$.

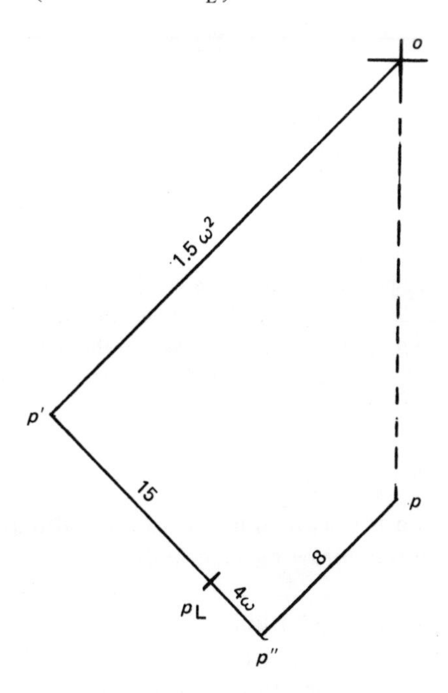

A sketch is given here of the acceleration diagram. The steps are:

(i) Locate the starting point O. Draw the centripetal vector (P_L relative to O) $op' = 1.5\omega^2$ parallel to OP in the direction $P \rightarrow O$.

(ii) Draw the tangential vector (P_L relative to O) $p'p_L = 15$ m s^{-2} perpendicular to OP (in accordance with clockwise angular acceleration).

(iii) Draw the Coriolis vector (P relative to P_L) $p_L p'' = 4\omega$ perpendicular to OP. The direction is the sliding velocity vector (radially outwards) turned through 90° in the direction of the angular velocity which is assumed clockwise).

(iv) Draw the sliding acceleration vector $p''p = 8$ m s^{-2}.

Since p is required to be vertically in line with o, we equate the total horizontal component of all four vectors to zero.

$$1.5\omega^2 \cancel{\cos 45°} - 15 \cancel{\cos 45°} - 4\omega \cancel{\cos 45°} - 8 \cancel{\cos 45°} = 0.$$

$$\therefore\ 1.5\omega^2 - 15 - 4\omega - 8 = 0.$$

$$\therefore\ \omega^2 - 2.667\omega - 15.33 = 0.$$

$$\therefore\ \omega = \tfrac{1}{2}(2.667 \pm \sqrt{(2.667)^2 - 4(-15.33)})$$

$$= 1.333 \pm 4.137$$

$$= \underline{5.470 \text{ rad s}^{-1}} \text{ and } \underline{-2.804 \text{ rad s}^{-1}}.$$

We do not need to draw the diagram accurately to determine the acceleration. It must be the resultant vertical component of the four vectors.

For $\omega = +5.470$ rad s^{-1}:

$$\begin{aligned} op &= \sin 45° \, (1.5\omega^2 + 15 + 4\omega - 8) \\ &= 0.7071\,[1.5\,(5.470)^2 + 15 + 4\,(5.470) - 8] \\ &= \underline{52.16 \text{ m s}^{-2}}. \end{aligned}$$

Example 1.3

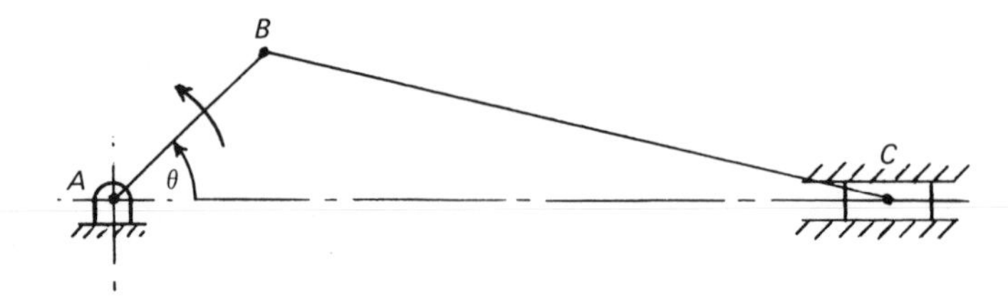

The diagram shows a simple crank and piston called the slider–crank mechanism. The crank AB is 0.1 m long and the connecting-rod BC is 0.3 m long. AB rotates anticlockwise at a constant speed of 5 rad s^{-1}. Sketch the velocity and acceleration diagrams, and hence determine the sliding velocity and acceleration of the piston at C for values of the crank angle θ of (a) 0°; (b) 90°; (c) 180°.

Solution 1.3

Attention is drawn to the wording of the question. If told to sketch, do not spend valuable time drawing accurately.

For all three positions:
Velocity of B relative to $A = \omega R = 5 \times 0.1 = 0.5$ m s^{-1}.
Centripetal acceleration of B relative to $A = \omega^2 R = 5^2 \times 0.1 = 2.5$ m s^{-2}.

Part (a): $\theta = 0°$.

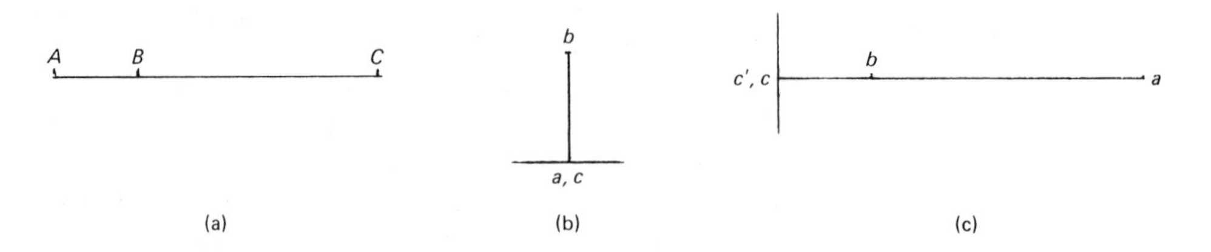

(a) (b) (c)

(a) is the configuration diagram, (b) the velocity diagram and (c) the acceleration diagram.

For drawing the velocity diagram (b) the steps are:

(i) Draw ab = 0.5 m s^{-1} vertically upwards (i.e. perpendicular to AB). This is consistent with anticlockwise rotation.

(ii) Draw a line through b perpendicular to BC. c must lie on this line.

(iii) Draw a horizontal line through a representing the sliding velocity of c.c must lie on this line.

Hence a and c are coincident.

To draw the acceleration diagram (c):

$$\text{Centripetal acceleration of } C \text{ relative to } B = \frac{v^2}{R} = \frac{(bc)^2}{BC} = \frac{(0.5)^2}{0.3} = 0.833 \text{ m s}^{-2}.$$

(i) Draw the centripetal vector ab = 2.5 m s^{-2} in direction $B \to A$.
(ii) Draw the centripetal vector bc' = 0.833 m s^{-2} in direction $C \to B$.
(iii) Draw the tangential vector $c'c$ perpendicular to BC. c must lie on this line.
(iv) Draw the sliding (horizontal) vector through a. c must lie on this line. Hence c and c' are coincident.

Answers: sliding velocity = 0; sliding acceleration = 3.333 m s^{-2}.

Part (b): $\theta = 90°$.

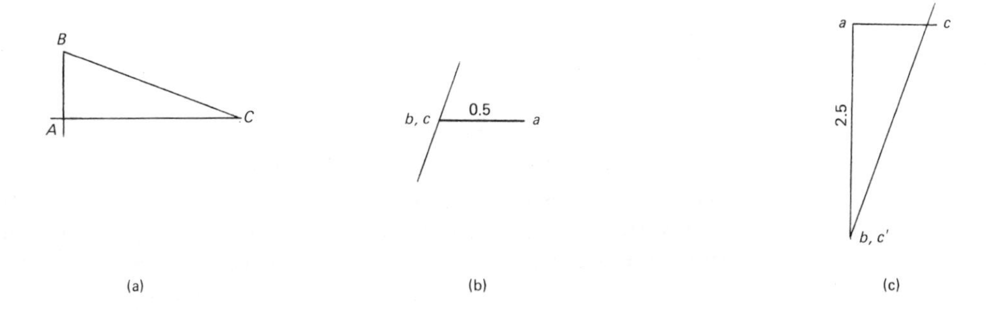

Velocity diagram (b):

(i) Draw ab = 0.5 m s^{-1} perpendicular to AB, i.e. horizontally.
(ii) Draw a line through b perpendicular to BC. c must lie on this line.
(iii) Draw a horizontal through a (sliding velocity). c must lie on this line.

Hence b and c are coincident.

Acceleration diagram (c):

Centripetal acceleration of C relative to B = 0 (bc = 0).

(i) Draw the centripetal vector ab = 2.5 m s^{-2} in the direction $B \to A$.
(ii) Draw the centripetal vector bc' (= 0). Hence b and c' are coincident.
(iii) Draw the tangential vector $c'c$ of unknown length through c'.
(iv) Draw the sliding vector ac horizontally through a to locate c.

Answers: sliding velocity = 0.5 m s^{-1}; sliding acceleration = 0.884 m s^{-2}.

Part (c) $\theta = 180°$.

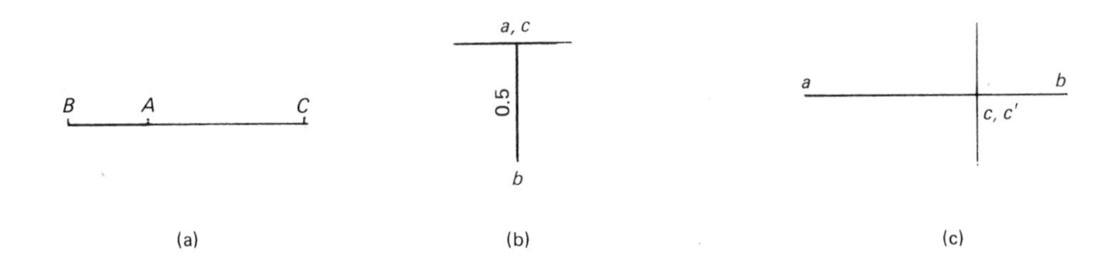

Velocity diagram (b):

(i) Draw ab = 0.5 m s^{-1} perpendicular to AB, downwards (i.e. anticlockwise rotation).
(ii) Draw line bc of unknown length perpendicular to BC.
(iii) Draw ac horizontally through a. Hence locate c, which is coincident with a.

Acceleration diagram (c):

Centripetal acceleration of C relative to $B = \frac{(bc)^2}{BC} = \frac{(0.5)^2}{0.3} = 0.833 \text{ m s}^{-2}$.

(i) **Draw the centripetal vector $ab = 2.5 \text{ m s}^{-2}$ in the direction $B \rightarrow A$.**
(ii) **Draw the centripetal vector $bc' = 0.833 \text{ m s}^{-2}$ in the direction $C \rightarrow B$.**
(iii) **Draw the tangential vector $c'c$ perpendicular to BC. c lies on this line.**
(iv) **Draw the sliding (horizontal) vector through a. c lies on this line.**

Hence c and c' are coincident.

Answers: sliding velocity = 0; sliding acceleration = 1.667 m s^{-2}.

Do not become over-confident by reason of the apparent simplicity of these diagrams. It is a strange but undoubted fact that students who become competent at drawing quite complicated velocity and acceleration diagrams suddenly begin to make mistakes when faced with 'special cases' such as these, in which one or more of the vectors is zero, resulting in a simplified diagram. If you have any doubt about the procedure in such cases, just strictly follow the procedure set out in the Fact Sheet.

As a follow-up to this example, you should try Problem 1.5 at the end of this chapter.

Example 1.4

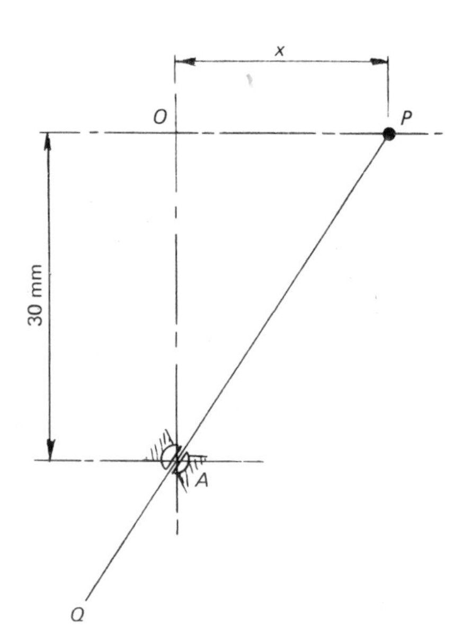

A rod PQ passes through a swivel-block A which is fixed in position 30 mm below a fixed point O (see diagram). P is constrained to move along a horizontal line such that the displacement x from O after t seconds is given by

$$x = 15t - \tfrac{1}{2}t^3 \text{ mm.}$$

Either graphically or analytically, determine the angular velocity and acceleration of PQ when $t = 1$ second.

Solution 1.4

We first determine the displacement, velocity and acceleration of P.

$$x = 15t - \tfrac{1}{2}t^3 = 15(1) - \tfrac{1}{2}(1)^3 = 14.5 \text{ mm}.$$

$$\dot{x} = 15 - 3 \times \tfrac{1}{2}t^2 = 15 - 1.5(1)^2 = 13.5 \text{ mm s}^{-1}.$$

$$\ddot{x} = -1.5 \times 2t = -3 \text{ mm s}^{-2}.$$

Graphical Solution

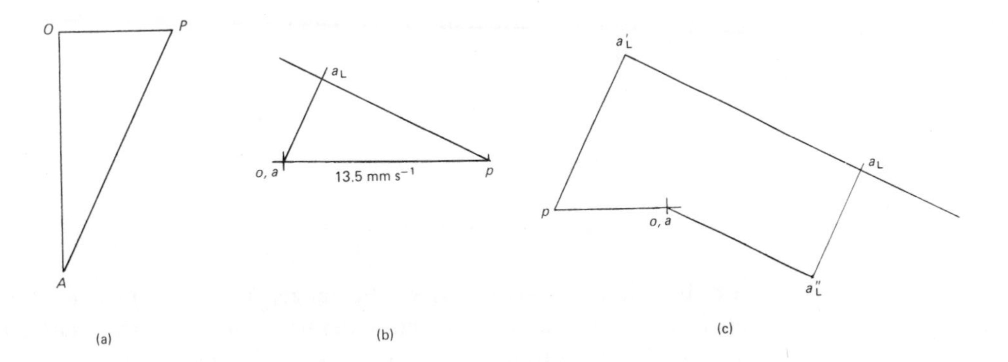

Configuration diagram (a):
The distance PA scales 33.3 mm. We define a point A_L on the link PQ (A-link) which is coincident with the swivel-block.

(This feature of mechanisms involving sliding blocks is explained more fully in Example 1.7.)

Velocity diagram (b):

(i) Locate the starting-point o, a. Draw op = 13.5 mm s^{-1} horizontally to the right.
(ii) Draw the sliding velocity aa_L parallel to AP_L of unknown length.
(iii) Draw the tangential velocity pa_L perpendicular to PA to locate a_L.

aa_L scales 5.87 mm s^{-1}
pa_L scales 12.15 mm s^{-1}.

$$\text{Angular velocity of } PA_L = \frac{v}{R} = \frac{pa_L}{PA_L} = \frac{12.15}{33.3} = \underline{0.3649 \text{ rad s}^{-1}}.$$

Acceleration diagram (c):

$$\text{Centripetal acceleration } A_L \text{ relative to } P = \frac{v^2}{R} = \frac{(pa_L)^2}{PA_L}$$

$$= \frac{(12.15)^2}{33.3} = 4.433 \text{ mm s}^{-2}.$$

$$\text{Coriolis acceleration } A_L \text{ relative to } A = 2v\omega = 2(aa_L)\left(\frac{pa_L}{PA_L}\right)$$

$$= 2 \times 5.87\left(\frac{12.15}{33.3}\right) = 4.28 \text{ mm s}^{-2}.$$

(i) **Draw the linear acceleration vector (P relative to O) op = –3 mm s^{-2} (i.e. right to left).**
(ii) **Draw the centripetal vector (A_L relative to P) pa'_L = 4.433 mm s^{-2} parallel to A_LP in the direction $A_L \to P$.**
(iii) **Draw the tangential vector (A_L relative to P) a'_La_L of unknown length perpendicular to A_LP.**
(iv) **Draw the Coriolis vector (A_L relative to A) aa''_L = 4.28 mm s^{-2} perpendicular to AP. Direction: velocity vector a_L relative to a, turned through 90° in the direction of rotation of AP, i.e. clockwise.**
(v) **Draw the sliding vector (A_L relative to A) a''_La_L to locate a_L.**

This completes the diagram.

The tangential vector a'_La_L scales 6.98 mm s^{-2}.

$$\therefore \text{The angular acceleration of } PQ = \frac{a}{R} = \frac{6.98}{33.3} = \underline{0.2096 \text{ rad s}^{-2}}.$$

The direction is anticlockwise: the tangential component of P relative to A_L is 'up–left'. Thus P accelerates anticlockwise about A. (Similarly, considering the tangential acceleration of A_L relative to P, this is 'down–right', which is also consistent with anticlockwise angular acceleration.)

Analytical Solution

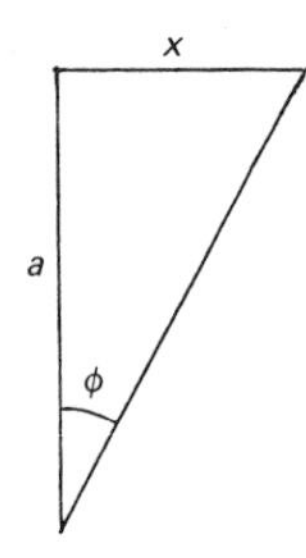

From the diagram,

$$\tan\phi = \frac{x}{a}.$$

Differentiating implicitly with respect to t,

$$\sec^2\phi\dot{\phi} = \frac{\dot{x}}{a}.$$

$$\dot{\phi}\,(1 + \tan^2\phi) = \frac{\dot{x}}{a}.$$

$$\therefore \dot{\phi}\left(1 + \frac{x^2}{a^2}\right) = \frac{\dot{x}}{a}.$$

$$\therefore \dot{\phi} = \dot{x}\left(\frac{a}{a^2 + x^2}\right). \qquad (1)$$

Differentiate again with respect to t, remembering that $\dot{x}$ is not constant.

$$\ddot{\phi} = \dot{x}a(-1)\,(a^2 + x^2)^{-2}(2x)\,(\dot{x}) + \left(\frac{a}{a^2 + x^2}\right)\ddot{x}$$

$$= -(\dot{x})^2\left(\frac{2ax}{(a^2 + x^2)^2}\right) + \ddot{x}\left(\frac{a}{a^2 + x^2}\right). \qquad (2)$$

Substituting in equation 1,

$$\dot{\phi} = 13.5\left(\frac{30}{(30)^2 + (14.5)^2}\right) = \underline{0.3648 \text{ rad s}^{-1}}$$

and in equation 2,

$$\ddot{\phi} = -(13.5)^2\left(\frac{2 \times 30 \times 14.5}{((30)^2 + (14.5)^2)^2}\right) + (-3)\left(\frac{30}{(30)^2 + (14.5)^2}\right)$$

$$= -0.1286 - 0.0811$$

$$= \underline{-0.2097 \text{ rad s}^{-2}}.$$

Example 1.5

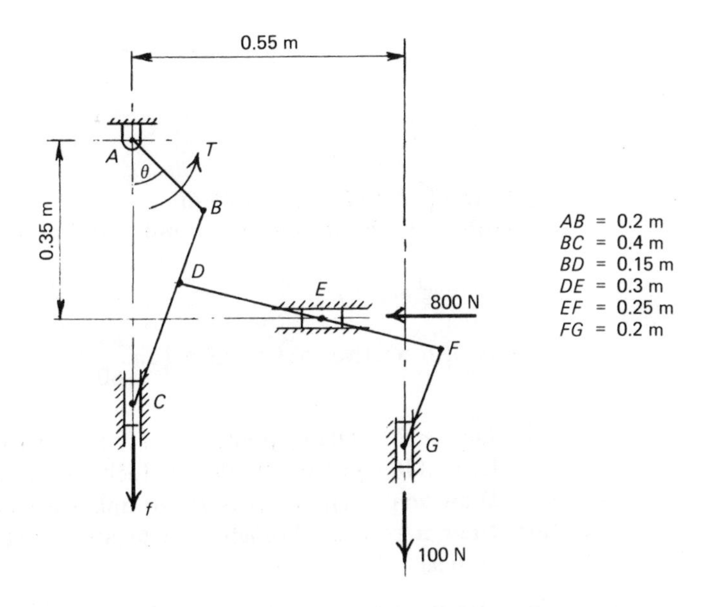

A machine consists of a crank AB turning at 45 rev min^{-1} anticlockwise and three links, BC, DF and FG, arranged as shown in the diagram to operate three sliders at C, E and G. AB is subjected to a torque = 80 N m, and forces of f, 800 N and 100 N act respectively at C, E and G in the directions shown. For the configuration shown, when the angle θ at A is 45°, determine the value of the force f at C. The machine may be assumed to have an efficiency of 1.

Solution 1.5

This question does not ask for the velocity diagram to be drawn, but it should be clear that this is necessary, in order to calculate the power output at the points *C*, *E* and *G*.

Configuration diagram (a) (p. 14):

(i) Fix point *A*. Draw vertically through *A* and 0.55 m to the right of *A* to locate the sliders at *C* and *G*. Draw a horizontal 0.35 m below *A* to locate the slider at *E*.
(ii) Draw *AB* = 0.2 m at 45°.
(iii) With centre *B*, draw an arc of radius 0.4 m to locate *C*.
(iv) Measure *BD* = 0.15 m along *BC* to locate *D*.
(v) With centre *D*, draw an arc of radius 0.3 m to locate *E*.

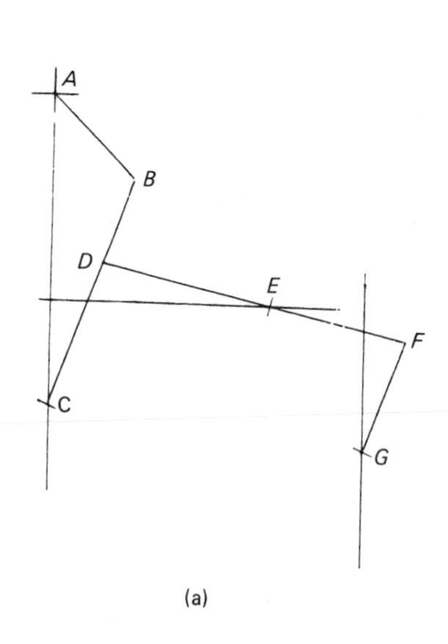

(a)

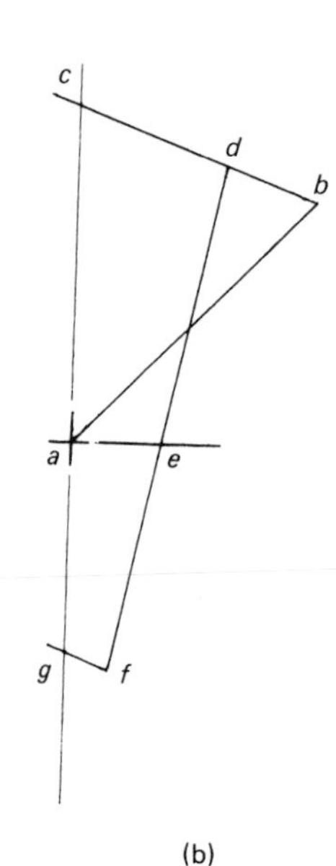

(b)

(vi) **Extend DE to F; $EF = 0.25$ m.**
(vii) **With centre F, draw an arc of radius 0.2 m to locate G.**

Velocity diagram (b):

Velocity of B relative to $A = \omega R = \left(2\pi \frac{45}{60}\right) \times 0.2 = 0.942 \text{ m s}^{-1}$.

(i) **Locate the starting-point a. Draw the vector $ab = 0.942 \text{ m s}^{-1}$ perpendicular to AB, consistent with anticlockwise rotation (i.e. 'up–right').**
(ii) **Draw bc perpendicular to BC of unknown length.**
(iii) **Draw a vertical through a to locate c. (The velocity of c relative to A is vertical.)**
(iv) **Measure bc and divide at d in the ratio $\frac{bd}{dc} = \frac{BD}{DC}$.**

bc scales 0.714 m s^{-1}. $\therefore bd = 0.714 \times \frac{0.15}{0.4} = 0.268 \text{ m s}^{-1}$.

(v) **Draw de of unknown length perpendicular to DE.**
(vi) **Draw a horizontal through a to locate e.**
(vii) **Extend de to f in the ratio $\frac{de}{ef} = \frac{DE}{EF}$.**

de scales 0.77 m s^{-1}. $\therefore ef = 0.77 \times \frac{0.25}{0.3} = 0.642 \text{ m s}^{-1}$.

(viii) **Draw fg perpendicular to FG of unknown length.**
(ix) **Draw vertical through a to locate g.**

This completes the diagram. Scaling from it:

$ac = 0.92 \text{ m s}^{-1}$ upwards.
$ae = 0.232 \text{ m s}^{-1}$ to right.
$ag = 0.58 \text{ m s}^{-1}$ downwards.

The force f at point C, and the force of 800 N at E, are both seen to be in the opposite direction to the respective velocities; thus, work is done *by* the machine

at these two points. But at G the force is in the same direction as the velocity, so that at this point work is done *on* the machine; or, putting it another way, the work done by the machine at this point is negative.

$$T\omega = \Sigma\ (\text{force} \times \text{velocity}).$$

$$80\left(2\pi\ \frac{45}{60}\right) = f(ac) + 800(ae) - 100(ag).$$

$$\therefore\ 377 = f \times 0.92 + 800 \times 0.232 - 100 \times 0.58.$$

$$\therefore f = \frac{377 - 800 \times 0.232 + 100 \times 0.58}{0.92} = \underline{271.1\ \text{N.}}$$

Example 1.6

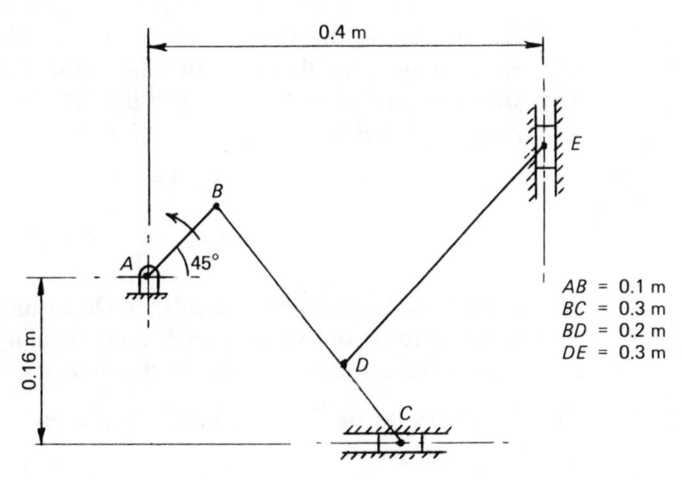

The diagram shows details of a mechanism comprising three links. The crank AB turns about a fixed point A anticlockwise at a constant speed of 4 rad s^{-1}. The end C of BC is constrained by a horizontal slide. Link DE is pinned to BC and end E is constrained by a vertical slide. Dimensions are as shown. Determine the velocities and accelerations of the sliders at C and E for the configuration shown, when AB makes an angle of 45° with the horizontal.

Solution 1.6

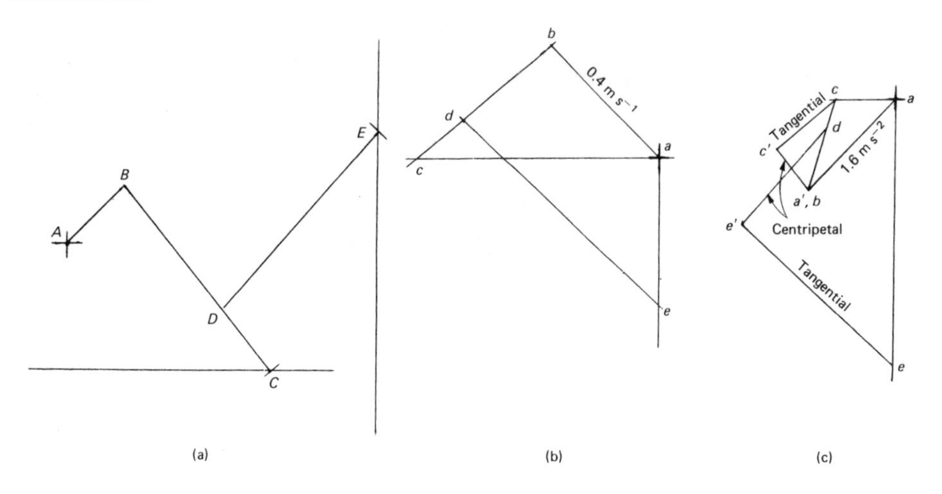

Configuration diagram (a) (p. 15):

Locate A. Measure 0.4 m to the right and 0.16 m below *A* and draw vertical and horizontal lines for the two slides. Draw *AB* 0.1 m long at 45°. With centre *B*, draw an arc of radius 0.3 m to intersect the horizontal slide; hence locate *C*. Draw *BC*. Measure 0.2 m from *B* to locate *D*. With centre *D*, draw an arc of radius 0.3 m to intersect the vertical slide; hence locate *E*.

Velocity diagram (b):
Calculate the velocity of B relative to A.

$$v = \omega R = 4 \times 0.1 = 0.4 \text{ m s}^{-1}.$$

(i) Locate the starting-point *a*. Draw *ab* = 0.4 m s^{-1} perpendicular to *AB*.
(ii) Draw *bc* through *b* perpendicular to *BC*, of unknown length.
(iii) Relative to *A*, the velocity of *C* must be along the direction of the slide. Hence draw a line through *a* parallel to the slide at *C*; this locates *c*.
(iv) Divide *bc* at *d* such that *bd* : *dc* = *BD* : *DC*. Vector *b-c-d* is then the velocity image of link *BCD*.

$$bc \text{ scales } 0.44 \text{ m s}^{-1}.$$

$$\therefore bd = 0.44 \times \frac{0.2}{0.3} = 0.293 \text{ m s}^{-1}.$$

(v) Draw *de* through *d* perpendicular to *DE* of unknown length.
(vi) Relative to *A*, the velocity of *E* must be along the direction of slide. Hence draw a line through *a* parallel to the slide at *E*; this locates *e*.

This completes the diagram. Scaling from it gives:

$bc = 0.44 \text{ m s}^{-1}$.
$de = 0.69 \text{ m s}^{-1}$.
Centripetal accelerations:
Centripetal acceleration of B relative to $A = \omega^2 R = 4^2 \times 0.1 = 1.6 \text{ m s}^{-2}$.

Centripetal acceleration of C relative to $B = \frac{v^2}{R} = \frac{(bc)^2}{BC} = \frac{(0.44)^2}{0.3} = 0.645 \text{ m s}^{-2}$.

Centripetal acceleration of E relative to $D = \frac{v^2}{R} = \frac{(de)^2}{DE} = \frac{(0.69)^2}{0.3} = 1.587 \text{ m s}^{-2}$.

Acceleration diagram (c):

(i) Locate the starting-point *a*. Draw the centripetal vector (*B* relative to *A*):

$$aa_1 = 1.6 \text{ m s}^{-2} \text{ parallel to } AB \text{ in the direction } B \rightarrow A.$$

(ii) There is no tangential component of *B* relative to *A* (*AB* has *constant* speed of 4 rad s^{-1}): thus $a_1 b = 0$. Hence *b* is coincident with a_1.
(iii) Draw the centripetal vector (*C* relative to *B*) bc_1 = 0.645 m s^{-2} parallel to *BC* in direction $C \rightarrow B$.
(iv) Draw the tangential vector (*C* relative to *B*) $c_1 c$ of unknown length perpendicular to *BC*.
(v) Acceleration of *C* relative to *A* must be along the direction of slide. Hence draw a line through *a* parallel to slide at *C* to locate *c*.
(vi) Join *bc*. Divide *bc* at *d* in the ratio *bc* : *cd* = *BC* : *CD*. Vector *b-c-d* is the acceleration image of link *BCD*.
(vii) Draw the centripetal vector (*E* relative to *D*) de_1 = 1.587 m s^{-2} parallel to *DE* in direction $E \rightarrow D$.

(viii) Draw the tangential vector (E relative to D) $e_1 e$ of unknown length perpendicular to DE.

(ix) The acceleration of E relative to A must be along direction of slide. Hence draw line through a parallel to slide at E to locate e.

This completes the diagram. Scaling from it gives:

Acceleration of slider C (ac) = $\underline{0.75 \text{ m s}^{-2}}$.
Acceleration of slider E (ae) = $\underline{3.3 \text{ m s}^{-2}}$.

and from the velocity diagram:
Velocity of slider C = $\underline{0.62 \text{ m s}^{-1}}$.
Velocity of slider E = $\underline{0.37 \text{ m s}^{-1}}$.

For this sort of exercise you need reasonably good instruments (hard pencil, compasses, setsquare, rule and protractor) and you need to use them with reasonable skill. (For example, you should be able to draw lines parallel to, and perpendicular to other lines, as in Fact Sheet (i).) You should use a reasonably big scale, within the limits of a sheet of A4 paper. For this example, the configuration diagram should be to a scale not less than 2 cm $\equiv$ 0.1 m, the velocity diagram to a scale not less than 1 cm $\equiv$ 0.1 m s^{-1} and the acceleration diagram to a scale not less than 2 cm $\equiv$ 1 m s^{-2}. Too small a scale will result in loss of accuracy, while too large a scale might result in your running out of paper half way through the solution. Examiners sometimes recommend suitable scales. Sometimes, a question will include a part of the graphical solution, to ensure that a candidate has a fair chance of completing the question in the allotted time. If, however, you are solving a question yourself from the beginning, it is good practice to make a sketch of the velocity and acceleration diagrams before beginning the accurate drawing. This helps you to arrange the drawing in the available space, and also helps you to decide on a suitable scale. Remember that however accurate your work may be, an exact solution is impossible, and if your answers are within ± 5 per cent of the 'book' answers this may be considered sufficiently accurate.

Two almost trivial suggestions may be made to encourage greater accuracy with graphical work. First, if you have to draw a line through a point along the edge of a setsquare, it is more accurate to place the pencil on the point and to slide the edge up to the pencil, than to set the edge where you think it coincides with the point, and draw the line there. Second, when scaling the length of a vector from a diagram, rather than set a scale directly to the vector, and read off directly, it is more accurate to set dividers to the vector and transfer the dividers to the scale, to read the length.

With careful attention to detail of this sort, the error of ± 5 per cent can be reduced to as low as ± 1 per cent. But in the context of an examination, you need to economise on time, and although greater accuracy might earn you one or two extra marks, it may well be at the cost of more extra time than you can perhaps afford.

Example 1.7

The diagram shows details of a quick-return mechanism comprising two links AB and CD. A and C are fixed points. End B of the link AB is attached by a pin to a slider-block which thereby can turn relative to AB and also can slide along CD as AB turns.

Determine the angular velocity and angular acceleration of the link CD, and the sliding velocity and acceleration of the slider-block relative to link CD, when AB turns clockwise at a constant speed of 4 rad s^{-1} and at the instant when it makes an angle of θ = 60° to the vertical.

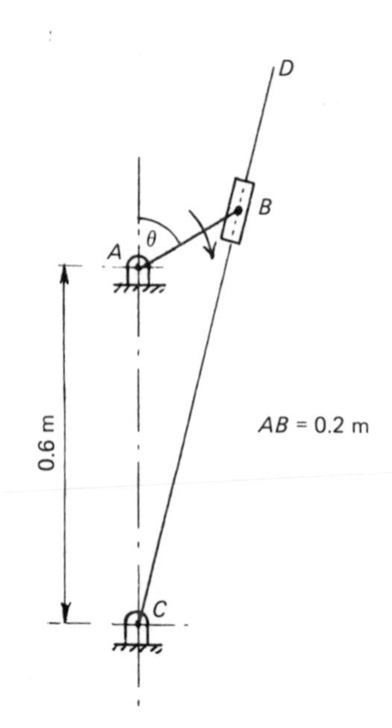

Solution 1.7

The slider-block sliding along link *CD* which is itself turning alerts us to the presence of the Coriolis component. Also, whenever a mechanism includes a block sliding along a link, a point of notation must be clearly understood. Although the centreline of the pin in the slider-block is labelled *B* in the diagram above, we have to recognise that we have to deal here not with one point, but with two. One is the pin centreline itself, and the second is the corresponding point on the link *CD*. This is not separately labelled in the diagram because the two points are coincident in space. But they are not coincident on the velocity diagram, as there is clearly a relative velocity between them. For this reason, we will use the notation b_b (*b*-block) and b_L (*b*-link) when drawing the velocity and acceleration diagrams.

The next diagrams, (a) to (c), show the complete graphical solution. The comments follow.

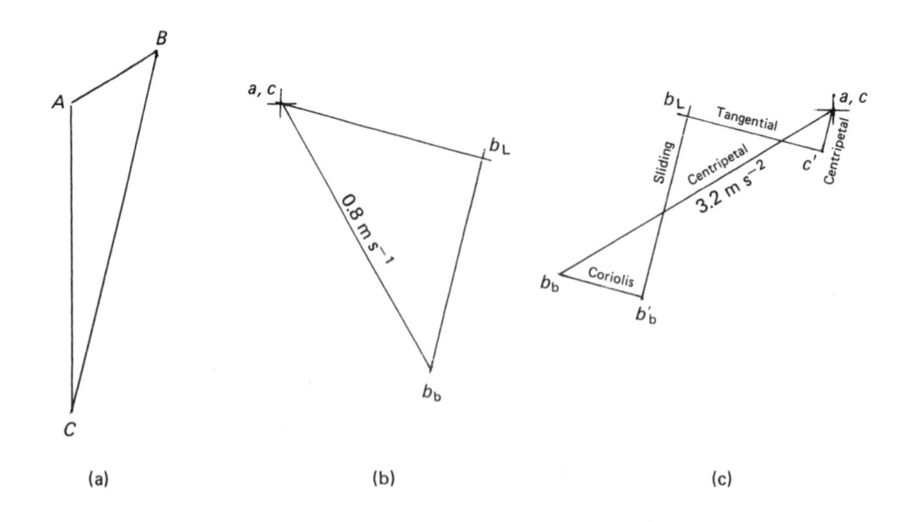

The configuration diagram (a) is simple and requires no comment.

Velocity diagram (b):
Calculate the velocity of B relative to A.

$$v = \omega R = 4 \times 0.2 = 0.8 \text{ m s}^{-1}.$$

(i) Locate the starting-point a. The point is also labelled c as there is no relative velocity of C to A, both points being fixed.

(ii) Draw $ab_b = 0.8 \text{ m s}^{-1}$ perpendicular to AB and in accordance with the clockwise rotation of AB.

(iii) Draw $b_b b_L$ parallel to CD of unknown length. This vector is the velocity of the point on link CD corresponding to the slider, relative to the slider – the sliding velocity.

Step 3 requires additional explanation. The velocity of the link relative to the slider is required. If you find this difficult to understand, it may help to think of a relative velocity as the velocity of one point, *assuming the other point to be fixed*. Now if the block was fixed, the link CD could move only up, or down, in the direction it was pointing at the instant, i.e. the direction C-D. Similarly, if the link CD was fixed then the block could only move up, or down, along the link.

(iv) Draw cb_L perpendicular to CD to locate b_L.

Step 4 completes the diagram. Scaling from it gives:

$cb_L = 0.551 \text{ m s}^{-1}$.
$b_b b_L = 0.577 \text{ m s}^{-1}$.

Scaling from the configuration diagram gives:

$CB = 0.72$ m.

Centripetal and Coriolis accelerations:

Centripetal acceleration of B relative to $A = \omega^2 R = (4)^2 \times 0.2 = 3.2 \text{ m s}^{-2}$.

Centripetal acceleration of B relative to $C = \dfrac{v^2}{R} = \dfrac{(cb_L)^2}{CB} = \dfrac{(0.551)^2}{0.72} = 0.422 \text{ m s}^{-2}$

Coriolis acceleration of B relative to $B = 2v\omega = 2(b_b b_L)\dfrac{cb_L}{CB_L} = 2 \times 0.577 \times \dfrac{0.551}{0.72}$

$$= 0.883 \text{ m s}^{-2}.$$

Acceleration diagram (c):

(i) Locate the starting-point (a, c). Draw the centripetal vector (B relative to A).

$ab_b = 3.2 \text{ m s}^{-2}$ parallel to AB in direction $B \to A$.

(ii) Draw the Coriolis component vector (B_L relative to B_b) $b_b b_b' = 0.883 \text{ m s}^{-2}$. The rule for direction (Fact Sheet (g)) is: '90° in advance of the sliding velocity vector, in the direction of rotation'. The sliding velocity vector here is $b_b b_L$, *not* $b_L b_b$, because we already have point b_b on the diagram, and require to draw the Coriolis component of b_L relative to b_b. The sliding velocity vector $b_b b_L$, *read in that order*, is 'up-right' (at about 1 o'clock). The required 'direction of rotation' is that of the link CD. Again referring to the velocity diagram, the velocity vector cb_L is 'down-right' (at about 4 o'clock). This shows the direction of rotation of CD to be clockwise. Although this may be fairly obvious in this case, you should learn to 'read' the velocity diagram, as described, since cases may arise later on where the direction of rotation is not obvious.

Thus the required direction of the Coriolis vector is that of the sliding vector $b_b b_L$ turned *clockwise* through 90°, that is perpendicular to CB at about 4 o'clock.

(iii) Draw the sliding acceleration vector (B_L relative to B_b) parallel to CD through b_b' of unknown length. (Point b_L must lie on this line.)

(iv) Draw the centripetal vector (B_L relative to C) $cc' = 0.422$ m s^{-2} parallel to CB_L in the direction $B_L \rightarrow C$.

(v) Draw the tangential vector (B_L relative to C) $c'b_L$ perpendicular to CB through c'; hence locate b_L.

This completes the diagram. Scaling from it gives:

$b_b' b_L = 1.83$ m s^{-2}.
$b_L c' = 1.41$ m s^{-2}.

From the velocity diagram:

$$\text{Angular velocity of } CD = \frac{v}{R} = \frac{b_L c}{B_L C} = \frac{0.551}{0.72} = \underline{0.765 \text{ rad s}^{-1}}.$$

Sliding velocity of block $= b_b b_L = \underline{0.577 \text{ m s}^{-1}}$.

From the acceleration diagram:

$$\text{Angular acceleration of } CD = \frac{\text{Tangential acceleration}}{\text{Radius}} = \frac{c'b_L}{CB} = \frac{1.41}{0.72} = \underline{1.96 \text{ m s}^{-2}}.$$

Sliding acceleration of block $= b_L b_b' = \underline{1.83 \text{ m s}^{-2}}$.

This mechanism can also be solved analytically. See Example 1.8.

Example 1.8

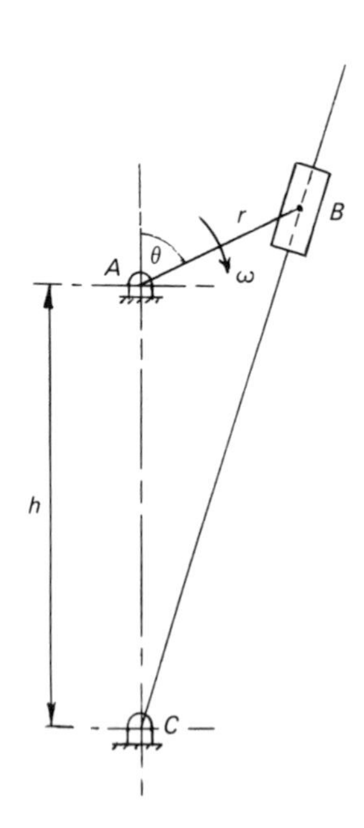

A simple quick-return mechanism is shown in the diagram. The crank AB has a length r and the distance between the crank pivot A and the link pivot C is h. The crank rotates at constant velocity ω in a clockwise direction. Derive expressions for the angular velocity and angular acceleration of the link CD in terms of the displacement angle θ of crank AB. Calculate values for $\theta = 60°$, given $r = 0.2$ m and $h = 0.6$ m.

Solution 1.8

To determine angular velocity and acceleration of a link, we first require an expression for the angular displacement. This may then be differentiated twice with respect to time. The next diagram shows the elements of the mechanism; the required angular displacement is shown as ϕ.

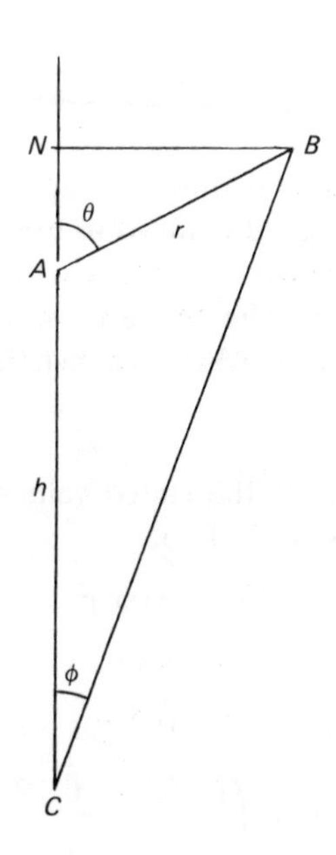

From the diagram,

$BN = r \sin\theta = (h + r\cos\theta)\tan\phi.$

$$\therefore \tan\phi = \frac{r\sin\theta}{h + r\cos\theta}.$$

We differentiate 'implicitly', using the 'quotient' formula for the right-hand side.

[k] $$(\sec^2\phi)\frac{d\phi}{dt} = \left(\frac{(h + r\cos\theta)(r\cos\theta) - (r\sin\theta)(-r\sin\theta)}{(h + r\cos\theta)^2}\right)\frac{d\theta}{dt}.$$

Replace $\sec^2\phi$ by $(1 + \tan^2\phi)$. Use the notation $\frac{d\phi}{dt} = \dot{\phi}$ and recall that $\frac{d\theta}{dt} = \omega$.

$$1 + \left(\frac{r^2\sin^2\theta}{(h + r\cos\theta)^2}\right)\dot{\phi} = \left(\frac{hr\cos\theta + r^2\cos^2\theta + r^2\sin^2\theta}{(h + r\cos\theta)^2}\right)\omega.$$

Recalling that $\sin^2\theta + \cos^2\theta = 1$,

$((h + r\cos\theta)^2 + r^2\sin^2\theta)\,\dot{\phi} = (hr\cos\theta + r^2)\,\omega.$

$$\therefore \dot{\phi} = \omega\left(\frac{hr\cos\theta + r^2}{h^2 + 2hr\cos\theta + r^2\cos^2\theta + r^2\sin^2\theta}\right)$$

$$\dot{\phi} = \omega\left(\frac{hr\cos\theta + r^2}{h^2 + r^2 + 2hr\cos\theta}\right). \quad (1)$$

This is the expression required. We substitute the stated values:

$$\dot{\phi} = 4\left(\frac{0.6 \times 0.2\cos 60^\circ + (0.2)^2}{(0.6)^2 + (0.2)^2 + 2 \times 0.6 \times 0.2\cos 60^\circ}\right) = \underline{0.769 \text{ rad s}^{-1}}.$$

You may have observed that the data are the same as for Example 1.7 in which an answer of 0.765 rad s^{-1} was obtained.

Equation 1 may be differentiated directly to obtain the angular acceleration. We must remember that when we have differentiated the right-hand side with respect to θ we must again multiply this by $d\theta/dt = \omega$. Again using the quotient formula,

$$\ddot{\phi} = \omega^2 \left(\frac{(h^2 + r^2 + 2hr \cos\theta)(-hr \sin\theta) - (hr \cos\theta + r^2)(-2hr \sin\theta)}{(h^2 + r^2 + 2hr \cos\theta)^2} \right).$$

We could spend quite a long time rearranging this rather formidable expression, but if we read the question then we see that we have done what is required. If the answer is required in some particular form, it is the responsibility of the questioner to state this. The above is 'an expression for the angular acceleration' and in an examination, working against time, you do not have to do any more than you are asked.

Substituting the stated values (and noting again that this is an analytical solution for Example 1.7),

$h^2 + r^2 + 2hr \cos\theta = (0.6)^2 + (0.2)^2 + 2 \times 0.6 \times 0.2 \cos 60° = 0.52.$

$hr \sin\theta = 0.6 \times 0.2 \sin 60° = 0.104.$

$hr \cos\theta + r^2 = 0.6 \times 0.2 \cos 60° + (0.2)^2 = 0.10.$

$$\therefore \ddot{\phi} = (4)^2 \left(\frac{(0.52)(-0.104) - (0.10)(-2 \times 0.104)}{(0.52)^2} \right) = \underline{-1.969 \text{ rad s}^{-2}}.$$

The graphical answer (Example 1.7) was 1.96 rad s^{-2}.

In comparing this example with Example 1.7 we see clearly that the graphical solution is simpler. But a graphical solution will yield values of velocity and acceleration for one configuration only, and if values for other configurations are required then the whole procedure must be repeated. Whereas the expressions for angular velocity and acceleration just derived are, of course, valid for any angle θ. By substituting in them, graphs could be drawn of $\dot{\phi}$ and $\ddot{\phi}$ for the complete cycle of the mechanism.

Example 1.9

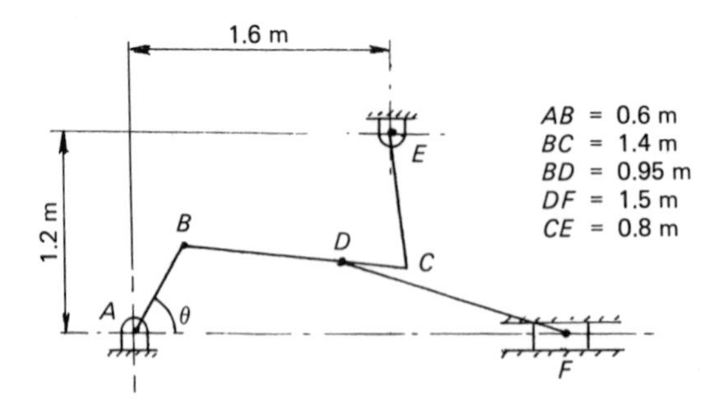

The mechanism shown in the diagram comprises four links, *AB*, *BC*, *CE* and *DF*; *DF* is attached by a pin to a point *D* on *BC*, and is constrained by the horizontal slide at *F*. The dimensions are shown. Determine the acceleration of *F* in the slide when link *AB* is turning at a constant angular velocity and the velocity of sliding of *F* is 2 m s^{-1} from right to left, at the instant that *AB* makes an angle $\theta = 60°$ to the horizontal. Find also the angular accelerations of *BC*, *CE* and *DF*.

Solution 1.9

Questions of this type have trapped many examinees. The temptation is to begin the velocity diagram with the sliding vector *af*, as this is stated, both as regards magnitude and direction. If you try this then you will find that the diagram cannot be completed.

It is very useful to remember that once you have a configuration diagram you can always draw the velocity diagram, *even if you do not know a single velocity*. In this example, we will begin the velocity diagram with the vector *ab*, representing the velocity of *B* relative to *A*. And because we don't know its magnitude, we will make this vector an arbitrary length – say, 5 cm (although of course, it will not have this value on the printed page of the book).

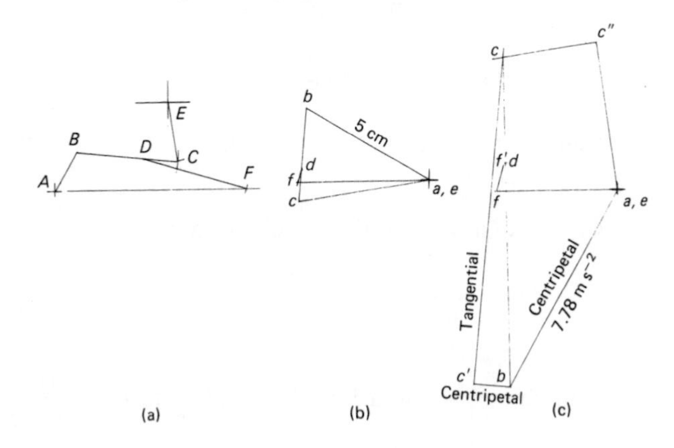

Configuration diagram (a):

Fix point *A*, and point *E*, 1.2 m above *A* and 1.6 m to the right. The slider at *F* is on the horizontal passing through *A*. Draw *AB* at 60°, length 0.6 m. With centre *B*, draw an arc of radius *BC*; with centre *E*, draw an arc of radius *EC* to locate *C*. Join *BC* and measure off *BD* along it. With centre *D*, draw an arc radius *DF* to intersect the slider line and thus locate *F*.

Velocity diagram (b):

(i) Locate the starting-point *a*. This is also point *e*. Draw *ab* = 5 cm (arbitrary) perpendicular to *AB*, assuming anticlockwise rotation. (This is obviously consistent with the information that *F* slides from right to left.)

(ii) Draw *bc* through *b* perpendicular to *BC* of unknown length: *c* lies on this line.

(iii) Draw *ec* through *e* perpendicular to *EC* to locate *c*.

(iv) Measure *bc* and calculate the location of *d* for the velocity image:

$$bc \text{ scales } 3.10 \text{ cm},$$

$$\therefore bd = bc\left(\frac{BD}{BC}\right) = 3.10\left(\frac{0.95}{1.4}\right) = 2.10 \text{ cm}.$$

(v) Draw *df* through *d* perpendicular to *DF* of unknown length.

(vi) Draw a horizontal through (*a*, *e*) to intersect the above, to locate *f*.

This completes the diagram.

af scales 4.63 cm; *ab* is 5 cm; *bc* scales 3.10 cm; *ce* scales 4.67 cm; *df* scales 0.44 cm.

From the information that the slider velocity is 2 m s^{-1} we can now put a scale on the diagram:

$$2 \text{ m s}^{-1} \equiv 4.63 \text{ cm}.$$

The various velocities may now be calculated:

$$ab = 5\left(\frac{2}{4.63}\right) = 2.16 \text{ m s}^{-1}; \qquad bc = 3.10\left(\frac{2}{4.63}\right) = 1.34 \text{ m s}^{-1};$$

$$ce = 4.67\left(\frac{2}{4.63}\right) = 2.02 \text{ m s}^{-1}; \qquad df = 0.44\left(\frac{2}{4.63}\right) = 0.19 \text{ m s}^{-1}.$$

Observe that the complete diagram was drawn before we made use of the information that the slider speed was 2 m s^{-1}.

Centripetal accelerations:

$$\text{Centripetal acceleration of } B \text{ relative to } A = \frac{v^2}{R} = \frac{(ab)^2}{AB} = \frac{(2.16)^2}{0.6} = 7.78 \text{ m s}^{-2}.$$

$$\text{Centripetal acceleration of } C \text{ relative to } B = \frac{(bc)^2}{BC} = \frac{(1.34)^2}{1.4} = 1.28 \text{ m s}^{-2}.$$

$$\text{Centripetal acceleration of } C \text{ relative to } E = \frac{(ce)^2}{CE} = \frac{(2.02)^2}{0.8} = 5.10 \text{ m s}^{-2}.$$

$$\text{Centripetal acceleration of } F \text{ relative to } D = \frac{(df)^2}{DF} = \frac{(0.19)^2}{1.5} = 0.02 \text{ m s}^{-2}.$$

Acceleration diagram (c):

(i) Locate the starting-point (a, e). Draw the centripetal vector (B relative to A) ab = 7.78 m s^{-2} in the direction $B \rightarrow A$.
(ii) Draw the centripetal vector (C relative to B) bc' = 1.28 m s^{-2} parallel to BC in the direction $C \rightarrow B$.
(iii) Draw the tangential vector (C relative to B) through c' of unknown length perpendicular to BC.
(iv) Draw the centripetal vector (C relative to E) ec'' = 5.10 m s^{-2} parallel to CE in the direction $C \rightarrow E$.
(v) Draw the tangential vector (C relative to E) $c''c$ perpendicular to CE to locate c.
(vi) Join bc.
(vii) Measure bc and divide proportionately to locate d. (Acceleration image.)

bc scales 11.3 m s^{-2}.

$$\therefore bd = bc\left(\frac{BD}{BC}\right) = 11.3\left(\frac{0.95}{1.4}\right) = 7.67 \text{ m s}^{-2}.$$

(viii) Draw the centripetal vector (F relative to D) df' = 0.02 m s^{-2} parallel to DF in the direction $F \rightarrow D$. This vector is so small that for practical purposes, f' is coincident with d.
(ix) Draw the tangential vector (F relative to D) $f'f$ of unknown length perpendicular to DF.
(x) Draw a horizontal through a to locate f. (The acceleration of F relative to the fixed point A must be in the direction of the slider.)

This completes the diagram. Scaling gives:

af = 4.32 m s^{-2}; $c'c$ = 10.23 m s^{-2}; $c''c$ = 3.37 m s^{-2};
$f'f$ = 1.02 m s^{-2}.

Sliding acceleration of $F = af = 4.32$ m s^{-2}, right to left.

$$\text{Angular acceleration} = \frac{\text{Tangential component}}{\text{Radius}} .$$

For BC, angular acceleration $= \dfrac{c'c}{BC} = \dfrac{10.23}{1.4} = 7.31$ rad s^{-2} anticlockwise.

For CE, angular acceleration $= \dfrac{c''c}{CE} = \dfrac{3.37}{0.8} = 4.21$ rad s^{-2} clockwise.

For DF, angular acceleration $= \dfrac{f'f}{DF} = \dfrac{1.02}{1.5} = 0.68$ rad s^{-2} .

A single example will suffice to show the direction of angular acceleration. The acceleration diagram shows that, relative to B, the tangential component of C is $c'c$, not cc', which is the tangential component of B relative to C. $c'c$ is directed almost upwards vertically; thus, C accelerates tangentially upwards relative to B. This is consistent with an anticlockwise angular acceleration.

Example 1.10

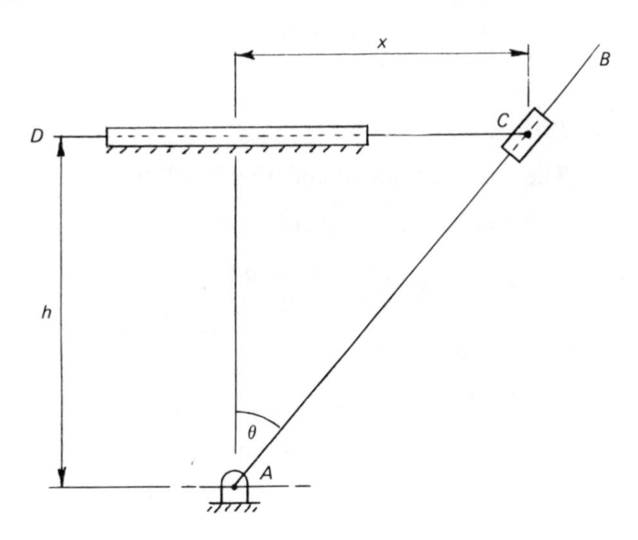

In the part of a mechanism shown in the diagram the rod CD is constrained to move horizontally only. It carries a swivelling slider at C which slides along the link AB which is pivoted at A.

(a) Derive expressions for the linear velocity and acceleration of CD when AB has an angular velocity ω clockwise and an angular acceleration α clockwise.

(b) Calculate the linear acceleration of CD when $\theta = 40°$, $h = 2$ m, $\omega = 1$ rad s^{-1} clockwise and $\alpha = 8$ rad s^{-2} anticlockwise.

(c) If CD moves with constant velocity and the angular velocity of AB is 1 rad s^{-1} clockwise, determine the angular acceleration of AB.

Solution 1.10

Part (a)

We require to express x as a function of θ and differentiate it twice with respect to time. (See Example 1.8.)

From the diagram,

$x = h \tan \theta$.

Using the notation $\dfrac{dx}{dt} = \dot{x}$,

$$\dot{x} = \frac{dx}{d\theta}\frac{d\theta}{dt} = \underline{(h \sec^2 \theta)\, \omega.} \qquad (1)$$

When differentiating the second time, we note that ω is not constant. Using the 'product' formula $d(uv) = v\, du + u\, dv$,

$$\ddot{x} = (\omega)(h)(2 \sec \theta)(\sec \theta \tan \theta)(\omega) + (h \sec^2 \theta)\, \ddot{\theta}$$

$$= \underline{\omega^2 (2h \sec^2 \theta \tan \theta) + \alpha h \sec^2 \theta.} \qquad (2)$$

Part (b)

Substituting in equation 2 and noting that α is negative,

$$\ddot{x} = (1)^2 (2 \times 2 \sec^2 40° \tan 40°) + (-8)(2 \sec^2 40°)$$

$$= 5.720 - 27.265$$

$$= \underline{-21.545 \text{ m s}^{-2}.}$$

The negative sign indicates an acceleration from right to left, since x is originally reckoned positive from left to right.

Part (c)

The left-hand side of equation 2 will be 0.

$$0 = \omega^2 (2h \sec^2 \theta \tan \theta) + \alpha h \sec^2 \theta.$$

$$\therefore \alpha = -\omega^2 \left(\frac{2h \sec^2 \theta \tan \theta}{h \sec^2 \theta}\right)$$

$$= -\omega^2\, 2 \tan \theta$$

$$= -(1)^2\, 2 \tan 40°$$

$$= \underline{-1.678 \text{ rad s}^{-2}\text{, i.e. anticlockwise.}}$$

Example 1.11

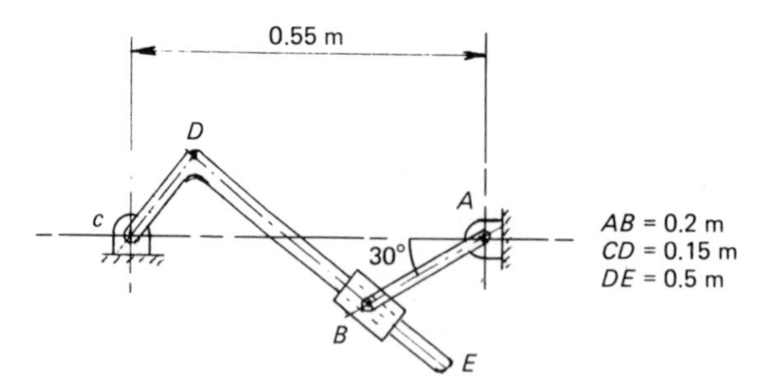

The diagram shows part of a machine. *CDE* is a single rigid right-angled link. A block can slide along the portion *DE*; this block is pinned to the crank *AB*, and the pin enables the block to turn relative to *AB*. At a certain instant, *AB* is at 30° below the horizontal, as shown, and is turning clockwise with an angular velocity of 4.5 rad s^{-1}. For this configuration determine the velocity of the point *E* and the angular velocity of *CDE*.

Solution 1.11

This is the sort of problem that some students believe an examiner sets out of sheer spite. It can be very distressing, if you have practised drawing velocity dia-

grams until you are absolutely confident that you can answer any examination question on the topic, only to find, when you get into the examination hall, that you run into difficulty just drawing the configuration diagram.

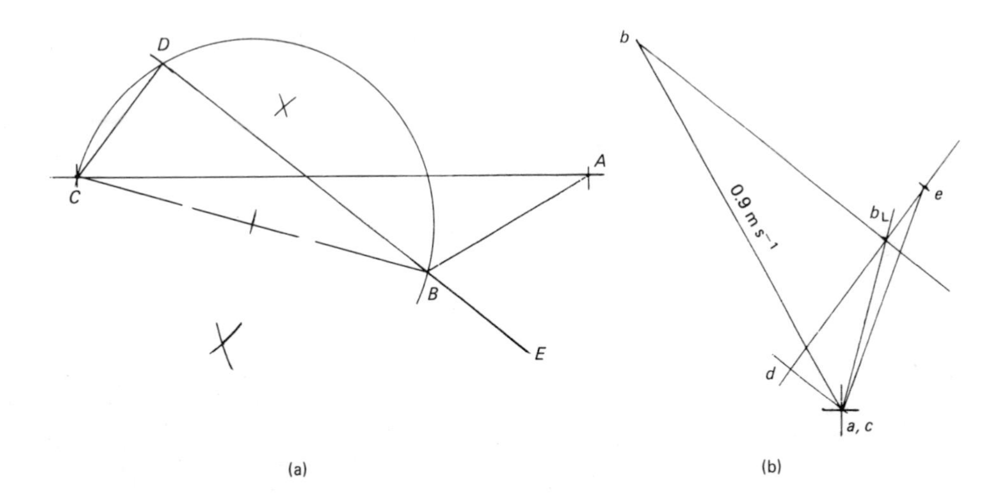

(a) (b)

Configuration diagram (a):

Recall the axiom of geometry that the angle in a semicircle is a right angle.

(i) Locate A and C. Draw AB.
(ii) Join BC. Draw a circle on BC as diameter.
(iii) With centre C and radius CD, draw an arc to cut the circle at D.
(iv) Join DB and extend to E.

As an alternative construction, after (i) above, you could draw a circle with centre C, radius CD, and construct the tangent from B to this circle.

Velocity diagram (b):
Velocity of B relative to $A = \omega R = 4.5 \times 0.2 = 0.9$ m s^{-1}.
Define point B_L coincident with B on link DE. (See Example 1.7.)

(i) Locate the starting-point (a, c). Draw ab = 0.9 m s^{-1} perpendicular to AB.
(ii) Draw bb_L of unknown length parallel to DE.
(iii) Draw cb_L perpendicular to CB; hence locate b_L.
(It is clear that B_L moves in a circular path about C.)
(iv) Draw cd of unknown length perpendicular to CD.
(v) Draw $b_L d$ perpendicular to $B_L D$; hence locate d.
(vi) Extend db_L to e in the ratio

$$\frac{db_L}{de} = \frac{DB_L}{DE}.$$

db_L scales 0.34 m s^{-1}; DB_L scales 0.358 m.

$$\therefore\ de = db_L \left(\frac{DE}{DB_L}\right) = 0.34 \times \frac{0.5}{0.358} = 0.475 \text{ m s}^{-1}.$$

(vii) Draw ae.

Velocity of point $E = ae$ which scales 0.497 m s^{-1}.
For the angular velocity of the link CDE choose any vector and divide by the

corresponding radius, e.g.

$$\omega_{CDE} = \frac{v}{R} = \frac{cb_{\mathrm{L}}}{CB} = \frac{0.367}{0.39} = \underline{0.941 \text{ rad s}^{-1}}.$$

The lesson of this exercise is that you should always be prepared for the unexpected.

Example 1.12

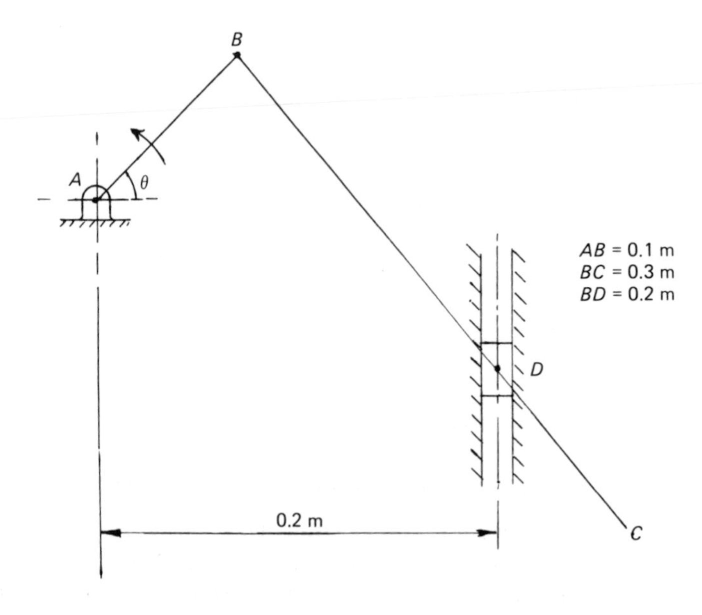

The diagram shows a simple mechanism comprising a crank AB connected by a pin to the link BC. At point D on BC, 0.2 m from B, a slider is attached by a pin; this slider is constrained to move in the vertical guide shown.

(a) Given that AB turns anticlockwise at a constant rate of 5 rad s^{-1}, determine the velocity of the slider D in the guide, the velocity of the end C of link BC, and the angular velocity of BC, when the angle θ is 45°.

(b) Given that the velocity of C is 2 m s^{-1} for the configuration shown, when angle θ is 45°, calculate the corresponding angular speed of AB.

Solution 1.12

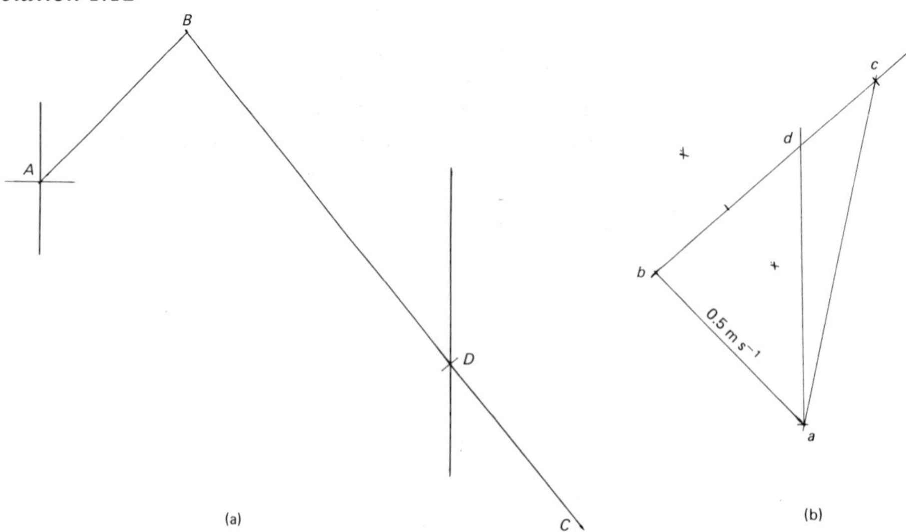

(a) (b)

(a) shows the configuration diagram and (b) the required velocity diagram. The point to note at the outset is one we have met before. When a question comprises two parts, never make the mistake of drawing a velocity diagram twice; it is not necessary.

Configuration

Locate A. Draw the vertical through D, 0.2 m to the right of A. Draw AB = 0.1 m at 45°. Draw an arc, centre B, radius 0.2 m, to intersect the vertical through D. Project BD to C.

Velocity diagram

$$\text{Velocity of } B \text{ relative to } A = \omega r$$
$$= 5 \times 0.1$$
$$= 0.5 \text{ m s}^{-1}.$$

Locate a. Draw ab = 0.5 m s^{-1} perpendicular to AB. Draw a line through b perpendicular to BD of unknown length. Draw a line through a vertical (i.e. parallel to slider) to locate d. Extend bd to c in the ratio $bd : dc = BD : DC$. (In this case, since DC is $\frac{1}{2}BD$, the simplest way is to bisect bd and project it by half its length.) Join ac.

ANSWERS

Part (a)

From the diagram, ad scales $\underline{0.65 \text{ m s}^{-1}}$.
ac scales $\underline{0.824 \text{ m s}^{-1}}$.

$$\omega_{BC} = \frac{v}{r}$$
$$= \frac{bd}{BD}$$
$$= \frac{0.46}{0.2}$$
$$= \underline{2.3 \text{ rad s}^{-1}}.$$

Part (b)

Since an angular velocity of AB of 5 rad s^{-1} produces a velocity of 0.824 m s^{-1} of C, the required angular velocity of AB is

$$\omega_{AB} = 5 \times \frac{2.0}{0.824}$$
$$= \underline{12.14 \text{ rad s}^{-1}}.$$

Example 1.13

The diagram shows a simple mechanism comprising a crank AB connected by a pin to a single rigid link BCD. End C of this link moves along a horizontal slider. The crank is attached to the link at B which is 0.2 m from C.

(a) Given that AB turns clockwise at 10 rad s^{-1}, determine the velocity of the slider C in the horizontal guide, the magnitude of the velocity of end D of the link BCD, and the angular velocity of this link, when AB is at 30° to the horizontal, as shown overleaf.

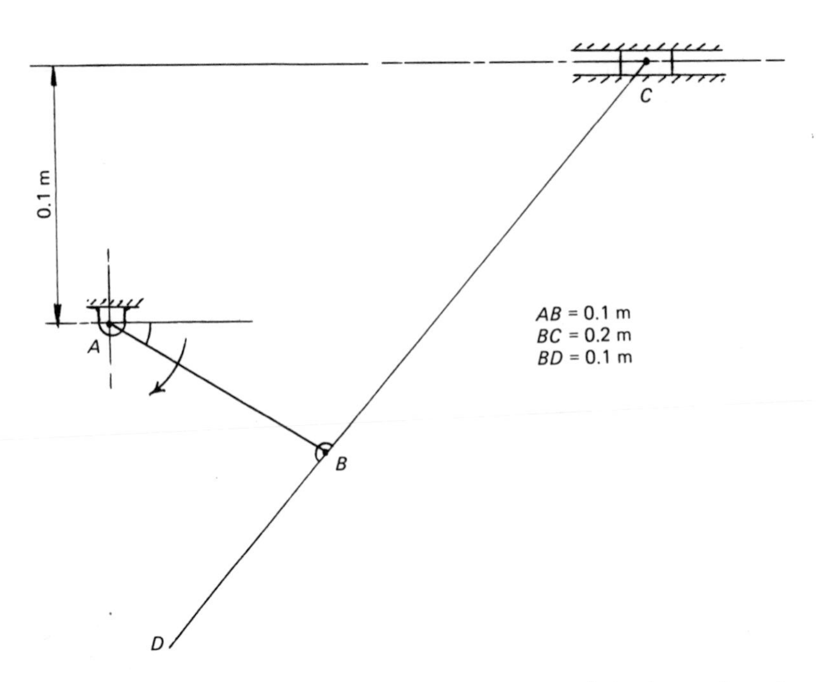

(b) Given that the magnitude of the velocity of D is 0.1 m s^{-1} for the configuration shown, determine the corresponding angular speed of AB. PCL

Solution 1.13

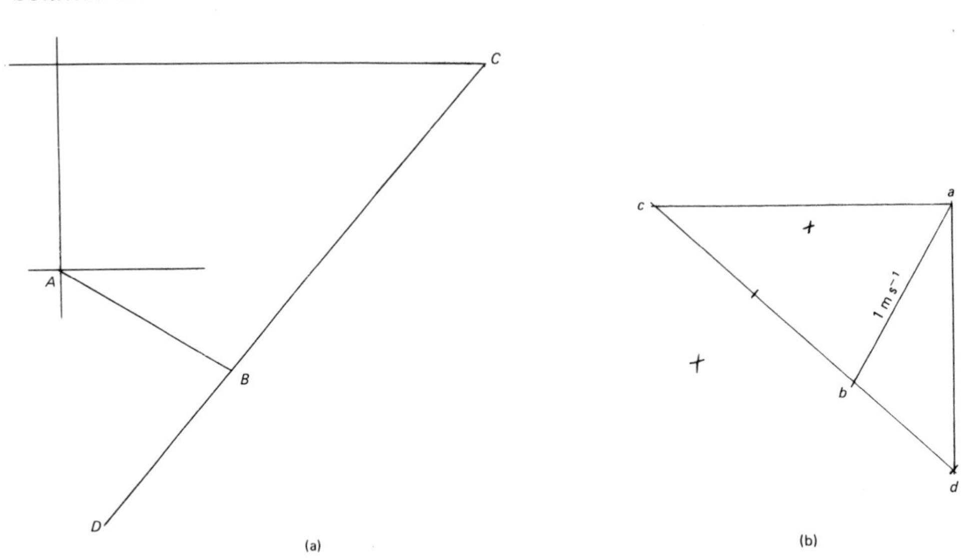

(a) and (b) are respectively the configuration and velocity diagrams. This is a further example of the principle that a single velocity diagram will suffice for the kinematic analysis of a mechanism for a particular configuration.

The configuration diagram needs little explanation. A convenient scale for A4 paper would be 5 cm ≡ 0.1 m.

Velocity diagram (b):

Velocity of B relative to A $= \omega r$

$= 10 \times 0.1$

$= 1$ m s^{-1}.

Locate a. Draw ab = 1 m s^{-1} perpendicular to AB. Draw line through b perpendicular to BC of unknown length. Draw line through a horizontal to locate c. Extend cb to locate d ($cb : bd = CB : BD$: velocity image.) In this case, since the ratio is 2 : 1, cb has been bisected, and half its length stepped off. Join ad.

ANSWERS

Part (a)

ac scales $\underline{1.48 \text{ m s}^{-1}}$.
ad scales $\underline{1.3 \text{ m s}^{-1}}$.

$$\omega_{BC} = \frac{bc}{BC} = \frac{1.32}{0.2} = \underline{6.6 \text{ rad s}^{-1}}.$$

Part (b)

By simple ratio:

$$\omega_{AB} = 10 \times \frac{0.1}{1.3}$$
$$= \underline{0.769 \text{ rad s}^{-1}}.$$

Example 1.14

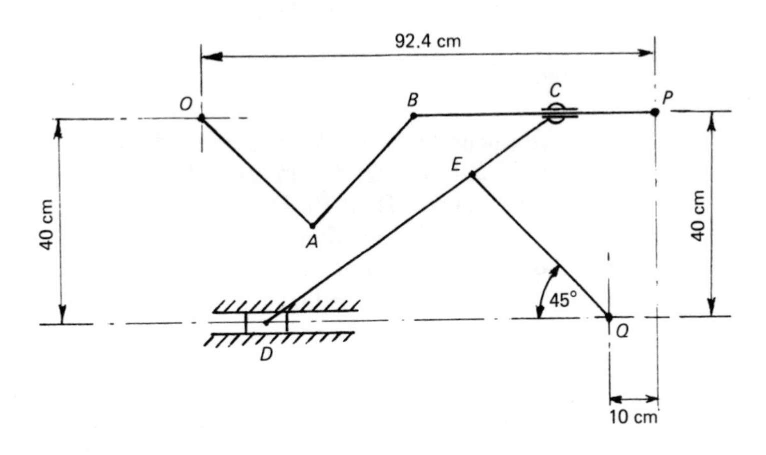

In the mechanism shown in the diagram the link OA rotates about O in the clockwise direction at 25 radians/second. The link PB pivots about P and is connected to link OA by the member AB. The link QE oscillates about Q and is pinned at E to the link CD, which has pinned to it at C a slotted block through which the link PB passes. The end D of link CD is pinned to a slider which constrains its motion to that along a horizontal line passing through Q.

The lengths of the members are: OA 30 cm; AB 30 cm; PB 50 cm; and QE 40 cm. The length DE is 50 cm and the length of the link CD is such that link PB is horizontal in the configuration shown in the diagram.

Determine for this configuration

(a) the velocity of the slider at D;
(b) the angular velocity of the link PB; and
(c) the speed of sliding of the link PB through the slotted block pinned to the link CD at C.

U. Lond. K.C.

Solution 1.14

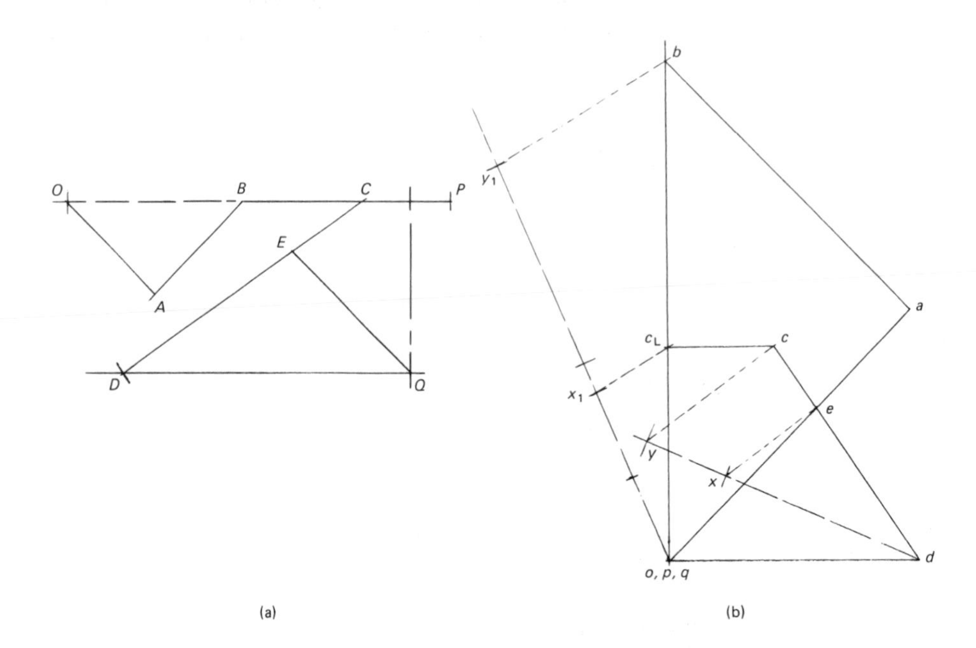

(a) (b)

The configuration and velocity diagrams are shown as (a) and (b) respectively. The dimension 92.4 cm is chosen deliberately to ensure that OA and AB will both be inclined at 45° to the horizontal.

Configuration

(A convenient scale would be one-tenth actual size.) Draw the two horizontals. through O-B-P and Q, 40 cm apart. Locate Q, P (10 cm to the right of Q), and O (92.4 cm to the left of P). Draw QE = 40 cm at 45°. With centre E and radius 50 cm, draw an arc to intersect the lower horizontal to locate D. Extend DE to cut the upper horizontal to locate C. Step off PB = 50 cm. Draw arcs, centres O and B, of length 30 cm to locate A. (Check that OA and AB are at 45°.)

Velocity

This is another of those problems where proceeding the obvious way will result in difficulties. It is a good example of the desirability of sketching the velocity diagram before you draw it accurately – partly to gain an impression of the overall shape, so that a convenient scale may be chosen, and partly (in this case particularly) to ensure that you *can* draw the diagram. If you begin your sketch by drawing ***oa*** (the obvious choice, since the speed of OA is given), you can draw ***oa***, ***ab***, ***pb***, and the velocity image of PC_LB (C_L being the point on ***PB*** coincident with the slider C) but you can't get any further. The place to begin is to draw the vector ***qe***, and since you do not know the magnitude of this velocity, you make the length arbitrary. On A4 paper a convenient length would be ***qe*** = 5 cm.

Locate o, p, q. Draw ***qe*** 5 cm long at 45°, i.e. perpendicular to ***QE***. Draw a line through e perpendicular to ***DE*** to intersect the horizontal through q to locate ***d***. Draw the velocity image ***dec***. Instead of measuring lengths off the configuration diagram, it is simpler to use 'similar triangles' construction.

Thus: Draw any line through ***d*** (the broken line in diagram (b)). Step off with compasses ***dx*** = ***DE*** and ***dy*** = ***DC*** along this line. Do not *measure* ***DE*** or ***DC***; just transfer the lengths directly with the compasses. Join *ex* (the dotted line in diagram (b)). Draw a line through *y* parallel to *ex* to intersect *de* produced at *c*.

Draw a horizontal through *c* of unknown length; c_L must lie on this line. Draw a line through *p* perpendicular to *PB* (i.e. vertical) to locate c_L. Draw the velocity image $pc_L b$. (This time, because *PC* is very short in comparison with pc_L, the length of *PC* is stepped off *twice* along the broken line to locate x_1, and, similarly, y_1 is *PB* again stepped off twice. Otherwise, the parallel dotted lines would intersect the vertical through *opq* at rather acute angles, and this would make for inaccuracy.) Draw a line through *b* perpendicular to *AB* of unknown length. Draw a line through *o* perpendicular to *OA* to intersect this and thus locate *a*.

ANSWERS

(a) Velocity of $A = \omega r = 25 \times 30 = 750$ cm s^{-1}.
But *oa* scales 8.25 cm.
Hence the scale of the velocity diagram is:

$8.25 \equiv 750$ cm s^{-1}.

Velocity of slider at *D* is *od* which scales 6.0 cm.

$$\therefore\ od = 750 \times \frac{6.0}{8.25} = \underline{545.5 \text{ cm s}^{-1}}.$$

(b) $$\omega_{PB} = \frac{pb}{PB} = \left(750 \times \frac{11.65}{8.25}\right)\frac{1}{50} = \underline{21.18 \text{ rad s}^{-1}}.$$

(c) Block sliding speed = cc_L which scales 2.55 cm.

$$\therefore\ cc_L = 750 \times \frac{2.55}{8.25} = \underline{231.8 \text{ cm s}^{-1}}.$$

The results above are taken directly from a graphically constructed velocity diagram, drawn according to the plan set out above. The diagram is reproduced as (b) on page 32, but the restrictions imposed upon the printers mean that it will appear in this book to a different scale from that to which it was originally drawn. Thus you will not be able to check these calculations against the diagram. But of course, in your interests, you should go through the complete procedure yourself and draw your own diagrams.

Example 1.15

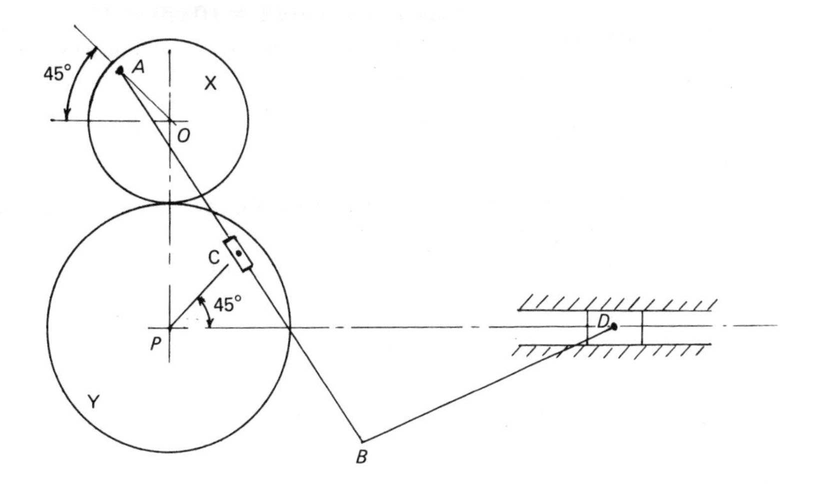

In the mechanism shown in the diagram the gear X rotates clockwise about the fixed centre O at 15 rad s^{-1} and meshes with gear Y, which turns about the fixed centre P. The link AB is pinned to gear X at A and slides through the slotted block C, which can turn on a pin carried by gear Y. The link BD is pinned at B to the link AB and at D to a slider, which is constrained to move along a horizontal straight line passing through P.

The pitch circle diameters of gears X and Y are respectively 16 cm and 24 cm. OA is 7 cm and PC is 10 cm. The lengths of the links AB and BD are respectively 44 cm and 28 cm.

For the configuration shown in the figure determine

(a) the velocity of the slider at D;

(b) the angular velocity of the link AB; and

(c) the speed of sliding of the link AB through the slotted block at C.

U. Lond. K.C.

Solution 1.15

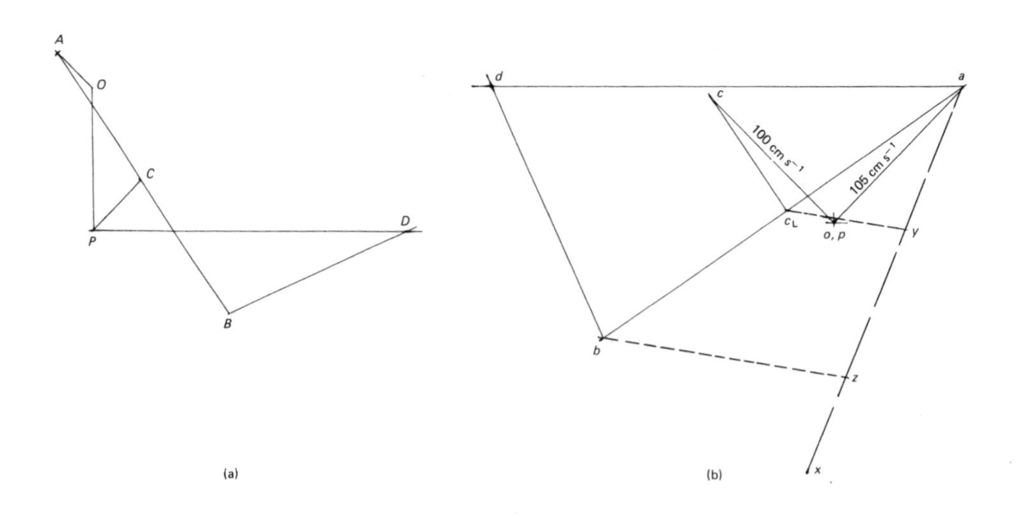

Configuration

A convenient scale for A4 paper is 1 cm ≡ 2 mm i.e. one-fifth full size.

Locate O. OP is the sum of the two pitch radii, i.e. $\frac{1}{2}(16 + 24) = 20$ cm. Hence locate P and draw the horizontal through P.

Draw $OA = 7$ cm at 45° and $PC = 10$ cm at 45°. Join AC and extend to B; $AB = 44$ cm. Centre B, draw arc radius $BD = 28$ cm to intersect the horizontal through p, to locate D. Join BD.

Velocity

As with all problems involving sliding blocks, always remember to allocate *two* points for velocity at the block; one point, c, for the block itself and another point, c_L, for the corresponding point on AB.

Velocity of $A = \omega_{OA} \times OA = 15 \times 7 = 105$ cm s^{-1}.

By ratio of the pitch circle radii:

$$\omega_{PC} = 15 \times \frac{8}{12} = 10.0 \text{ rad s}^{-1}.$$

Velocity of $C = 10.0 \times 10 = 100$ cm s^{-1}.

Locate point o, p. Draw oa = 105 cm s^{-1} perpendicular to OA. Draw pc = 100 cm s^{-1} perpendicular to PC. Draw line through a perpendicular to AC_LB of unknown length; c_L must lie on this line. Velocity of c_L relative to c must be parallel to AB. Draw the line through c parallel to AB; hence locate c_L. Draw the velocity image a–c_L–b. (Draw any line ax. Step off $ay = AC$ and $az = AB$ along it. Join yc_L. Draw parallel line through z; hence locate b.) Draw the line through b perpendicular to BD of unknown length. Draw the horizontal through a to locate d.

ANSWERS

(a) ad scales 270 cm s^{-1}.

(b) $\omega_{AB} = \dfrac{ab}{AB} = \dfrac{250}{44} = 5.68$ rad s^{-1}.

(c) cc_L scales 78 cm s^{-1}.

Example 1.16

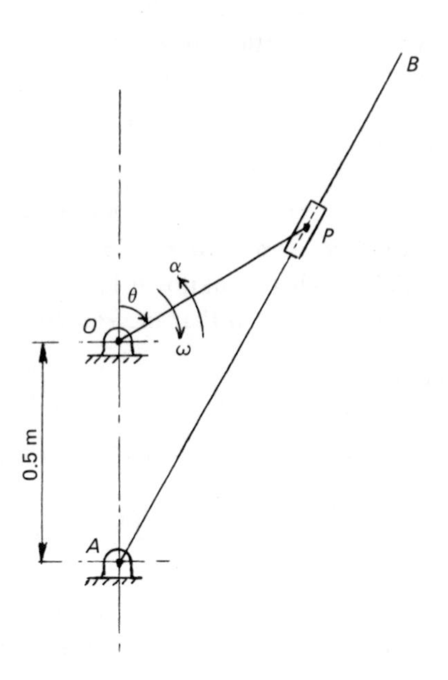

The diagram shows a quick-return mechanism. The crank OP has a length of 0.5 m and is attached to a slider by means of a swivel pin, so that the slider can turn relative to the crank. The link AB passes through the slider. At the instant shown, the angle θ is 60 degrees and OP has an angular velocity of 6 rad s^{-1} clockwise and an angular acceleration in an anti-clockwise direction. A scale drawing of the configuration is supplied.

(a) Draw the velocity diagram for the mechanism, to scale, on the same sheet as the configuration diagram, choosing your own scale. Use the diagram to calculate the angular velocity of the link AB.

(b) Make a *sketch only* of the acceleration diagram of the mechanism. You need calculate no values of acceleration, but your sketch should clearly show, by appropriate lettering, the directions of the vectors drawn. It should also show, either by numbering, or by explanatory notes, the order in which the vectors are drawn. The nature of each vector (e.g. 'centripetal', 'tangential', etc.) should also be indicated.

PCL

Solution 1.16

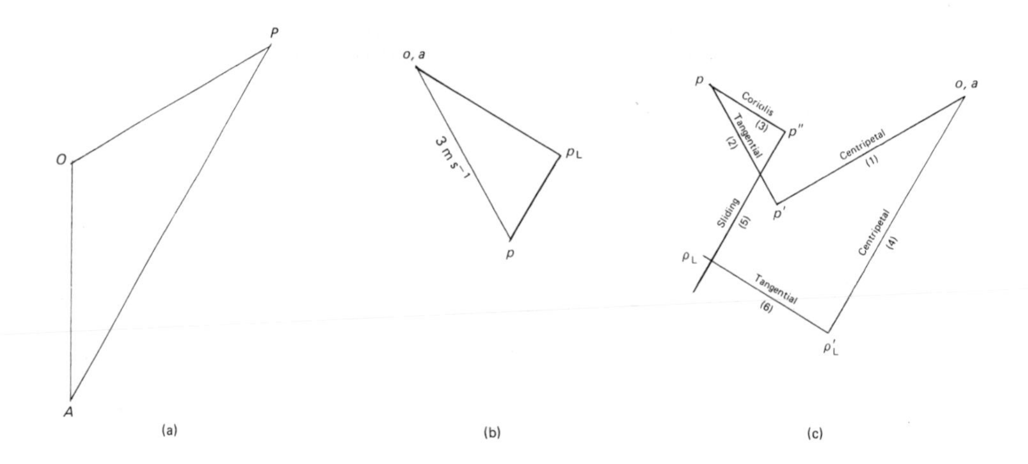

(a) (b) (c)

Once more recognising the special feature of sliding blocks on links, we allocate letter p to the block and letter p_L to the corresponding point on the link AB.

The configuration diagram is supplied. (This is a familiar procedure, designed to provide a question of reasonable length.)

Velocity

Velocity of $P = \omega_{OP} \times OP = 6 \times 0.5 = 3 \text{ m s}^{-1}$.
Locate o, a. Draw $op = 3 \text{ m s}^{-1}$ perpendicular to OP. Draw sliding vector pp_L parallel to AB of unknown length. Draw vector through a perpendicular to AB to locate p_L. It is seen that the velocity diagram is a 30–60–90–degree triangle.

ANSWER

ap_L scales, or is simply calculated to be, 2.598 m s^{-1}.

$$\omega_{AB} = \frac{ap_L}{AP} = \frac{2.598}{0.866} = \underline{3.00 \text{ rad s}^{-1}}.$$

(Again, the configuration diagram is seen to be a 30° isosceles triangle, and AP is simply calculated, eliminating the necessity of actually scaling it off the configuration diagram.)

Acceleration

Recall that all centripetal and Coriolis components could be calculated once the velocity diagram is drawn. In general, tangential components of acceleration cannot be calculated but in this case there is an exception, because the angular acceleration of OP is specified. Thus, given the value of this angular acceleration, the tangential component could be calculated.

The diagram is self-explanatory, in that the vectors are numbered in the order in which they are drawn. Given actual values, op', $p'p$, pp'' and ap'_L would be drawn to actual known lengths. Point p_L is then located by the intersection of vectors 5 and 6.

A fair proportion of the available marks would be allocated to the correct direction of the Coriolis vector. This is drawn from point p and is one of the two components of acceleration between p and p_L. The velocity vector pp_L points 'up-right'. We turn this vector through 90° in the direction of rotation of the link AB. Velocity vector ap_L shows this link to be turning clockwise. Hence the Coriolis vector pp'' is 'down-right' (or at about 4 o'clock).

Notice that you *could* calculate the magnitudes of the centripetal acceleration components of OP and AB. But *do not do so*. This would take time, and would earn no marks.

Example 1.17

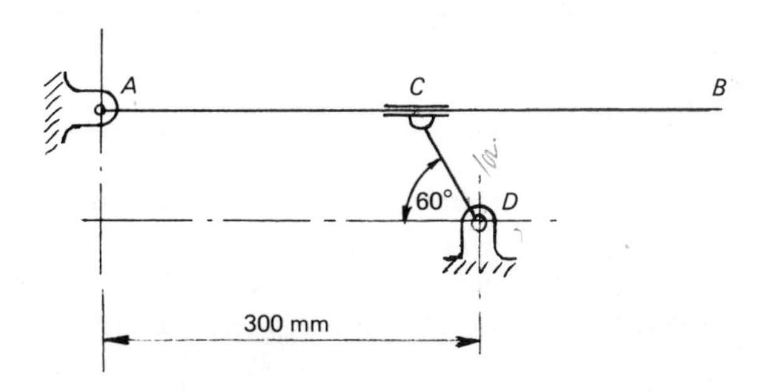

The mechanism shown consists of a link AB freely pinned at A, a crank CD freely pinned at D, and a slider attached to the end of the crank and through which the link AB passes. At the instant shown in the figure the crank is being driven at an angular speed of 10 rad s^{-1} anticlockwise and an angular acceleration of 100 rad s^{-2} clockwise.

Determine the velocity and acceleration of the slider along the link AB, the angular velocity and angular acceleration of the link AB, and the velocity and acceleration of B.

AB is 500 mm and CD is 100 mm.

U. Lond. K.C.

Solution 1.17

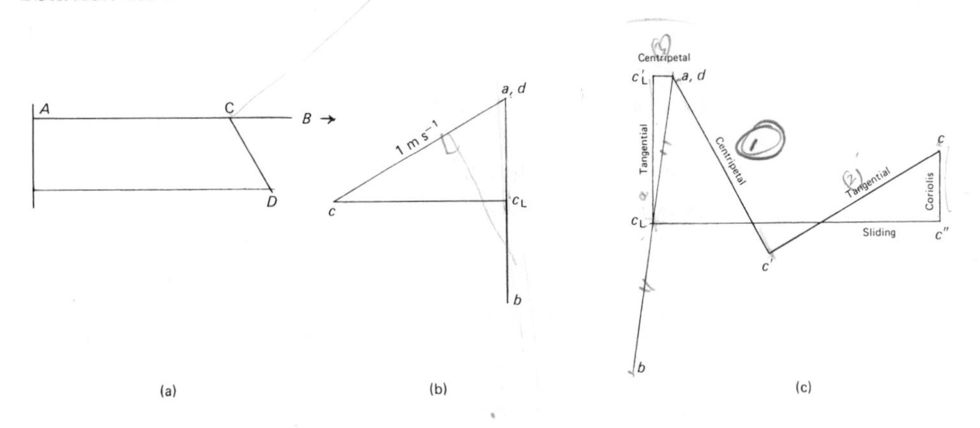

(a), (b) and (c) are respectively the configuration, velocity and acceleration diagrams.

Configuration

This is simple, but to save space, AB has not been drawn to its full length. A simple calculation shows that AC will be 250 mm; hence C is the mid-point of AB.

Velocity

We call 'c_L' the point on AB. Velocity of $C = \omega r = 10 \times 0.1 = 1$ m s^{-1}.

Draw $ac = 1$ m s^{-1} perpendicular to CD. Draw the horizontal through c and line through a perpendicular to AB to locate c_L. Complete the velocity image by extending ac_L to b.

Scaling, or calculating from the velocity diagram:

$ac_L = 0.5$ m s^{-1}.
$cc_L = 0.866$ m s^{-1}.

$$\omega_{AB} = \frac{ac_L}{AC} = \frac{0.5}{0.25} = 2.0 \text{ rad s}^{-1}.$$

Acceleration

Centripetal acceleration $CD = \omega^2 r = 10^2 \times 0.1 = 10$ m s^{-2}.
Tangential acceleration $CD = \alpha r = 100 \times 0.1 = 10$ m s^{-2}.
Centripetal acceleration $C_L A = v^2/r = (0.5)^2/0.25 = 1$ m s^{-2}.
Coriolis acceleration $CC_L = 2v\omega = 2 \times 0.866 \times 2 = 3.464$ m s^{-2}.

Locate the point a, d. Draw the centripetal vector $dc' = 10$ m s^{-2} in the direction $C \rightarrow D$. Draw the tangential vector $c'c$ perpendicular to CD 'up-right' (in accordance with clockwise angular acceleration). Draw the centripetal vector ac'_L = 1 m s^{-2} in the direction $C_L A$. The sliding velocity vector cc_L is left-to-right. Link AB is turning clockwise (the velocity vector ac_L is downwards). Hence the Coriolis vector cc'' is 'left-to-right turned clockwise through 90°', i.e. vertically down. Draw the sliding vector of unknown length horizontally (i.e. parallel to AB). Draw the tangential vector through A perpendicular to AB to locate c_L. Join ac_L. Extend it (acceleration image) to b. (Since C is the mid-point of AB, $ac_L = c_L b$.)

ANSWERS

From the velocity diagram:

Sliding velocity of slider is cc_L = 0.866 m s^{-1}.

$\omega_{AB} = 2.0$ rad s^{-1}.

Velocity of $B = ab$ = 1 m s^{-1}.

From acceleration diagram:

Sliding acceleration of slider is $c''c_L$ = 14.6 m s^{-2}.

$$\text{Angular acceleration of } AB = \frac{\text{tangential component } c'_L c_L}{AC} = \frac{7.1}{0.25} = 28.4 \text{ rad s}^{-2}.$$

Acceleration of b is ab: scales 14.7 m s^{-2}.

Example 1.18

The mechanism shown in the diagram consists of a link AB, pivoted to a fixed point at A and driven by a crank CD through a slider at C, and a link BE, the end E of which is attached to a slider constrained to move only in a horizontal direction.

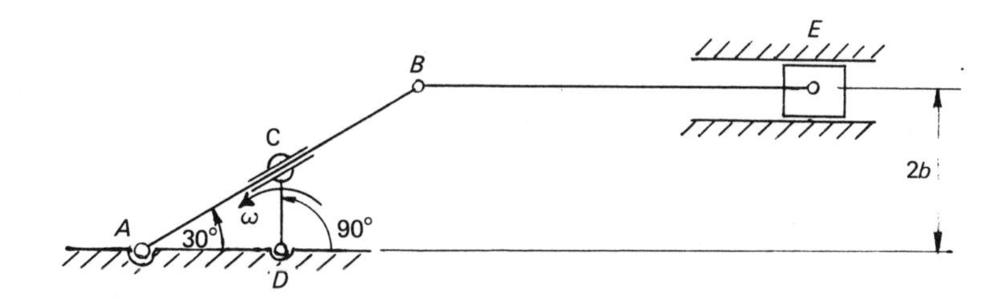

At a certain instant the mechanism is in the position shown and the crank CD is rotating about D with a constant angular velocity ω anticlockwise. Determine, in terms of b, the length of the crank, and ω, (a) the angular velocity and angular acceleration of the link AB, (b) the angular velocity of the link BE, and (c) the velocity and acceleration of the slider at E.

$AB = 4b, AD = \sqrt{3b}, CD = b, BE = 5b.$

U. Lond. K.C.

Solution 1.18

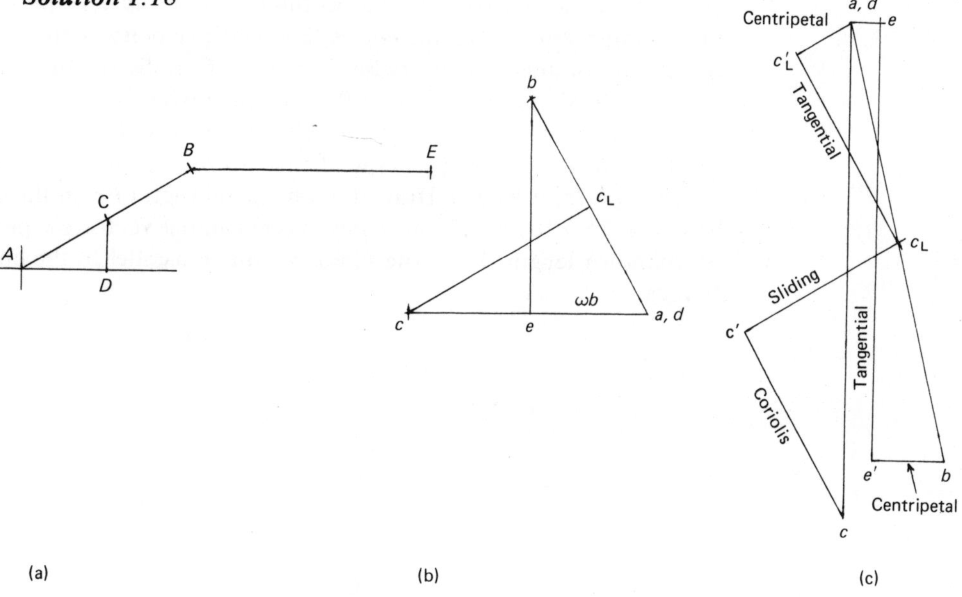

The three diagrams for the solution are shown here. In such a case, the configuration diagram is hardly necessary, as all members are either horizontal, or inclined at 30°; however, it is included for the sake of completeness.

Velocity

Velocity of $C = \omega r = \omega b$.

Locate the point a, d. Draw the vector dc perpendicular to DC. An arbitrary length of 5 cm is a reasonable scale. Draw the vector cc_L parallel to AB of unknown length. Draw vector ac_L perpendicular to AB to locate point c_L. Draw the velocity image of AC_LB. (The simple trigonometry of the configuration should show you that AC_L is $2b$, without having to scale.) Draw vector be perpendicular to BE of unknown length. Draw horizontal vector through a, d to locate e.

Scaling from the velocity diagram,

$$ac_L = \tfrac{1}{2}\omega b; \quad cc_L = \tfrac{1}{2}\sqrt{3\omega b}; \quad \omega_{AB} = \frac{ac_L}{AC} = \frac{\tfrac{1}{2}\omega b}{2b} = \tfrac{1}{4}\omega; \quad be = \tfrac{1}{2}\sqrt{3}\omega b_L.$$

Acceleration

Centripetal acceleration $CD = \omega^2 b$.
Tangential acceleration $CD = 0$.

Centripetal acceleration $C_L A = v^2/r = \dfrac{(ac_L)^2}{AC} = \dfrac{(\frac{1}{2}\omega b)^2}{2b} = \frac{1}{8}\omega^2 b$.

Centripetal acceleration $EB = v^2/r = \dfrac{(be)^2}{BE} = \dfrac{(\frac{1}{2}\sqrt{3}\omega b)^2}{5b} = 0.15\omega^2 b$.

Coriolis acceleration $CC_L = 2v\omega = 2(cc_L)\,\omega_{AB} = 2 \times \frac{1}{2}\sqrt{3}\omega b \times \frac{1}{4}\omega = 0.433\omega^2 b$.

Locate the point a, d. Draw the centripetal vector dc in the direction $C \rightarrow D$ (i.e. vertically down). (An arbitrary length of 10 cm is suitable for A4 paper.) Draw the centripetal vector ac'_L in the direction $C_L \rightarrow A$. (Length is according to the scale adopted, i.e. $\omega^2 b \equiv 10$ cm; hence ac'_L is $\frac{1}{8} \times 10 = 1.25$ cm.) Draw the Coriolis vector cc' of length (0.433×10) in the direction 'up–left' perpendicular to AC_L. (The sliding velocity vector cc_L is 'up–right'; velocity vector ac_L shows the link AC to be rotating anticlockwise; hence the Coriolis vector is in the direction of the velocity vector cc_L rotated 90° anticlockwise, i.e. 'up–left' as stated.) Draw the sliding acceleration vector through c' parallel to AC of unknown length. Draw the tangential vector $c'_L c_L$ perpendicular to AC to locate c_L. Join ac_L. Draw the image $ac_L b$ ($c_L b = ac_L$). Draw the centripetal vector be' in the direction $E \rightarrow B$ of length $(0.15 \times 10) = 1.5$ cm. Draw the tangential vector $e'e$ perpendicular to BE of unknown length. Draw the sliding vector ae parallel to the horizontal slide at E to locate e.

ANSWERS

Part (a)

Scaling from the acceleration diagram gives $c'_L c_L = 0.44\omega^2 b$.

$\omega_{AB} = \frac{1}{4}\omega$.

$$\alpha_{AB} = \frac{c'_L c_L}{AC} = \frac{0.44\omega^2 b}{2b} = \underline{0.22\omega^2}.$$

Part (b)

$$\omega_{BE} = \frac{be}{BE} = \frac{\frac{1}{2}\sqrt{3}\omega b}{5b} = \underline{0.1732\omega}.$$

Part (c)

From velocity diagram, ae scales $\frac{1}{2}\omega b$.
From acceleration diagram, ae scales $0.062\omega^2 b$.
$\therefore$ Velocity of slider at $E = \underline{\frac{1}{2}\omega b}$.
Acceleration of slider at $E = \underline{0.062\omega^2 b}$.

Example 1.19

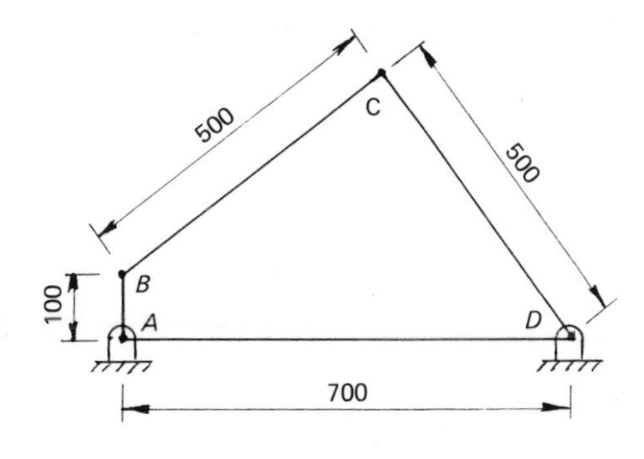

All dimensions in mm

In the four-bar link mechanism shown in the diagram, AB rotates at a uniform speed of 5 rad s^{-1} and CD oscillates about D. Determine the angular velocity and angular acceleration of the link CD for the position shown.

U. Lond. U.C.

Solution 1.19

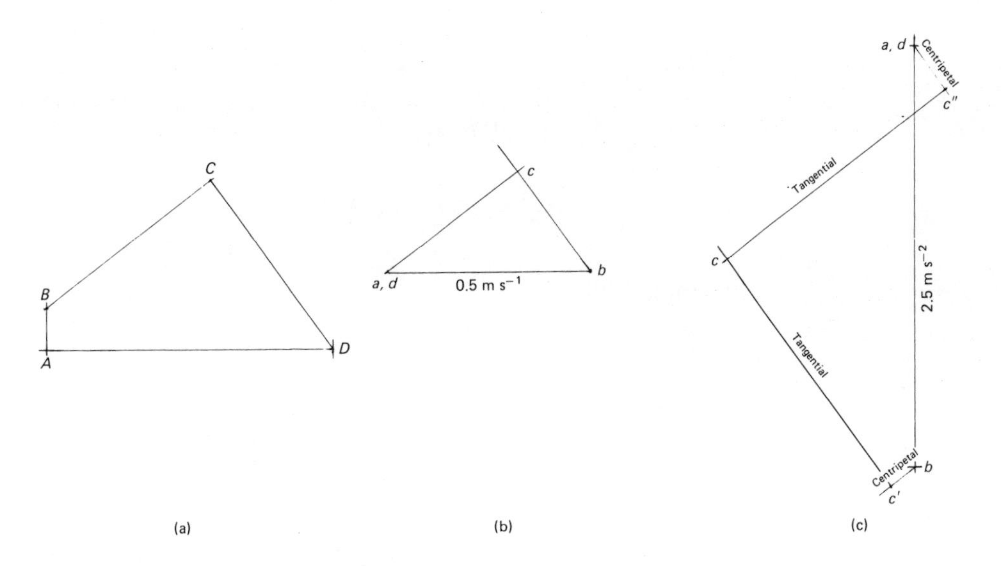

(a) (b) (c)

The mechanism is an example of the simple four-bar chain. Although the figure does not so specify, it is clear that the link AB is vertical, and it is necessary to assume this in order to draw the configuration diagram. The configuration is so chosen to make the angle at C a right angle.

Configuration

Draw AD horizontal and AB vertical. Locate C by drawing arcs radius 500 mm from B and D.

Velocity

Velocity of $B = \omega r = 5 \times 0.1 = 0.5$ m s^{-1}.
A suitable scale is 1 cm $\equiv$ 0.1 m s^{-1}.

Locate point a, d. Draw ab perpendicular to AB length 0.5 m s^{-1}. Draw bc perpendicular to BC of unknown length. Draw dc perpendicular to DC to locate c.

ac scales 0.4 m s^{-1}.
bc scales 0.3 m s^{-1}.

Acceleration

Centripetal acceleration of B relative to $A = \omega^2 r = 5^2 \times 0.1 = 2.5$ m s^{-2}.

Centripetal acceleration of C relative to $B = v^2/r = \dfrac{(bc)^2}{BC} = \dfrac{(0.3)^2}{0.5} = 0.18$ m s^{-2}.

Centripetal acceleration of C relative to $D = v^2/r = \dfrac{(cd)^2}{CD} = \dfrac{(0.4)^2}{0.5} = 0.32$ m s^{-2}.

Locate the point a, d. Draw the centripetal vector ab = 2.5 m s^{-2} in the direction $B \rightarrow A$. A scale of 4 cm $\equiv$ 1 m s^{-2} is suitable. Draw the centripetal vector bc' = 0.18 m s^{-2} in the direction $C \rightarrow B$. Draw the centripetal vector dc'' = 0.32 m s^{-2} in the direction $C \rightarrow D$. Draw the tangential vector $c''c$ perpendicular to CD of unknown length. Draw the tangential vector $c'c$ perpendicular to BC to locate c.

$c''c$ scales 1.66 m s^{-2}.

ANSWERS

$$\omega_{CD} = \frac{dc}{DC} = \frac{0.4}{0.5} = \underline{0.8 \text{ rad s}^{-1}.}$$

$$\alpha_{CD} = \frac{c''c}{CD} = \frac{1.66}{0.5} = \underline{3.32 \text{ rad s}^{-2}.}$$

Example 1.20

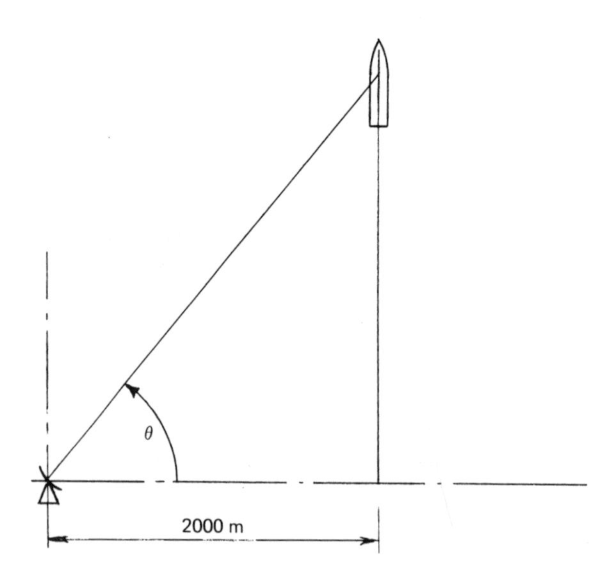

A rocket is fired vertically from the surface of the earth and is tracked at a station 2000 m from the point of firing as shown in the diagram.

At a certain instant, readings of $\theta = 60°$, $\dot{\theta} = 0.035$ rad s^{-1} and $\ddot{\theta} = 0.0055$ rad s^{-2} are recorded at the tracking station. Determine for this instant the height, velocity and acceleration of the rocket.

U. Lond. K.C.

Solution 1.20

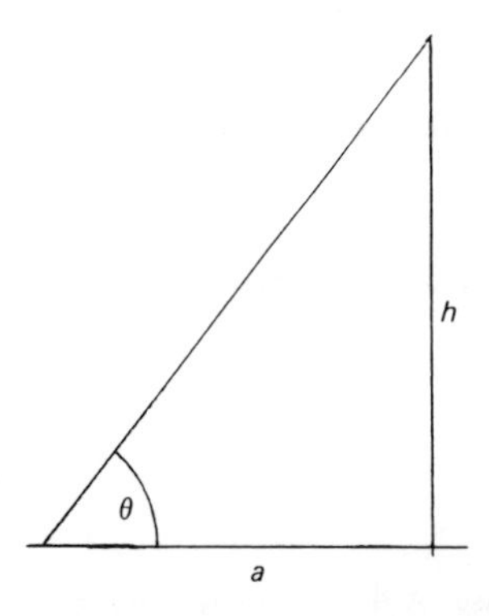

Calling the vertical height h and the horizontal distance from station to firing-point a, we see from the diagram that

$$h = a \tan\theta. \tag{1}$$

Velocity and acceleration may be determined by direct differentiation. Using the notation $dh/dt = \dot{h}$ and $d^2h/dt^2 = \ddot{h}$,

$$\dot{h} = a \sec^2\theta\,(\dot{\theta}), \tag{2}$$

(noting carefully that we must differentiate with respect to t; thus $\dot{h} = (dh/d\theta \times d\theta/dt)$).

Using the 'product' formula $d(u, v) = u\,dv + v\,du$,

$$\ddot{h} = a\,(\sec^2\theta)\,\ddot{\theta} + a\dot{\theta}\,(2\sec\theta)(\sec\theta\tan\theta)\,\dot{\theta}$$

$$= a\ddot{\theta}\sec^2\theta + a(\dot{\theta})^2(2\sec^2\theta\tan\theta). \tag{3}$$

Substituting given data in equation 1,

$$h = 2000 \tan 60°$$

$$= \underline{3464.1 \text{ m.}}$$

and in equation 2,

$$\dot{h} = 2000 \sec^2 60° \times 0.035$$

$$= \underline{280 \text{ m s}^{-1}.}$$

and in equation 3,

$$\ddot{h} = 2000\,[0.0055 \sec^2 60° + 2(0.035)^2 \sec^2 60° \tan 60°]$$

$$= \underline{77.95 \text{ m s}^{-2}.}$$

Example 1.21

The diagram shows a mechanism in which the light straight rod *BCD* moves horizontally in frictionless guides and the rod *OE* rotates about a fixed axis *O*. When $\angle NOC = 30°$ (as shown) the angular velocity of *OE* is 2 rad s^{-1} clockwise and its angular acceleration is 8 rad s^{-2}

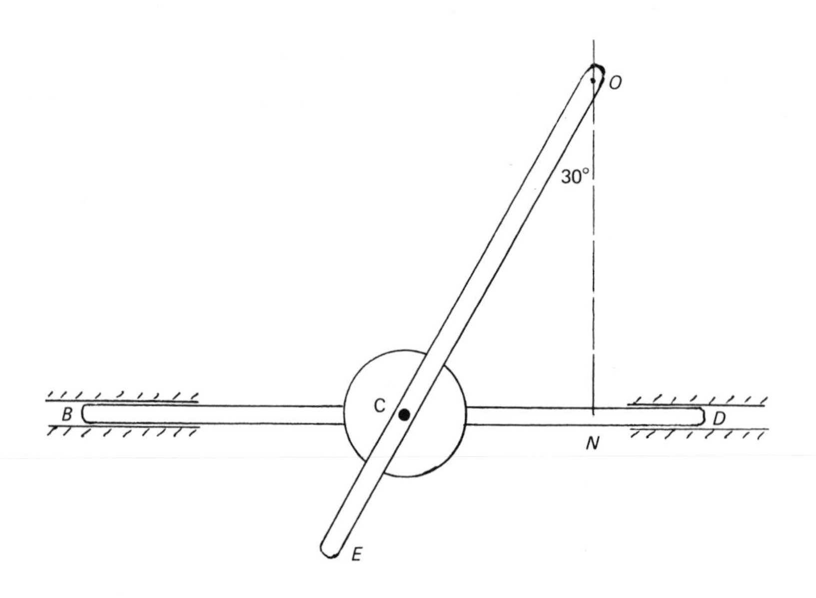

clockwise. A disc is freely pinned to BCD at C and the rod OE slides through the disc without friction. Given that ON = 0.26 m and OE = 0.4 m determine the velocity and acceleration of the rod BCD.

U. Lond. U.C.

Solution 1.21

The description and drawing of the device suggests solution by drawing configuration, velocity and acceleration diagrams. The problem may be solved in this way, but with some simple mechanisms an analytical solution is simpler, and this is such a case. (See also Examples 1.4, 1.8 and 1.10.) If we call 'h' the constant distance ON and 'x' the displacement of the point C from N, we may represent the essentials of the device as in the diagram, and we may write

$$x = h \tan \theta .$$

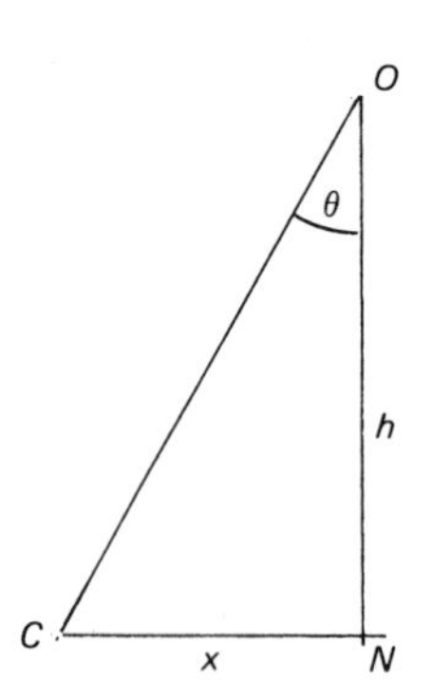

The analysis is seen to be identical to the previous example (Example 1.20) and the procedure is the same.

Differentiating twice with respect to t,

$$\dot{x} = h \sec^2 \theta \, (\dot{\theta}). \tag{1}$$

$$\ddot{x} = h \sec^2 \theta \, (\ddot{\theta}) + h\dot{\theta} \, (2 \sec \theta \sec \theta \tan \theta) \, (\dot{\theta})$$

$$= h \, [\ddot{\theta} \sec^2 \theta + (\dot{\theta})^2 \, (2 \sec^2 \theta \tan \theta)] . \tag{2}$$

Substituting in equation 1,

$$\dot{x} = 0.26 \sec^2 30° \times 2 = \underline{0.693 \text{ m s}^{-1}}$$

and in equation 2,

$$\ddot{x} = 0.26(8 \sec^2 30° + 2^2 \times 2 \sec^2 30° \tan 30°) = \underline{4.375 \text{ m s}^{-2}}.$$

Example 1.22

An atomiser consists of a horizontal disc of radius R with radial ribs on its upper face. The disc rotates about a vertical axis through its centre with an angular velocity ω. Liquid of density ρ flows out of a vertical pipe on to the centre of the disc and forms a fine mist on ejection from the periphery.

Derive an expression for the torque required to drive the disc in terms of the given quantities and the volume flow rate Q.

Obtain also an expression for the gain in kinetic energy of unit mass of the fluid as it passes across the disc.

U. Lond. U.C.

Solution 1.22

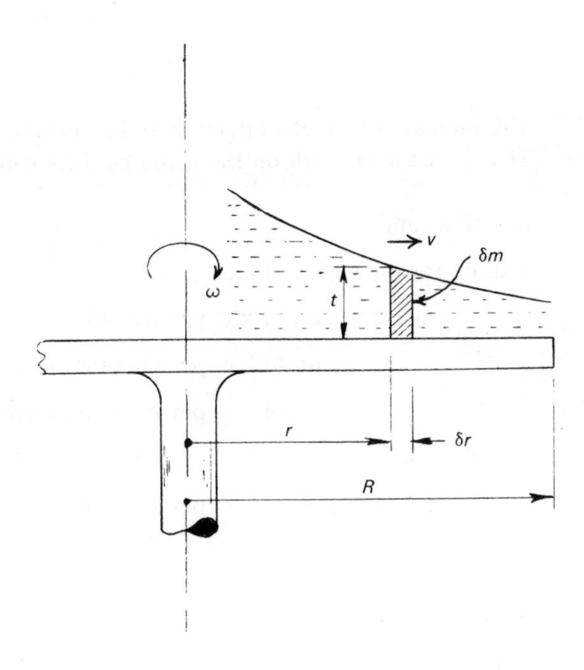

The diagram shows a cross-section of the spinning plate with the fluid flowing outwards. We will consider an element of fluid comprising a thin ring, of inner radius r, thickness δr and height t at that point. The mass of this element, δm, is its volume multiplied by its density.

$$\therefore \; \delta m = 2\pi r \times \delta r \times t \times \rho.$$

Because the fluid is moving outwards across the plate, while the plate is turning, a Coriolis acceleration is present.

The radial velocity, v, is related to the flow rate Q. At radius r, the circumferential area across which the fluid flows outwards is $2\pi rt$.

$$\therefore \; Q = 2\pi rt \times v.$$

$$\therefore\ v = \frac{Q}{2\pi rt}.$$

Coriolis acceleration = $2v\omega$.

For a mass δm the corresponding force exerted on the fluid by the plate, δF, is

$$\delta F = 2v\omega\ \delta m$$

$$= \frac{2Q\omega}{2\pi rt} 2\pi r\ \delta r t\rho$$

$$= 2Q\omega\rho\ \delta r.$$

The corresponding torque due to the element, δT, is

$$\delta T = 2Q\omega\rho\ \delta r \times r.$$

The total torque T for all such elements, is therefore

$$T = \Sigma(2Q\omega\rho r\ \delta r)$$

$$= 2Q\omega\rho \int_0^R (r)\ \mathrm{d}r$$

$$= 2Q\omega\rho\ (\tfrac{1}{2}R^2).$$

$$T = \underline{Q\omega\rho R^2}.$$

The increase of kinetic energy may be determined from a simple energy equation. The torque does work on the spinning fluid which thereby gains k.e.

In one second:

w.d. by torque = gain of k.e.

$\therefore\ T\omega$ = gain of k.e. per second

= gain of k.e. per kilogram × kilogram per second

= gain of k.e. per kilogram × $Q\rho$.

$\therefore\ Q\omega^2\rho R^2$ = gain of k.e. per kilogram × $Q\rho$

$\underline{\omega^2 R^2 = \text{gain of k.e. per kilogram.}}$

Example 1.23

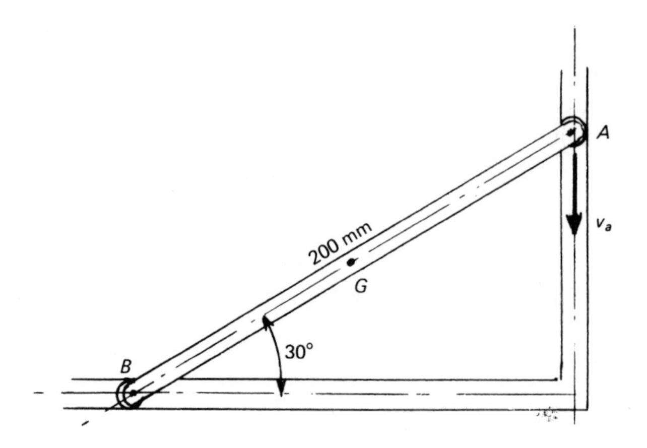

A link AB is guided by the vertical and horizontal slots, as shown in the diagram. End A is moving at constant velocity of 4 m s^{-1}. Find, by any method,

(a) the velocity of end B and of the mid-point G;

(b) the angular velocity of AB, and

(c) the acceleration of B.

Thames Poly.

Solution 1.23

Earlier examples have shown that with very simple mechanisms the kinematics may be determined by direct differentiation of displacement instead of by drawing velocity and acceleration diagrams. However, in this particular case, a diagrammatic solution probably offers the simpler solution, principally because the question also requires the velocity of the mid-point of the rod. Also, we will see that the resulting diagrams are so simple that accurate drawings are not required; simple sketches suffice.

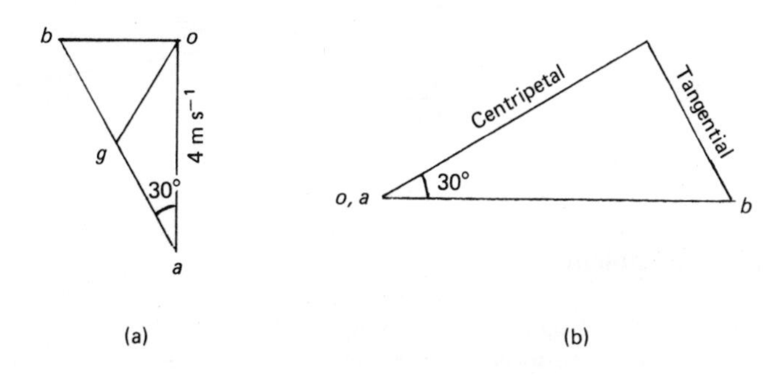

Diagrams (a) and (b) show respectively the velocity and acceleration diagrams, merely as sketches.

Part (a)

From the velocity diagram:

$$ob = 4 \tan 30° = \underline{2.309 \text{ m s}^{-1}}.$$

$$ab = 4 \sec 30° = 4.619 \text{ m s}^{-1}.$$

$$og = \tfrac{1}{2}ab = \underline{2.309 \text{ m s}^{-1}}.$$

(A circle drawn on ab as diameter would pass through o and would have g as centre; hence gb and go would be equal radii.)

Part (b)

$$\omega_{AB} = \frac{ab}{AB} = \frac{4.619}{0.2} = \underline{23.095 \text{ rad s}^{-1}}.$$

Part (c)

Centripetal acceleration of B relative to $A = \dfrac{v^2}{r}$

$$= \frac{(ab)^2}{AB} = \frac{(4.619)^2}{0.2} = 106.7 \text{ m s}^{-2}.$$

In diagram (b), o and a are coincident, as point A moves with *constant* velocity. ab must be horizontal. Hence

$$ab = ab \sec 30° = 106.7 \sec 30° = \underline{123.2 \text{ m s}^{-2}}.$$

1.3 Problems

Problem 1.1

The diagram shows a slider–crank mechanism. The crank AB is 30 mm long, the connecting-rod BC is 120 mm long, and the crank rotates at a constant speed of 750 rev min^{-1}. Determine the piston velocity and acceleration, the angular velocity and acceleration of the connecting-rod and the velocity and acceleration of G, the mass centre of the connecting-rod, which is 40 mm from the crank-pin end.

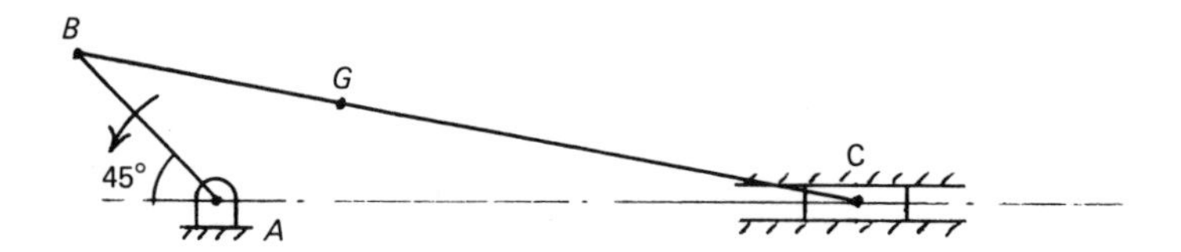

Problem 1.2

The diagram shows a simple mechanism for a press. If the crank AB turns clockwise at a uniform speed of 30 rev min^{-1}, determine the velocity of the slider at E when the angle θ is (a) 60°; (b) 180°. Determine the maximum force that the slider could overcome in both positions, if the maximum torque transmitted to the crank OA is 80 N m.

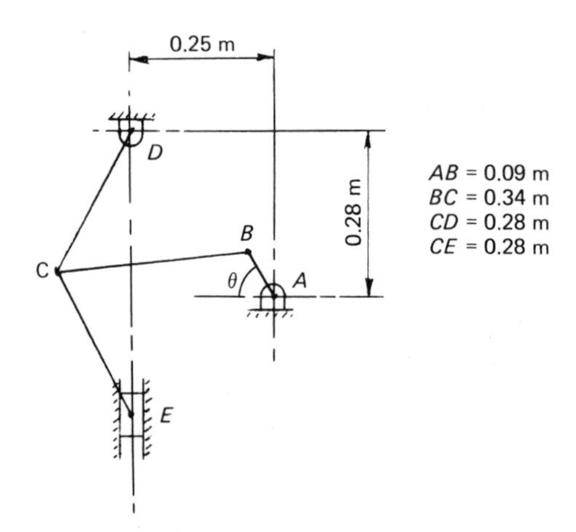

Problem 1.3

The mechanism shown in the diagram is called the elliptic trammel. Determine, either graphically or analytically, the velocity and acceleration of the end B in the vertical slide when end A has a displacement x from O of 0.4 m, a velocity left-to-right of 0.2 m s^{-1} and an acceleration right-to-left of 0.8 m s^{-2}. The length of AB is 1 m.

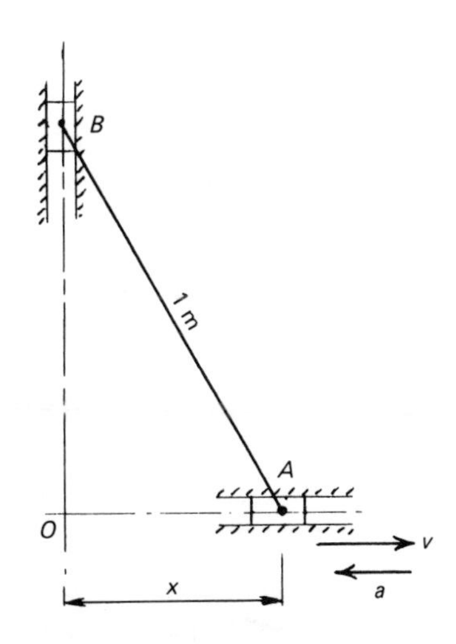

Problem 1.4

The diagram shows the configuration of a simple slider–crank mechanism. The length of the crank *AB* is *r*, and that of the connecting-rod *BC* is *kr*. Derive expressions for the angular velocity and angular acceleration of the connecting-rod when the crank turns with a constant angular velocity ω.

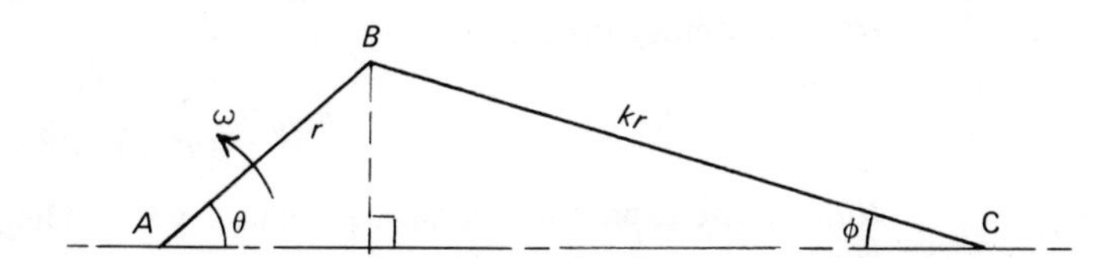

Problem 1.5

In the diagram is shown a simple mechanism for operating a pump. The piston at *F* is operated by a crank *AB*, connecting-rod *BC*, swinging beam *CDE*, and pump rod *EF*. For the particular configuration shown, *AB* and *CDE* are both horizontal. Determine the velocity of the piston when *AB* turns clockwise at a constant speed of 4 rad s^{-1}.

Hint: It is not necessary to draw accurate diagrams; sketches will be found sufficient.

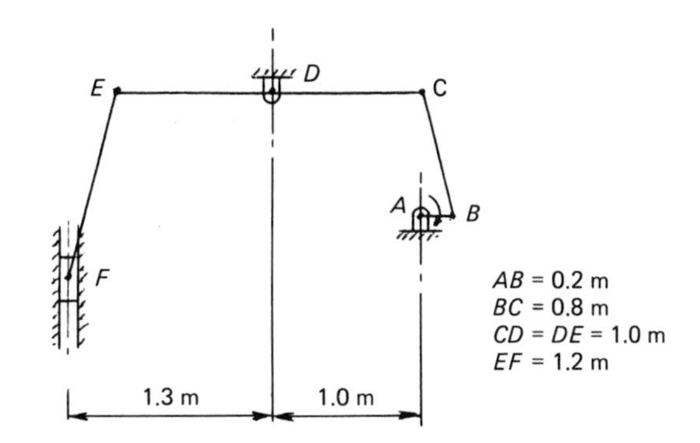

Problem 1.6

The diagram shows the outline of a crank, connecting-rod and piston mechanism called the oscillating-cylinder mechanism, or the trunnion engine. The crank, of length r, turns about the fixed point O. The cylinder is arranged to pivot about a transverse pin, or trunnion, so that as the crank turns the cylinder rocks up and down. The trunnion centre lies on the horizontal through the crank centre, O. The crank turns with constant angular velocity ω. Derive expressions for the angular velocity and acceleration of the connecting-rod.

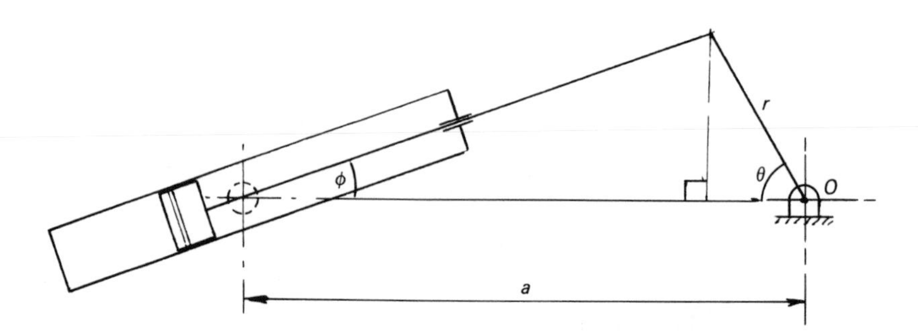

Problem 1.7

The end B of a light rigid rod AB is constrained to move in a vertical guide, the rod passing through a swivel at C which is fixed in position but permits the rod to turn. Prove that when the end B has displacement y and velocity $\dot{y}$ relative to C, the Coriolis acceleration at the swivel is given by the expression

$$a_{\text{Cor}} = \frac{(\dot{y})^2\, 2hy}{(h^2 + y^2)^{3/2}}\,.$$

Hint: Derive expression for angular velocity $\dot{\phi}$ and sliding velocity at the swivel, $\dot{x}$ and calculate Coriolis component from $2\dot{x}\dot{\phi}$.

Displacement x will be distance $BC - h$.

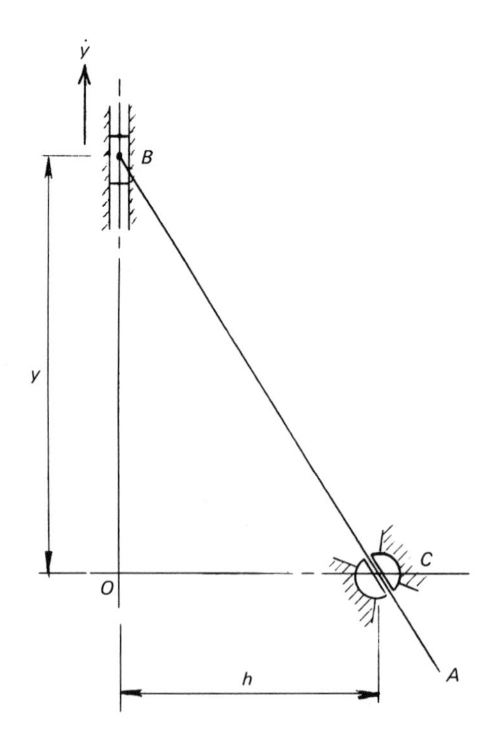

Problem 1.8

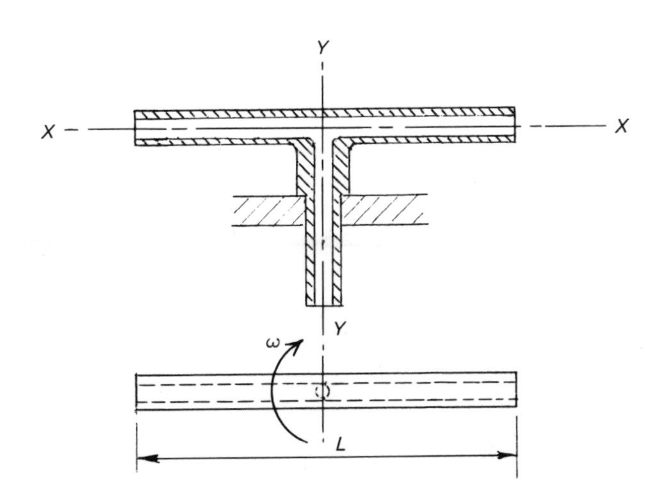

Two views of a tee-shaped pipe are shown in the diagram. The horizontal limb of the tee is of uniform circular-section bore and total length L and the pipe is designed to rotate clockwise about the vertical axis Y-Y. While the pipe is rotating, water is pumped up the vertical limb at a constant rate of Q m^3 s^{-1} and the flow divides equally along the horizontal limbs. Prove that under these conditions a torque T will be necessary to sustain a uniform velocity of rotation ω given by the expression:

$$T = \tfrac{1}{4}\rho Q \omega L^2$$

where ρ is the density of the water. Show by a simple diagram that this torque will be in the same direction as that of rotation.

If the pipe length L is 8 m, the bore diameter is 75 mm and the speed of rotation is 4 rev min^{-1}, calculate the torque required if water is pumped through at a rate of 20 000 kg per hour.

Hint: Consider half-flow through one limb only. Consider element of water at radius r, thickness δr, and calculate Coriolis force, and hence torque due to this. Integrate (0 to $\tfrac{1}{2}L$) and double (for both limbs).

Problem 1.9

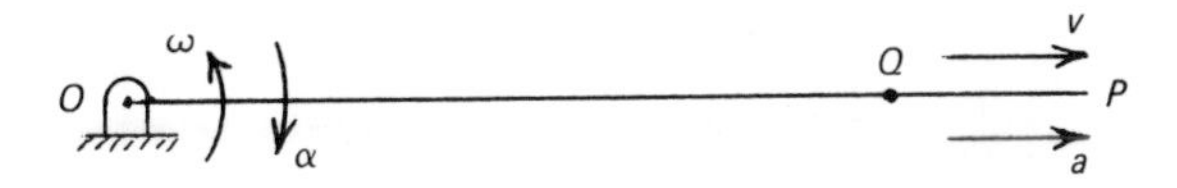

A rod OP is turning with angular velocity ω and constant angular acceleration $\alpha = 0.4$ rad s^{-2}. A point Q moves along OP with relative velocity $v = 0.2$ m s^{-1} and constant acceleration $a = 0.4$ m s^{-2}, all velocities and accelerations in the direction shown in the diagram. At the instant shown in the diagram when OP is horizontal, the total acceleration of Q relative to O is zero. Determine the magnitude only of the total acceleration of Q relative to O one second later.

Hints: Sketching the first acceleration diagram will enable ω and radius OQ to be calculated. (Answer: $\omega = 0.7368$ rad s^{-1}; $OQ = 0.7368$ m.) Use elementary kinematic equations to calculate the new angular velocity of OP (Answer: 0.3368 rad s^{-1}), the new sliding velocity of Q (Answer: 0.6 m s^{-1}) and the new radius OQ' (Answer: 1.1368 m). Calculate the centripetal, tangential and Coriolis components and evaluate the resultant.

Problem 1.10

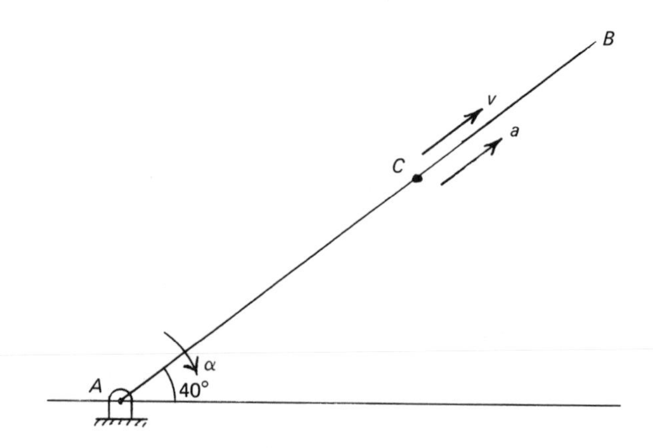

A rod AB turns with angular velocity ω and angular acceleration $\alpha = 2.5$ rad s^{-2} clockwise. A point C on the rod at a radius of 2 m has velocity $v = 1.5$ m s^{-1} outwards, relative to the rod, and an acceleration $a = 3.2$ m s^{-2} outwards. Calculate two values of ω such that at the instant shown, when AB is 40° to the horizontal, the total acceleration of C relative to A is vertical. Calculate this total acceleration in both cases.

Hint: See Example 1.2.

1.4 Answers to Problems

1.1 1.3 m s^{-1}; 134 m s^{-2}; 14.11 rad s^{-1}; 1072 rad s^{-2}; 1.9 m s^{-1}; 158 m s^{-2}.

1.2 (a) 0.305 m s^{-1}; 824 N. (b) 0; infinite force.

1.3 $\dot{y} = 0.0873$ m s^{-1} downwards; $\ddot{y} = 0.2971$ m s^{-2} upwards.

1.4 $\dot{\phi} = \omega\left(\dfrac{\cos\theta}{\sqrt{k^2 - \sin^2\theta}}\right)$; $\ddot{\phi} = (\omega)^2\left(\dfrac{\sin\theta\,(1 - k^2)}{(k^2 - \sin^2\theta)^{3/2}}\right)$.

1.5 0.8 m s^{-1}.

1.6 $\dot{\phi} = \omega\left(\dfrac{ar\cos\theta - r^2}{a^2 + r^2 - 2ar\cos\theta}\right)$;

$$\ddot{\phi} = (\omega)^2\left(\frac{ar\sin\theta\,(r^2 - a^2)}{(a^2 + r^2 - 2ar\cos\theta)^2}\right).$$

1.8 37.23 N m.

1.9 0.2757 m s^{-2}.

1.10 ω = 1.394 rad s^{-1} anticlockwise, and 2.652 rad s^{-1} clockwise, a = 1.068 m s^{-2} and 16.91 m s^{-2}, both downwards.

2 Kinetics of Rigid Bodies

Pure translation of rigid bodies: vehicles on flat and inclined surfaces. Pure rotation in a plane. Combined translation and rotation.

2.1 The Fact Sheet

(a) General Plane Motion of a Rigid Body

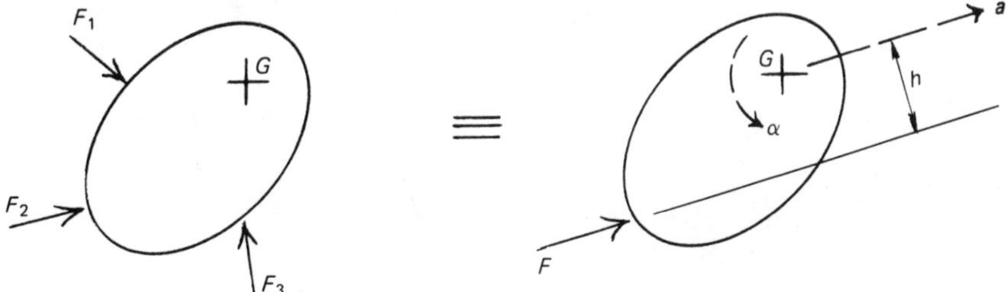

Fig. 2.1

When a rigid body is subjected to a number of forces having a resultant force F (Fig. 2.1) then the motion will consist of the following:

(i) A linear acceleration a of the mass centre G of the body, having the same direction as F, and given by

$$F = ma.$$

(ii) An angular acceleration α given by

$$Fh = I_g\alpha$$

where I_g is the moment of inertia of the body about an axis through G, the mass centre perpendicular to the plane of motion, and h is the perpendicular distance from G to the line of action of F.

(b) Special Case 1: Pure Translation of a Rigid Body

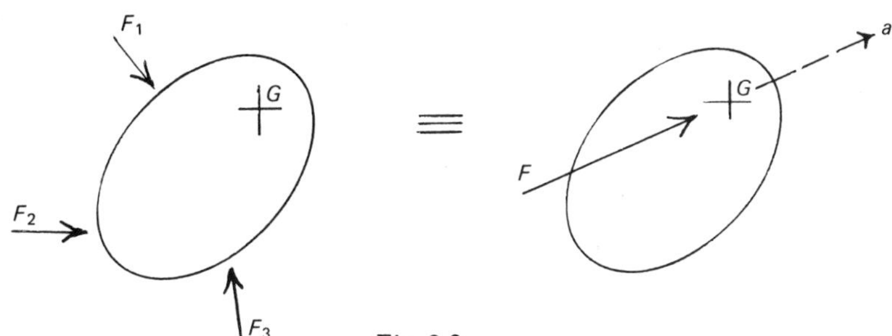

Fig. 2.2

When a rigid body having mass m moves with a linear acceleration a in a known direction (Fig. 2.2) and the angular acceleration is zero, then

(i) the resultant force component ΣF in the direction of a is given by

$$\Sigma F = ma;$$

(ii) the resultant force component in the direction perpendicular to the acceleration will be zero;

(iii) the resultant moment of all forces acting on the body about the mass centre G will be zero.

(c) Special Case 2: Pure Rotation of a Rigid Body

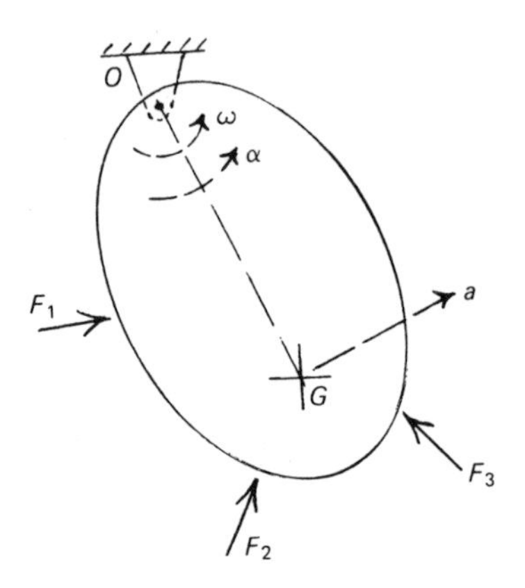

Fig. 2.3

When a rigid body subjected to a number of forces rotates about a fixed point O with an angular velocity ω and an angular acceleration α (Fig. 2.3) the following equations apply.

(i) $\Sigma M_o = I_o \alpha$

where ΣM_o is the resultant moment of all the forces about the turning-point O, and I_o is the moment of inertia of the body about an axis through G perpendicular to the plane of rotation.

(ii) $\Sigma M_g = I_g \alpha$

where ΣM_g is the resultant moment of all the forces about the mass centre G, and I_g is the moment of inertia of the body about an axis through G perpendicular to the plane of rotation.

(iii) $\Sigma F_t = ma$

where ΣF_t is the resultant component of all the forces in the direction of the linear acceleration a of the mass centre G, i.e. tangential to the circle of rotation of G about O.

(iv) $\Sigma F_r = m\omega^2 (OG)$

where ΣF_r is the resultant component of all the forces in the radial direction $G \rightarrow O$.

(v) Kinematically:

$$a = \alpha(OG).$$

(d) Moments of Inertia

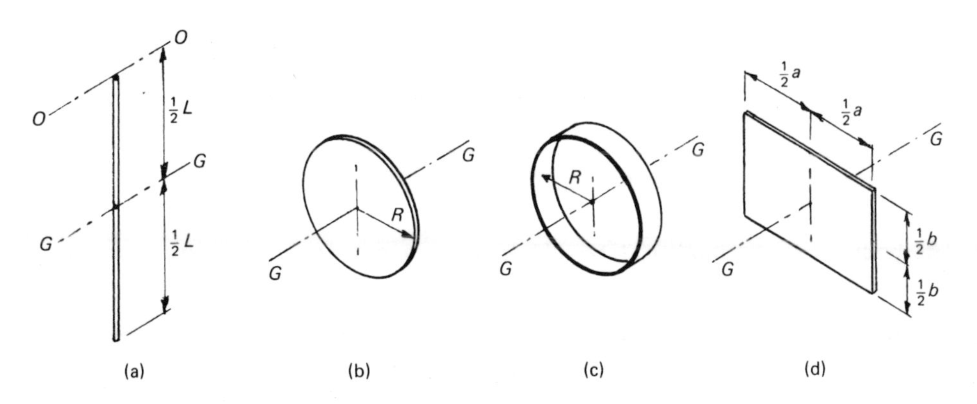

Fig. 2.4

Uniform thin bar of mass m, length L (Fig. 2.4(a)):

$$I_g = \tfrac{1}{12}mL^2; \qquad I_o = \tfrac{1}{3}mL^2.$$

Uniform circular disc of mass m, radius R (Fig. 2.4(b)):

$$I_g = \tfrac{1}{2}mR^2.$$

Uniform thin ring of mass m, radius R (Fig. 2.4(c)):

$$I_g = mR^2.$$

Uniform rectangular lamina mass m, dimensions a, b (Fig. 2.4(d)):

$$I_g = \tfrac{1}{12}m(a^2 + b^2).$$

The parallel-axis theorem:

$$I_o = I_g + mh^2,$$

where I_g is the moment of inertia about an axis through the mass centre G;
I_o is the moment of inertia about any parallel axis through a point O;
h is the perpendicular distance between the two axes.

(e) Energy and Work

Potential energy = mgh

where h is the vertical height of the mass centre of a body above an arbitrary datum.

Kinetic energy of translation = $\frac{1}{2}mv^2$.

Kinetic energy of rotation = $\frac{1}{2}I\omega^2$.

Work done by a torque $M = M\theta$

where θ is the angle turned through in radians.

(f) D'Alembert's Principle

(i) When a body of mass m has a linear acceleration of its mass centre of magnitude a, a 'reversed effective force' may be added to the free-body diagram.

The magnitude of this force is (ma).
Its line of action is parallel to a and passes through the mass centre.
Its direction is opposite to that of a.

(ii) When a body of moment of inertia I_g with respect to an axis through its mass centre G has an angular acceleration α, a 'reversed effective moment' may be added to the free-body diagram.

The magnitude of this moment is ($I_g \alpha$).
Its direction is opposite to that of α.

The body, with the modifications above, may then be treated *as if in static equilibrium*; thus:

Resultant force in any direction = 0.
Resultant moment about any point = 0.

2.2 Worked Examples

Example 2.1

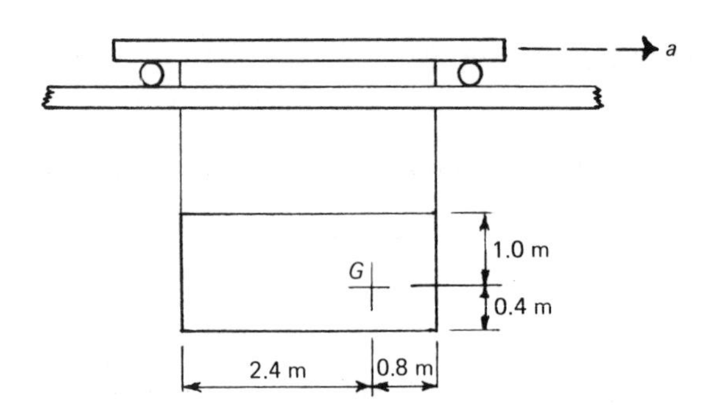

A loaded crate is 3.2 m wide and 1.4 m deep. Its mass is 2200 kg and the mass centre is located as shown in the diagram. It hangs from the beam of a crane by two initially vertical wires of equal length attached to the two upper edges of the crate. Calculate the tension in the wires, and the angle they will make with the vertical, when the crane accelerates uniformly to the right with an acceleration a = 1.2 m s^{-2}.

Solution 2.1

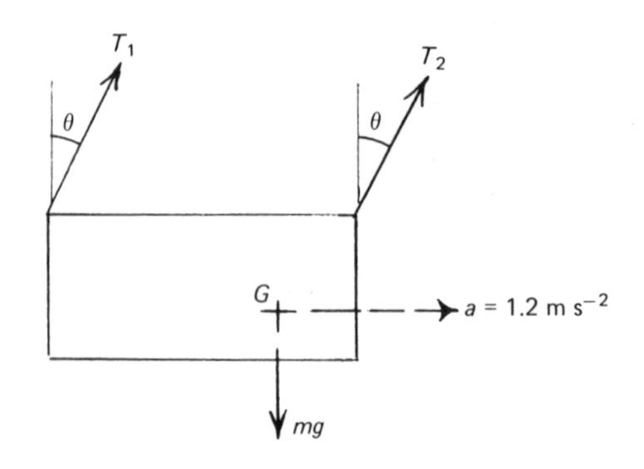

[b] From the free-body diagram shown at the bottom of the previous page it is clear that both wires are inclined at the same angle θ to the vertical.

Vertical equilibrium:

$$(T_1 + T_2) \cos \theta = mg = 2200 \times 9.81 = 21\,582 \text{ N}.$$

Horizontal equation of motion: $\Sigma F = ma$;

$$(T_1 + T_2) \sin \theta = ma = 2200 \times 1.2 = 2640 \text{ N}.$$

We may eliminate $(T_1 + T_2)$ by dividing the second equation by the first:

$$\frac{(T_1 + T_2) \sin \theta}{(T_1 + T_2) \cos \theta} = \frac{2640}{21\,582}.$$

$$\therefore \tan \theta = 0.1223.$$

$$\therefore \theta = \underline{6.974^\circ}.$$

Substituting in the first equation,

$$T_1 + T_2 = \frac{21\,582}{\cos 6.974^\circ} = 21\,743 \text{ N}. \qquad (1)$$

No rotation. $\therefore$ moments about $G = 0$. Resolving T_1 and T_2 into vertical and horizontal components,

$$(T_1 + T_2) \sin \theta \times 1.0 + T_1 \cos \theta \times 2.4 - T_2 \cos \theta \times 0.8 = 0.$$

Using equation 1 to substitute for T_2,

$$21\,743 \sin 6.974^\circ \times 1.0 + T_1 \cos 6.974^\circ \times 2.4 - (21\,743 - T_1) \cos 6.974^\circ \times 0.8 = 0.$$

$$T_1 \cos 6.974^\circ (2.4 + 0.8) = 21\,743 (\cos 6.974^\circ \times 0.8 - \sin 6.974^\circ \times 1.0).$$

$$\therefore T_1 = \frac{21\,743 (0.8 - \tan 6.974^\circ \times 1.0)}{3.2}$$

$$= \underline{4605 \text{ N}}.$$

From equation 1, $T_2 = 21\,743 - 4605$

$$= \underline{17\,138 \text{ N}}.$$

Example 2.2

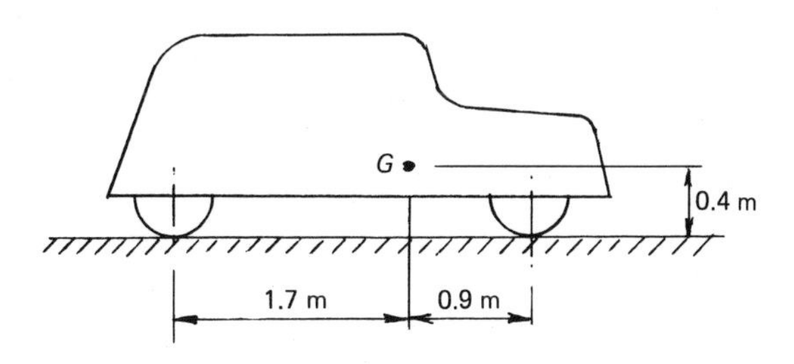

The diagram shows some principal dimensions of a car of total effective mass 1200 kg which is travelling up a slope of $\sin^{-1}$ (0.1).

(a) Calculate the magnitude of the front and rear wheel reaction forces when the car has a forward acceleration of 1.6 m s^{-2}.

(b) If while travelling at 20 m s^{-1} the car is suddenly braked, bringing it to rest with uniform retardation in a distance of 35 m, calculate the magnitude of the front and rear wheel reaction forces.

Solution 2.2

Part (a)

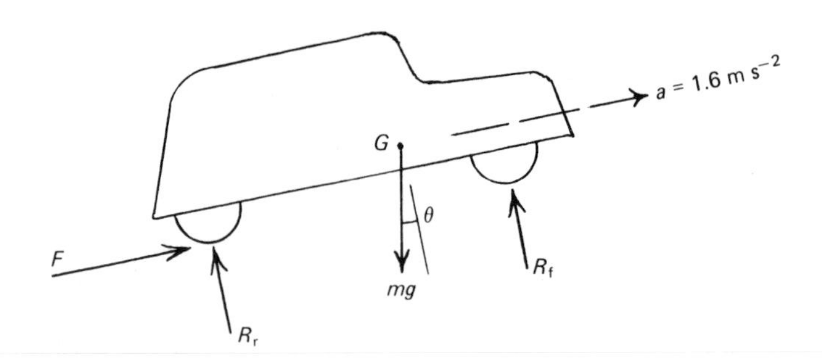

[b]

Shown here is the free-body diagram. *F* is the 'tractive force', or driving force provided by the engine. This is a reaction force between road and driving wheels, and so must act tangentially to the wheels, i.e. along the road surface. Although there are two wheels at front and rear, the wheel reaction forces are shown as a single force in each case; this will be the sum of the forces on the two wheels.

Equation of translation ($\Sigma F = ma$):

$$F - mg \sin \theta = ma$$

$$\therefore \; F = ma + mg \sin \theta$$

$$= 1200 \times 1.6 + 1200 \times 9.81 \times 0.1$$

$$= 3097.2 \text{ N.}$$

Equation of equilibrium perpendicular to the track:

$$R_r + R_f = mg \cos \theta$$

$$= 1200 \times 9.81 \cos (\sin^{-1} (0.1))$$

$$= 11\,713 \text{ N.}$$

No rotation. $\therefore$ moments of forces about $G = 0$.

$$R_r \times 1.7 = R_f \times 0.9 + F \times 0.4.$$

$$(11\,713 - R_f)\, 1.7 = R_f \times 0.9 + 3097.2 \times 0.4.$$

$$\therefore \; R_f = \frac{11\,713 \times 1.7 - 3097.2 \times 0.4}{0.9 + 1.7} = 7182.0 \text{ N.}$$

$$\therefore \; R_r = 11\,713 - 7182 = 4531 \text{ N.}$$

Part (b)

The retardation is calculated as follows:

$$v^2 = u^2 + 2ax.$$

$$a = \frac{v^2 - u^2}{2x}$$

$$= \frac{0 - (20)^2}{2 \times 35}$$

$$= -5.714 \text{ m s}^{-2}, \qquad \text{i.e. backwards.}$$

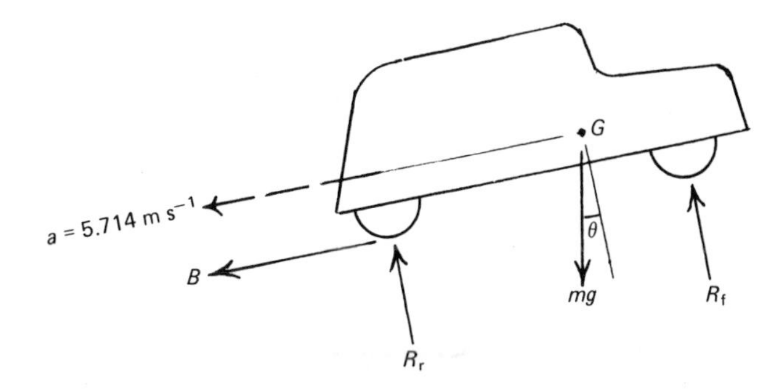

In the free-body diagram the tractive force is replaced by a braking force B acting against the direction of motion.

Equation of translation ($\Sigma F = ma$):

$$B + mg \sin \theta = ma.$$

$$\therefore B = ma - mg \sin \theta$$

$$= 1200 \times 5.714 - 1200 \times 9.81 \times 0.1$$

$$= 5679.6 \text{ N}.$$

Equation of equilibrium perpendicular to the track (exactly as part (a)):

$$R_r + R_f = 11\,713.$$

No rotation. $\therefore$ moments of forces about $G = 0$.

$$\therefore\ R_r \times 1.7 + B \times 0.4 = R_f \times 0.9.$$

$$\therefore\ (11\,713 - R_f)\ 1.7 = R_f \times 0.9 - 5679.6 \times 0.4.$$

$$\therefore\ R_f = \frac{11\,713 \times 1.7 + 5679.6 \times 0.4}{0.9 + 1.7}$$

$$= \underline{8532.3 \text{ N.}}$$

$$\therefore\ R_r = 11\,713 - R_f$$

$$= \underline{3180.7 \text{ N.}}$$

Notice how forward acceleration reduces the weight on the front wheels and adds it to the rear wheels, while braking has the opposite effect.

Example 2.3

A car has a mass of 1175 kg. The track width (i.e. distance between inner and outer wheels) is 2.62 m. The mass centre is 0.54 m above the ground and mid-way between the wheels. The car travels at a constant speed around a circular track of mean radius 26 m, the track being banked inwards at an angle of 20°.

(a) Calculate the limiting speed for the car to begin to tip outwards, i.e. for the inner-wheel reaction force to be zero. Assume that the car does not slip sideways.

(b) Calculate the limiting speed for the car to side-slip outwards, assuming a coefficient of friction between tyres and road of 0.7.

Solution 2.3

[b]

Part (a)

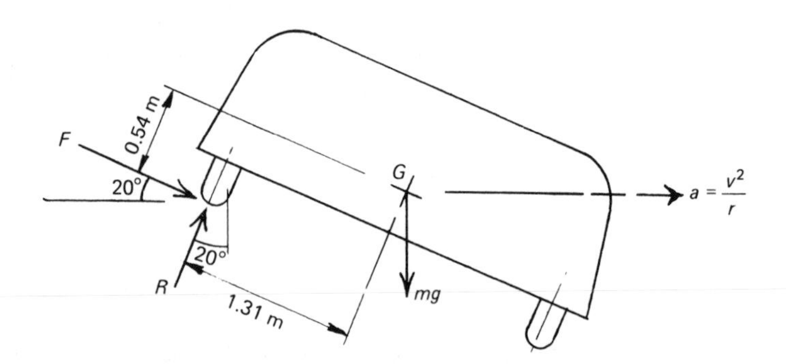

The free-body diagram is shown here. Because the car is about to tip outwards the inner reaction force is zero.

Equation of translation; horizontally ($\Sigma F = ma$):

$$F \cos 20° + R \sin 20° = \frac{mv^2}{r}. \quad (1)$$

Equation of equilibrium: vertically,

$$mg + F \sin 20° = R \cos 20°. \quad (2)$$

No rotation: hence moments of forces about $G = 0$;

$$F \times 0.54 = R \times 1.31. \quad (3)$$

Dividing equation 1 by equation 2 and at the same time substituting for R from equation 3,

$$\frac{\not{F} \cos 20° + \not{F}\left(\frac{0.54}{1.31}\right) \sin 20°}{\not{F}\left(\frac{0.54}{1.31}\right) \cos 20° - \not{F} \sin 20°} = \frac{mv^2}{rmg}.$$

$$\therefore \frac{1 + 0.4122 \tan 20°}{0.4122 - \tan 20°} = \frac{v^2}{26 \times 9.81}.$$

$$\therefore v = \sqrt{23.84 \times 26 \times 9.81} = \underline{77.98 \text{ m s}^{-1}}.$$

Part (b)

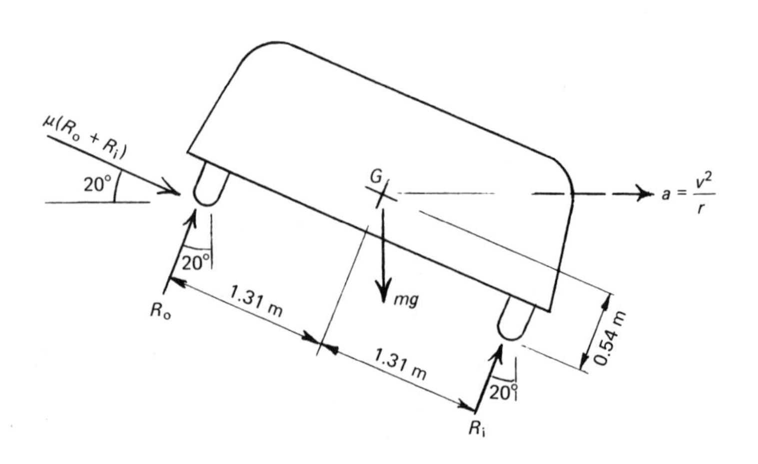

The free-body diagram is shown. Because slipping is imminent, we show the friction force as the product of normal reaction force ($R_i + R_o$) and friction coefficient μ.

Equation of translation: horizontally,

$$\mu (R_o + R_i) \cos 20° + (R_o + R_i) \sin 20° = \frac{mv^2}{r}. \quad (4)$$

Vertical equilibrium:

$$mg + \mu (R_o + R_i) \sin 20° = (R_o + R_i) \cos 20°. \quad (5)$$

Rearranging equation 5 and dividing,

$$\frac{\mu \cancel{(R_o + R_i)} \cos 20° + \cancel{(R_o + R_i)} \sin 20°}{\cancel{(R_o + R_i)} \cos 20° - \mu \cancel{(R_o + R_i)} \sin 20°} = \frac{\cancel{m}v^2}{r\cancel{m}g}.$$

$$\therefore \frac{0.7 + \tan 20°}{1 - 0.7 \tan 20°} = \frac{v^2}{26g}.$$

$$\therefore v = \sqrt{1.428 \times 26 \times 9.81} = \underline{19.08 \text{ m s}^{-1}}.$$

Example 2.4

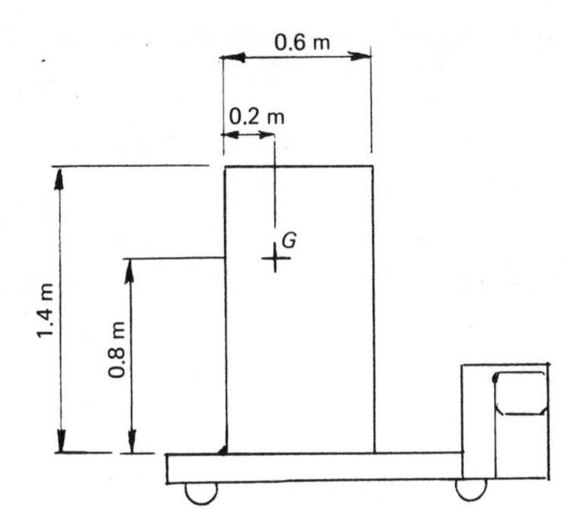

A crate rests on the flat floor of a truck. It has a mass of 780 kg and is 0.6 m wide by 1.4 m deep. The mass centre G is located as shown in the diagram. The crate is prevented from slipping backwards by a stop. Calculate the least forward acceleration of the truck to cause the crate to start tipping backwards: (a) if the truck travels along a straight horizontal road; (b) if it travels up a slope of 10°.

Solution 2.4

When the crate is about to tip up onto its rear edge, the reaction force of the truck floor will then act at this point. Also at this point, the stop will exert a forward force F on the crate. Notice also that the crate is mounted on the truck, and thus the acceleration of the crate will be that of the truck, both as to magnitude and as to direction. Free-body diagrams are shown overleaf as diagrams (a) and (b).

[b]

Part (a)

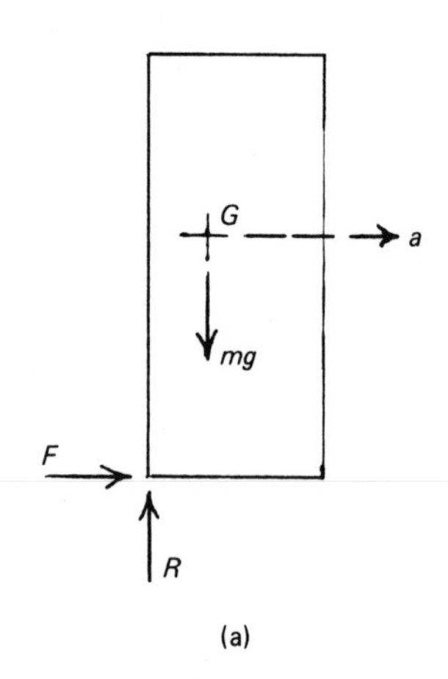

(a)

Equation of translation ($\Sigma F = ma$):

$$F = ma = 780a. \tag{1}$$

Equation of vertical equilibrium:

$$R = mg = 780g.$$

No rotation. $\therefore$ moments about $G = 0$.

$$\therefore\ F \times 0.8 = R \times 0.2.$$

$$780a \times 0.8 = 780g \times 0.2.$$

$$\therefore\ a = \underline{2.453 \text{ m s}^{-2}}.$$

Part (b)

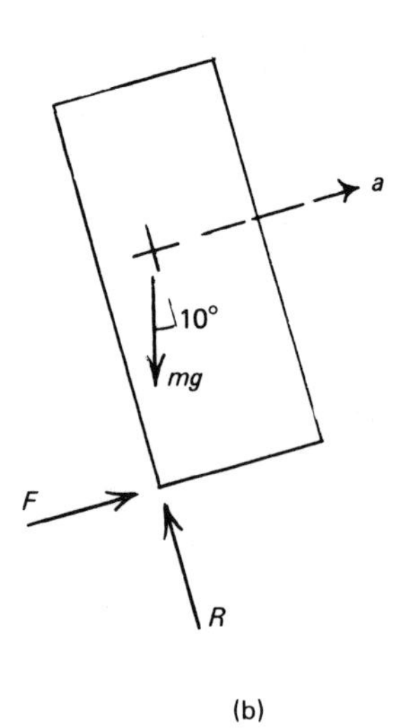

(b)

Equation of translation:

$$F - mg \sin \theta = ma.$$

$$\therefore\ F = 780\ (a + g \sin 10°).$$

Equation of equilibrium perpendicular to the direction of acceleration:

$$R = mg \cos \theta = 780g \cos 10° = 7536.$$

No rotation; moments of forces about $G = 0$:

$$F \times 0.8 = R \times 0.2.$$

$$780\,(a + g \sin 10°)\, 0.8 = 7536 \times 0.2.$$

$$\therefore\ a = \frac{7536 \times 0.2}{780 \times 0.8} - 9.81 \sin 10°$$

$$= 2.415 - 1.703$$

$$= \underline{0.712 \text{ m s}^{-2}}.$$

Example 2.5

A thin bar of uniform cross-section, of length L and mass m, is pivoted at one end to swing in a vertical plane. It is held with the free end level with the pivot, and released from this position. Obtain expressions for the angular velocity and acceleration of the bar, and derive expressions for the components of the pivot reaction force, after the bar has turned through an angle θ. Evaluate the total pivot reaction force in terms of (mg) when $\theta = 0°$ and $90°$.

Solution 2.5

[c]

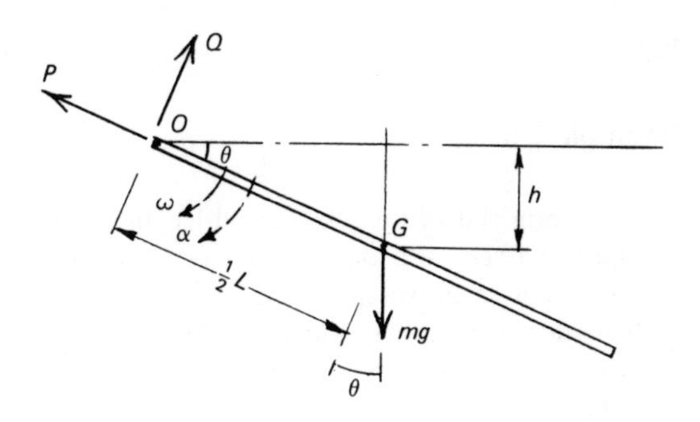

The diagram shows the free-body diagram when the bar has turned through an angle θ. The pivot reaction force is shown as components P and Q, respectively parallel to and perpendicular to the bar. This gives a simpler analysis than by assuming horizontal and vertical components.

Determination of ω may be obtained from an energy equation.

Loss of p.e. of bar = gain of k.e.

[e] $$mgh = \tfrac{1}{2} I_o \omega^2.$$

(The suffix 'o' refers to the moment of inertia of the bar with respect to a transverse axis through the pivot O.)

[d] $$mg\,(\tfrac{1}{2}L \sin \theta) = \tfrac{1}{2}(\tfrac{1}{3}mL^2)\,\omega^2.$$

$$\therefore\ \omega^2 = 3\,\frac{g}{L} \sin \theta.$$

$$\therefore\ \omega = \underline{\sqrt{3\,\frac{g}{L} \sin \theta}}.$$

Equation of angular motion ($\Sigma M = I\alpha$):
Taking moments of forces about O,

$$mg \cos\theta \times \tfrac{1}{2}L = I_o \alpha = \tfrac{1}{3}mL^2 \alpha.$$

$$\therefore\ \alpha = \frac{3}{2}\frac{g}{L}\cos\theta.$$

(We could have obtained this expression also by differentiating ω.)

Equation of angular motion ($\Sigma M = I\alpha$):
Now taking moments of forces about G,

[d] $$Q\tfrac{1}{2}L = I_g \alpha = \left(\tfrac{1}{12}mL^2\right)\left(\frac{3}{2}\frac{g}{L}\cos\theta\right).$$

$$\therefore\ Q = \tfrac{1}{4}mg\cos\theta.$$

Equation of translation ($\Sigma F = ma$) in direction $G \rightarrow O$ **(note that G has centripetal acceleration in this direction)**:

$$P - mg\sin\theta = m\omega^2 r = m\ 3\left(\frac{g}{L}\sin\theta\right)\tfrac{1}{2}L.$$

$$\therefore\ P = mg\sin\theta + \frac{3}{2}mg\sin\theta$$

$$= 2.5\ mg\sin\theta.$$

When $\theta = 0$, $P = 0$; $Q = \tfrac{1}{4}mg.$

When $\theta = 90°$, $P = 2.5\ mg$; $Q = 0.$

Example 2.6

A uniform rod of mass 16 kg and length 1.5 m is freely pivoted at one end so that it can swing in a vertical plane. The free end is raised until it is vertically above the pivot, and is then given an initial small and negligible displacement so that it swings downwards. Show that when the bar has swung through an angle of 70.53° then the vertical component of the pivot reaction force will be zero. Calculate the horizontal component of the force at this instant.

Solution 2.6

[e]

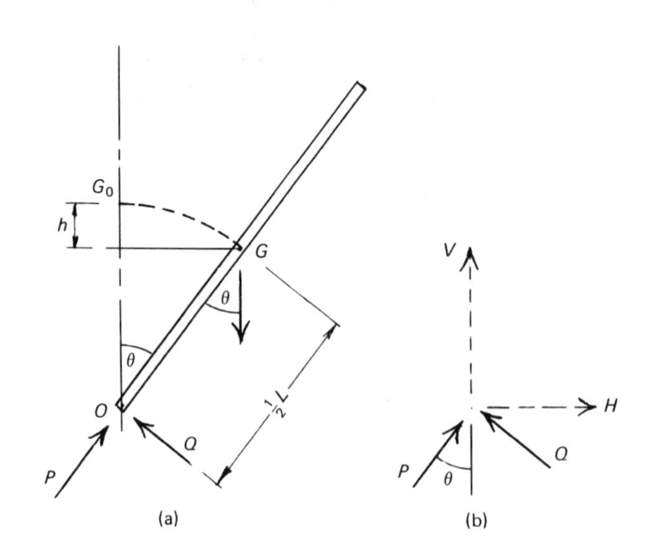

(a) is the free-body diagram. We will retain algebraic symbols, calling the bar mass m and its length L. As in Example 2.5, it is simpler to analyse using the force components P and Q, along and at right angles to the bar, respectively, rather than horizontal and vertical components.

Energy equation: Loss of p.e. = Gain of k.e.

[e] $$mgh = \tfrac{1}{2} I_o \omega^2 .$$

[d] $$mg\tfrac{1}{2}L\,(1 - \cos\theta) = \tfrac{1}{2}(\tfrac{1}{3}mL^2)\,\omega^2 .$$

$$\therefore\ \omega^2 = 3\,\frac{g}{L}\,(1 - \cos\theta).$$

As an alternative to writing an equation of motion, we can derive the acceleration α by differentiating this with respect to time t.

$$2\omega\,(\dot\omega) = 3\,\frac{g}{L}\,(\sin\theta)\,\dot\theta .$$

$$2\omega\alpha = 3\,\frac{g}{L}\,(\sin\theta)\,\omega .$$

$$\therefore\ \alpha = \frac{3}{2}\,\frac{g}{L}\,\sin\theta .$$

Equation of translation ($\Sigma F = ma$) in direction $G \rightarrow O$:

$$mg\cos\theta - P = m\omega^2 r = m\omega^2\,(\tfrac{1}{2}L) = \tfrac{1}{2}Lm \times 3\,\frac{g}{L}\,(1 - \cos\theta).$$

$$\therefore\ P = mg\cos\theta - \frac{3}{2}mg\,(1 - \cos\theta).$$

Substituting the values given,

$$P = 16g\cos 70.53^\circ - 1.5 \times 16g\,(1 - \cos 70.53^\circ)$$
$$= -104.64\ \text{N}.$$

Observe that the directions of P and Q were assumed when the free-body diagram was drawn; it happens that for this value of θ, P is in the opposite direction to that assumed.

Equation of rotation ($\Sigma M = I\alpha$):
Taking moments about G,

[c] $$Q \times \tfrac{1}{2}L = I_g\alpha = (\tfrac{1}{12}mL^2)\,\alpha$$

$$\therefore\ Q = \tfrac{1}{6}mL\alpha = \tfrac{1}{6}mL\left(\frac{3}{2}\,\frac{g}{L}\,\sin\theta\right) = \tfrac{1}{4}mg\sin\theta .$$

Substituting values,

$$Q = \tfrac{1}{4} \times 16 \times 9.81\sin 70.53^\circ = 37.00\ \text{N}.$$

The vertical and horizontal components V and H may be found by resolving P and Q appropriately: see diagram (b) on page 64. We may continue to assume the same direction of P, even though we now know this to be wrong, provided that we substitute the negative value obtained for it.

$$V = P\cos\theta + Q\sin\theta$$
$$= -104.64\cos 70.53^\circ + 37.00\sin 70.53^\circ$$

$= \underline{0.0063 \text{ N}},$

which is clearly negligible.

$$H = P \sin \theta - Q \cos \theta$$

$$= -104.64 \sin 70.53° - 37.00 \cos 70.53°$$

$$= \underline{-110.99 \text{ N}}.$$

You should, as always, note carefully the wording of this question. It does not say, 'Find the value of θ for which the vertical component of reaction force will be zero.' This would be a different, and a more difficult problem.

Example 2.7

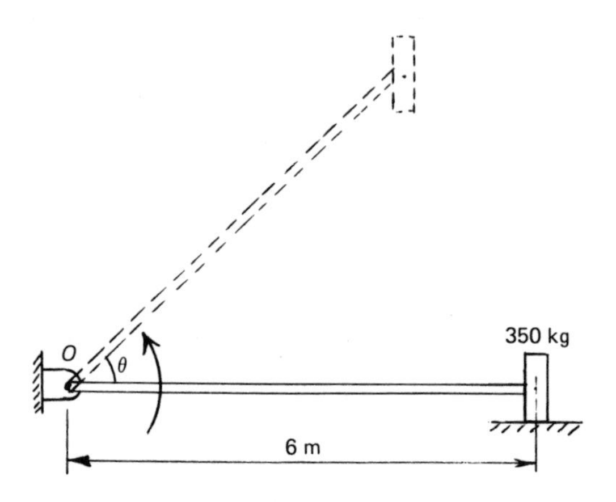

A simple elevator, used for servicing street lamps, consists essentially of a uniform bar of mass 220 kg and length 6 m, with a carrier at the free end, which when loaded, may be treated as a concentrated mass of 350 kg. A constant torque T = 28 kN m is applied at the pivot end to raise the assembly.

(a) Determine the angular acceleration of the beam when it is elevated an angle θ above the horizontal.

(b) At what value of θ should the torque be removed in order that the beam comes to rest when exactly vertical?

(c) Assuming the torque to be removed as prescribed in (b), determine the maximum angular velocity and maximum angular acceleration of the beam.

Solution 2.7

[c]

Part (a)

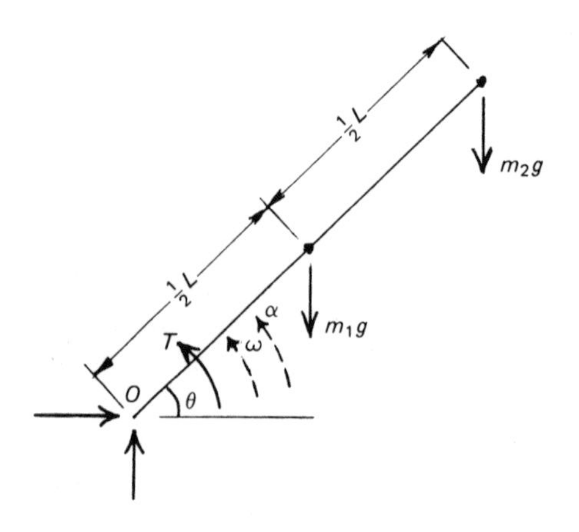

The free-body diagram is shown on page 66. We call the mass of the beam m_1 and and the mass at the end m_2.

Equation of angular motion ($\Sigma M = I\alpha$):
Taking moments about O,

[d] $$T - m_1 g \tfrac{1}{2}L \cos\theta - m_2 gL \cos\theta = I_o\alpha = (\tfrac{1}{3}m_1 L^2 + m_2 L^2)\,\alpha.$$

α may be determined directly from this equation. Rearranging and substituting values:

$$\alpha = \frac{28\,000 - 6\cos\theta \times 9.81\,(\frac{1}{2} \times 220 + 350)}{6^2\,(\frac{1}{3} \times 220 + 350)}$$

$$= \underline{1.837 - 1.777\cos\theta.}$$

Part (b)

A simple energy equation is needed. Initially, the assembly is at rest with no potential energy. Finally, it is also at rest, having gained potential energy. This gain of potential energy must be the work input from the torque. Hence the equation

[e] $$T\theta = m_1 g \tfrac{1}{2}L + m_2 gL.$$

$$\therefore\ \theta = \frac{220g \times 3 + 350g \times 6}{28\,000} = 0.9670 \text{ rad.}$$

$$\therefore\ \theta = \underline{55.4^\circ.}$$

Part (c)

The answer to part (a) shows that angular acceleration increases as θ increases, but this is true for only as long as the torque is applied. Thus the maximum acceleration will be obtained by substituting in the answer to part (a) the value of θ obtained in part (b):

$$\alpha_{max} = 1.837 - 1.777\cos 55.4^\circ$$

$$= \underline{0.8279 \text{ rad s}^{-2}.}$$

A similar argument is advanced for velocity; this will increase only as long as the torque is applied. We may write an energy equation relating the initial state when k.e. and p.e. = 0 and the state at $\theta = 55.4^\circ$ when the system has gained both k.e. and p.e. due to the work done by the torque. The diagram here applies.

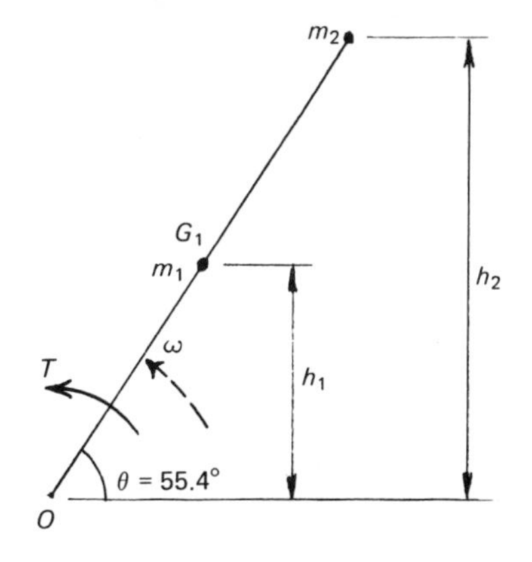

$$\text{Work done} = \text{Gain of p.e.} + \text{Gain of k.e.}$$

[e] $$T\theta = m_1 g h_1 + m_2 g h_2 + I_o \omega^2.$$

Substituting values,

$$28\,000 \times 0.9669 = 220g \times 3 \sin 55.4° + 350g \times 6 \sin 55.4°$$
$$+ \tfrac{1}{2}\omega^2 (\tfrac{1}{3} \times 220 \times 6^2 + 350 \times 6^2).$$

$$\therefore \omega^2 = \frac{(27\,073) - (\sin 55.4° \times 9.81)(660 + 2100)}{\tfrac{1}{2} \times 36 (\tfrac{1}{3} \times 220 + 350)}$$

$$= 0.628.$$

$$\therefore \underline{\omega = 0.7925 \text{ rad s}^{-1}}.$$

This answer could also have been calculated by equating the k.e. at $\theta = 55.4°$ to the gain of p.e. at $\theta = 90°$.

Example 2.8

A uniform bar of mass m and length L is pivoted at one end so that it can swing in a vertical plane. The free end is raised until it is level with the pivot, and it is then released from this position. Prove that at the instant of release, at a point on the bar distant y from the free end, there will be a bending moment M given by the expression

$$M = \frac{mgy^2}{4L}\left(1 - \frac{y}{L}\right)$$

and determine the maximum value of this bending moment.

You may assume that at the instant of release, the angular acceleration of the bar, α, is given by

$$\alpha = \frac{3}{2}\frac{g}{L}.$$

Solution 2.8

[c]

Bending moment, and also shear force, are both *internal* forces in a body. In order to determine them, we need not examine the whole body, only a part of it. These internal forces will then appear on the free-body diagram, as the forces exerted on the part by the rest of the body.

The free-body diagram is given here of a portion of the bar of length y measured from the free end at the instant of release.

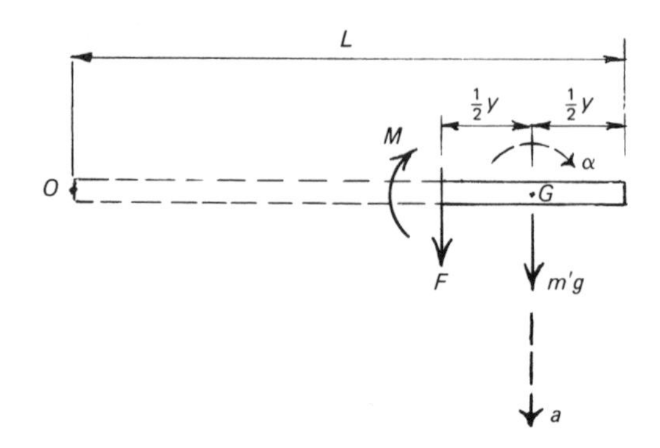

In examining the forces acting on this subsection of the bar, we first recognise the weight acting at the mass centre, G. Because the bar is of uniform section, G will be at the mid-point of the section, distant $\frac{1}{2}y$ from the end. The only other forces acting are those exerted by the remainder of the bar. These comprise a shear force, F, and a bending moment, M. (M actually comprises equal and opposite horizontal forces, but it is more convenient to show it as a moment.) There could also be a longitudinal (i.e. horizontal) tensile or compressive force, but since there is clearly no horizontal acceleration in this particular case, this force is absent.

The free-body diagram also shows a linear acceleration, a, of the mass centre, and an angular acceleration α. Since the subsection is an integral part of the whole bar, it is clear that α is the same as the angular acceleration of the bar.

We first kinematically relate a and α. Because G turns about the pivot O,

$$a = \alpha\left(L - \tfrac{1}{2}y\right) = \frac{3}{2}\frac{g}{L}\left(L - \tfrac{1}{2}y\right).$$

Equation of motion of translation ($\Sigma F = ma$):

$$F + m'g = m'a.$$

By proportion,

$$m' = m\frac{y}{L}.$$

$$\therefore\ F = m\frac{y}{L}(a - g)$$

$$= m\frac{y}{L}\left(\frac{3}{2}\frac{g}{L}\left(L - \tfrac{1}{2}y\right) - g\right)$$

$$= mg\frac{y}{L}\left(\frac{3}{2} - \frac{3y}{4L} - 1\right)$$

$$= mg\frac{y}{L}\left(\frac{1}{2} - \frac{3}{4}\frac{y}{L}\right).$$

Equation of motion of rotation ($\Sigma M = I\alpha$): taking moments about G:

$$M - F\tfrac{1}{2}y = I_g\alpha.$$

[d]

$$\therefore\ M = F\tfrac{1}{2}y + \left(\tfrac{1}{12}m'y^2\right)\left(\frac{3}{2}\frac{g}{L}\right)$$

$$= \frac{mgy^2}{2L}\left(\frac{1}{2} - \frac{3}{4}\frac{y}{L}\right) + \frac{1}{12}\left(\frac{my}{L}\right)y^2\frac{3}{2}\frac{g}{L}$$

$$= \frac{mgy^2}{4L}\left(1 - \frac{3}{2}\frac{y}{L} + \frac{1}{2}\frac{y}{L}\right)$$

$$\therefore\ M = \frac{mgy^2}{4L}\left(1 - \frac{y}{L}\right).$$

This moment varies with y, having values of 0 both when $y = 0$ and when $y = l$. To find the maximum value we differentiate with respect to y:

$$\frac{\mathrm{d}M}{\mathrm{d}y} = 0.$$

$$\therefore\ \frac{mg}{4L}\left(2y - \frac{3y^2}{L}\right) = 0.$$

$$\therefore y = \frac{2}{3}L.$$

$$\therefore M_{max} = \frac{mg}{4L}\left(\tfrac{2}{3}L\right)^2\left(1 - \tfrac{2}{3}\right)$$

$$= \frac{mgL}{27}.$$

The directions of M and F on the free-body diagram are actually positive bending moment and shear force, according to a widely accepted convention. But the correct answers would still be obtained by choosing the other directions for either or both M and F.

Example 2.9

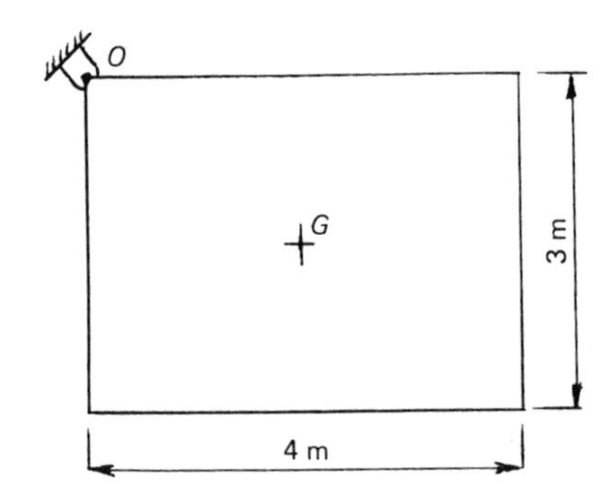

A uniform thin metal plate 3 m by 4 m and having a mass of 56 kg is hinged at one corner so that it can swing in a vertical plane about the fixed pivot O. It is released from rest in the position shown, with the long side horizontal. Determine its angular velocity and acceleration, and the magnitude of the force at the pivot, after it has turned through 90°.

The moment of inertia of a flat rectangular lamina of mass m, width a and breadth b about an axis through the centre perpendicular to the plane of the lamina is $\frac{1}{12}m(a^2 + b^2)$.

Solution 2.9

[c]

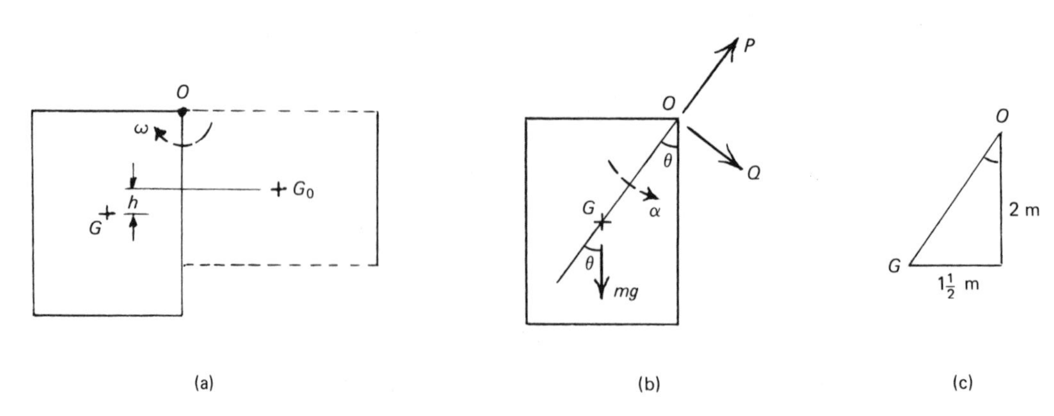

(a) (b) (c)

An energy equation is used to find ω. Diagram (a) shows the plate in the two positions.

$$\text{Loss of p.e.} = \text{Gain of k.e.}$$

[e] $$mgh = \tfrac{1}{2} I_o \omega^2 .$$

[d] $$\not{m}g(2 - 1.5) = \tfrac{1}{2}\omega^2 \, [\tfrac{1}{12}\not{m}(3^2 + 4^2) + \not{m}(OG)^2] .$$

From diagram (c),

$$(OG)^2 = 2^2 + 1\tfrac{1}{2}^2 = 6.25.$$

$$\therefore \; OG = 2.5 \; m.$$

$$\therefore \; \tfrac{1}{2}g = \tfrac{1}{2}\omega^2 \, (\tfrac{1}{12} \times 25 + 6.25).$$

$$\therefore \; \omega = \sqrt{\frac{9.81}{8.333}} = \underline{1.085 \text{ rad s}^{-1}} .$$

Equation of motion of rotation ($\Sigma M = I\alpha$):
Moments about O (diagram (b)):

$$\not{m}g \sin\theta \; OG = \alpha[\tfrac{1}{12}\not{m}(3^2 + 4^2) + \not{m}(OG)^2] .$$

From diagram (c),

$$\sin\theta = \frac{1\tfrac{1}{2}}{OG}$$

$$\therefore \; 1\tfrac{1}{2}g = \alpha\left(\frac{25}{12} + 6.25\right).$$

$$\therefore \; \alpha = \frac{1\tfrac{1}{2}g}{8.333} = \underline{1.766 \text{ rad s}^{-2}} .$$

Equation of translation ($\Sigma F = ma$) in direction $G \rightarrow O$ (diagram (b)):

$$P - mg\cos\theta = m\omega^2 (OG).$$

$$\therefore \; P = m\omega^2 (OG) + mg\cos\theta$$

$$= 56 \times (1.085)^2 \times 2.5 + 56 \times 9.81 \left(\frac{2}{2.5}\right)$$

$$= 604.4 \text{ N}.$$

Equation of translation ($\Sigma F = ma$) in direction perpendicular to OG:

$$mg\sin\theta + Q = ma = m(\alpha \times OG)$$

$$\therefore \; Q = m\alpha OG - mg\sin\theta$$

$$= 56 \times 1.766 \times 2.5 - 56 \times 9.81 \left(\frac{1.5}{2.5}\right)$$

$$= -82.38 \text{ N}.$$

(indicating that Q acts in the opposite direction to that assumed.)

Resultant pivot force $R = \sqrt{P^2 + Q^2}$.

$$R = \sqrt{(604.4)^2 + (-82.38)^2}$$

$$= \underline{610.0 \text{ N}}.$$

Example 2.10

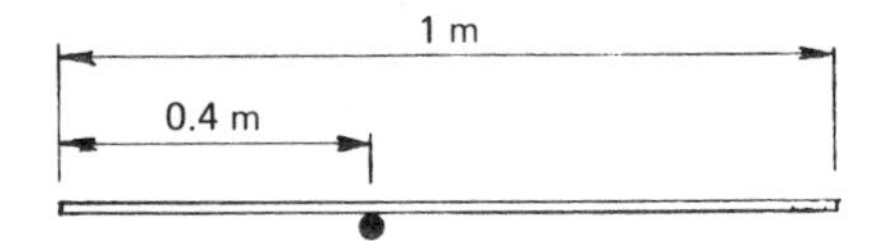

A uniform thin rod of mass 10 kg and length 1 m is held horizontally transversely across a peg 0.4 m from one end, as shown in the diagram. It is released from rest in this position. The coefficient of friction between rod and peg is 0.1. Calculate how far the rod will turn before it begins to slip on the peg, and determine its angular velocity and acceleration at that instant.

Solution 2.10

[c]

The free body diagram is as shown here. Since the rod is about to slip, the frictional force at the peg is μN.

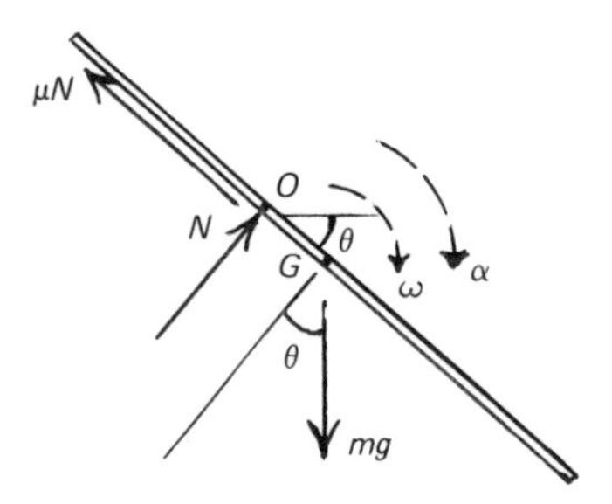

We may incorporate ω^2 into an energy equation. Loss of p.e. = gain of k.e.

$$mgh = \tfrac{1}{2} I_o \omega^2 .$$

It will probably be simpler in this example to substitute arithmetical values as we proceed. The manipulation of the equations will thereby be somewhat simpler.

[d]

$$10g\,(0.1 \sin\theta) = \tfrac{1}{2}\omega^2 \,[\tfrac{1}{12} \times 10 \times 1^2 + 10 \times (0.1)^2\,].$$

$$\therefore\ 9.81 \sin\theta = 0.4667\omega^2 .$$

$$\therefore\ \omega^2 = 21.022 \sin\theta. \qquad (1)$$

The equation of rotation about O ($\Sigma M = I\alpha$) will give an equation incorporating α.

$$mg \cos\theta \times 0.1 = I_o \alpha.$$

$$\therefore\ 10g \times 0.1 \cos\theta = \alpha\,[\tfrac{1}{12} \times 10 \times 1^2 + 10 \times (0.1)^2\,].$$

$$\therefore\ \alpha = 10.511 \cos\theta. \qquad (2)$$

The mass centre G has centripetal acceleration in the direction $G \rightarrow O$.

$$\mu N - mg \sin\theta = m\omega^2 r.$$

$$0.1N - 10g \sin\theta = 10 \times 0.1\,(21.022 \sin\theta).$$

$$\therefore\ 0.1N = \sin\theta\,(21.022 + 98.1).$$

$$\therefore\ N = 1191.2 \sin\theta. \qquad (3)$$

Equation of linear motion ($\Sigma F = ma$) transverse to the bar:

$$mg \cos \theta - N = ma = m (\alpha \times 0.1).$$

Substituting for N and α from equations 3 and 2,

$$10g \cos \theta - 1191.2 \sin \theta = 10 \times 0.1 (10.511 \cos \theta).$$

$$\cos \theta (98.1 - 10.511) = \sin \theta (1191.2).$$

$$\therefore \tan \theta = \frac{98.1 - 10.511}{1191.2} = 0.073\,53.$$

$$\therefore \underline{\theta = 4.205^\circ}.$$

Substituting in equation 1,

$$\omega^2 = 21.022 \sin 4.205^\circ = 1.5414.$$

$$\therefore \underline{\omega = 1.242 \text{ rad s}^{-1}}.$$

Substituting in equation 2,

$$\alpha = 10.511 \cos 4.205^\circ = \underline{10.483 \text{ rad s}^{-2}}.$$

You need to appreciate that this problem could be solved algebraically. It cannot be stressed too often that there are various ways of solving problems, and provided they all give the right answer, they are all right. But some methods are simpler than others. In this particular case, writing four equations to solve the four unknown quantities ω, α, N and θ is clearer if you do not include also the terms m, L and a (the distance from G to O). If, however, you are algebraically minded, you may care to attempt your own purely algebraic solution. You should be able to show that θ can be obtained from the expression

$$\tan \theta = \frac{\mu L^2}{L^2 + 36a^2}.$$

Example 2.11

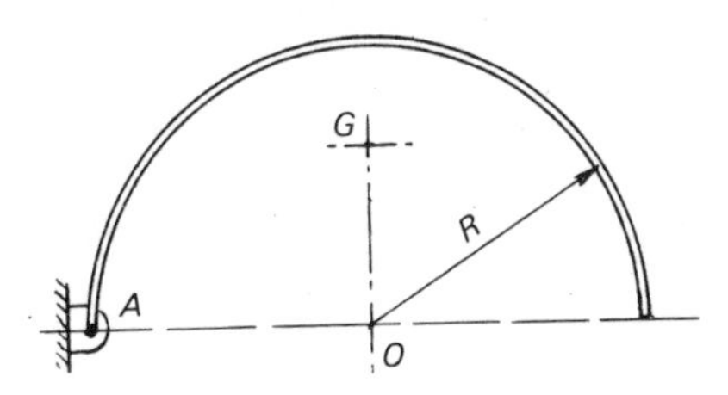

Half of a uniform thin circular ring is freely pivoted at one end A so that it can swing in a vertical plane. It has a mass of 2.6 kg and a radius of 0.14 m. It is initially held at rest with the centre of curvature O level with A and released from this position. Determine the angular velocity and acceleration of the half-ring after it has turned through 90°. Find also the magnitude of the pivot reaction force at this instant.

You may assume that the mass centre G of the half-ring is at a distance of $2R/\pi$ from O, the centre of curvature.

Solution 2.11

[c]

Angular velocity may be determined from an energy equation. We will require the moment of inertia of the half-ring about an axis through the pivot at A.

It should first be clear that the moment of inertia of the half-ring about an axis through O perpendicular to the plane of the ring is mR^2, since all the mass lies at the same distance R from this point. (An odd but very common error is to assume that because it is a half-ring, and not a full one, the moment of inertia is $\frac{1}{2}mR^2$.) To determine I_a we use the parallel-axis theorem, first determining I_g.

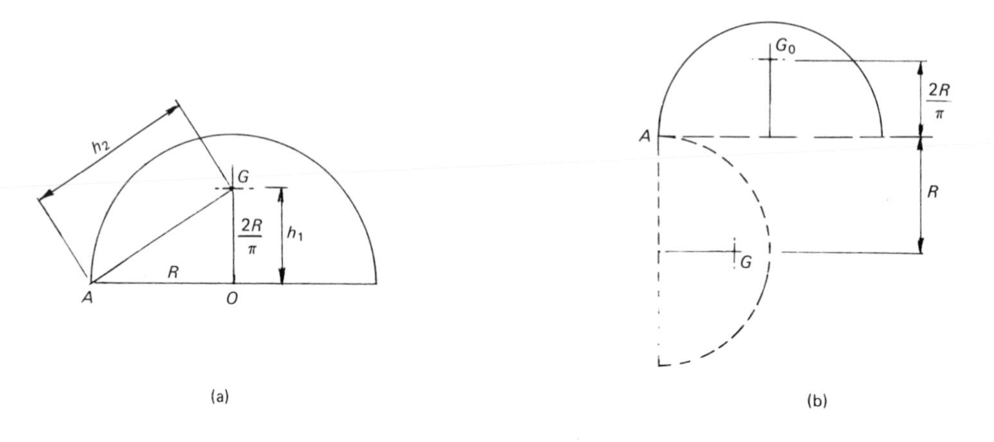

Referring to diagram (a),

[d]

$$I_g = I_o - mh_1^2 = mR^2 - m(OG)^2.$$

$$I_a = I_g + mh_2^2$$

$$= mR^2 - m(OG)^2 + m\,[(OG)^2 + (OA)^2]$$

$$= mR^2 + mR^2$$

$$= 2mR^2$$

$$= 2 \times 2.6 \times (0.14)^2.$$

$$\therefore\ I_a = 0.101\,92 \text{ kg m}^2.$$

Loss of p.e. = Gain of k.e. The mass centre drops a distance of $(R + 2R/\pi)$. See diagram (b).

[e]

$$mgh = \tfrac{1}{2}I_a\omega^2.$$

$$mg\left(R + \frac{2R}{\pi}\right) = \tfrac{1}{2}(2mR^2)\,\omega^2.$$

$$\therefore\ \omega^2 = \frac{2mgR}{2mR^2}\left(1 + \frac{2}{\pi}\right).$$

$$\therefore\ \omega^2 = \frac{9.81}{0.14}\left(1 + \frac{2}{\pi}\right) = 114.68.$$

$$\therefore\ \omega = \underline{10.709 \text{ rad s}^{-1}}.$$

In the free-body diagram, as with all problems of this kind, the force at the pivot is shown as two components, one having a line of action through the mass centre G, and the other at right angles to this.

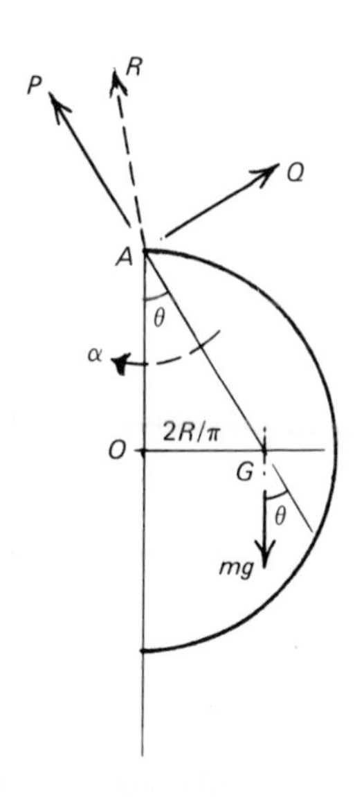

Equation of rotation ($\Sigma M = I\alpha$):
Taking moments about A, (mg) is the only force having a moment about A:

$$mg \times \frac{2R}{\pi} = I_a \alpha.$$

$$\therefore\ \alpha = \frac{2mgR}{\pi \times 2mR^2}$$

$$= \frac{9.81}{\pi \times 0.14}.$$

$$\therefore\ \alpha = \underline{22.304 \text{ rad s}^{-2}}.$$

Equation of translation ($\Sigma F = ma$) in direction $G \to A$:

$$P - mg\cos\theta = m\omega^2(AG).$$

$$\therefore\ P = m\omega^2(AG) + mg\cos\theta.$$

$$\theta = \tan^{-1}\left(\frac{2}{\pi}\right) = 32.48°.$$

$$AG = R\sec\theta = 0.14 \sec 32.48° = 0.1660 \text{ m}.$$

$$\therefore\ P = 1.8 \times 114.68 \times 0.1660 + 1.8 \times 9.81 \cos 32.48°.$$

$$\therefore\ P = 49.16 \text{ N}.$$

Equation of translation ($\Sigma F = ma$) in direction perpendicular to GA:

$$mg\sin\theta - Q = ma = m\alpha(AG).$$

$$\therefore\ Q = mg\sin\theta - m\alpha(AG)$$

$$= 1.8 \times 9.81 \sin 32.48° - 1.8 \times 22.304 \times 0.1660.$$

$$\therefore\ Q = 2.818 \text{ N}.$$

The resultant: $R = \sqrt{P^2 + Q^2}$

$$= \sqrt{(49.16)^2 + (2.818)^2}.$$

$$\therefore\ R = \underline{49.24 \text{ N}}.$$

Example 2.12

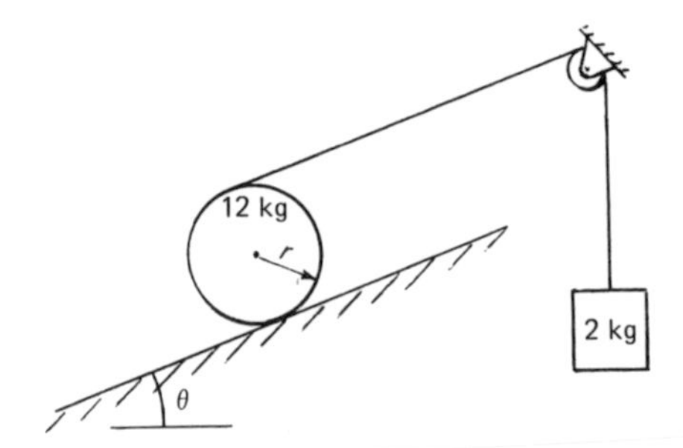

A uniform solid cylinder of mass 12 kg and radius r = 50 mm lies across a rough inclined plane of slope θ = 15°. A light cord is fixed to the cylinder, wrapped round it, and passed over a frictionless fixed pulley, a weight of mass 2 kg hanging freely from the end, as shown in the diagram. The cord between the cylinder and the pulley is parallel to the slope. The cylinder rolls on the inclined plane without slipping. Calculate the linear and angular acceleration of the cylinder and the acceleration of the hanging weight.

Solution 2.12

[a]

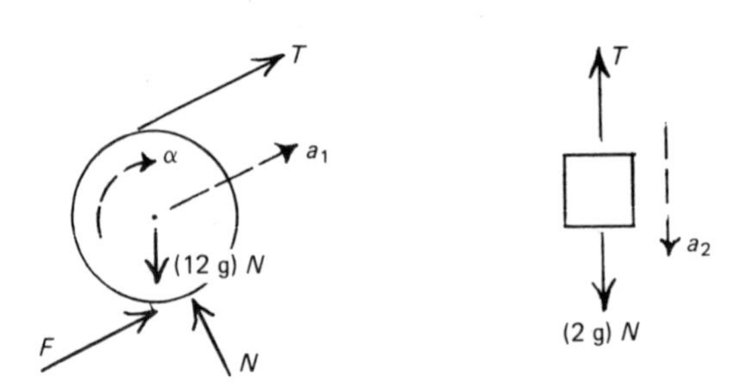

Shown here are the free-body diagrams of the cylinder and the hanging weight. We first relate kinematically the three accelerations, α, a_1 and a_2.

Because the cylinder rolls without slip,

$$a_1 = \alpha r.$$

a_2 will be the sum of the peripheral acceleration of the cylinder and the linear acceleration of its mass centre. Thus

$$a_2 = \alpha r + a_1 = 2\alpha r.$$

The friction force F is the plane reaction to the thrust of the cylinder against the plane as it rotates clockwise; hence the sense of this force is that of an anticlockwise rotation.

Equation of translation ($\Sigma F = ma$):

$$T + F - 12g \sin 15° = 12a_1 = 12\alpha r = 12 \times 0.05\alpha. \qquad (1)$$

Equation of rotation ($\Sigma M = I\alpha$):
Taking moments of forces about G:

[a] $$Tr - Fr = I_g\alpha = (\tfrac{1}{2}mr^2)\alpha.$$

$$\therefore\ T - F = \tfrac{1}{2}mr\alpha = \tfrac{1}{2} \times 12 \times 0.05\alpha. \qquad (2)$$

For the weight ($\Sigma F = ma$):

$$2g - T = 2a_2 = 2 \times 2\alpha r = 4 \times 0.05\alpha. \qquad (3)$$

Rearranging equation 1 and adding equation 2,

$$(T + F) + (T - F) = 0.6\alpha + 12g \sin 15° + 0.3\alpha.$$

$$\therefore\ 2T = 0.9\alpha + 12g \sin 15°.$$

Substitute for T in equation 3:

$$2g - \tfrac{1}{2} \times 0.9\alpha - 6g \sin 15° = 0.2\alpha.$$

$$\therefore\ \alpha\,(0.2 + 0.45) = 2g - 6g \sin 15°.$$

$$\therefore\ \alpha = \frac{2g - 6g \sin 15°}{0.65} = \underline{6.748 \text{ rad s}^{-2}}.$$

$$a_1 = \alpha r = 0.05 \times 6.748 = \underline{0.3374 \text{ m s}^{-2}}.$$

$$a_2 = 2\alpha r = \underline{0.6748 \text{ m s}^{-2}}.$$

Example 2.13

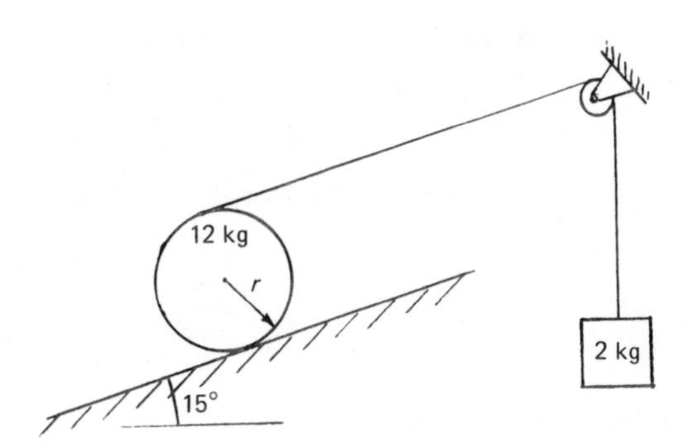

A uniform solid cylinder of mass 12 kg and radius r = 50 mm lies across a *smooth* inclined plane of slope 15°. A light cord is fixed to the cylinder, wrapped round it, and passed over a frictionless fixed pulley, a weight of mass 2 kg hanging freely from the end, as shown in the diagram. The cord between the cylinder and the pulley is parallel to the slope. Calculate the linear and angular acceleration of the cylinder and the acceleration of the hanging weight.

Solution 2.13

[a]

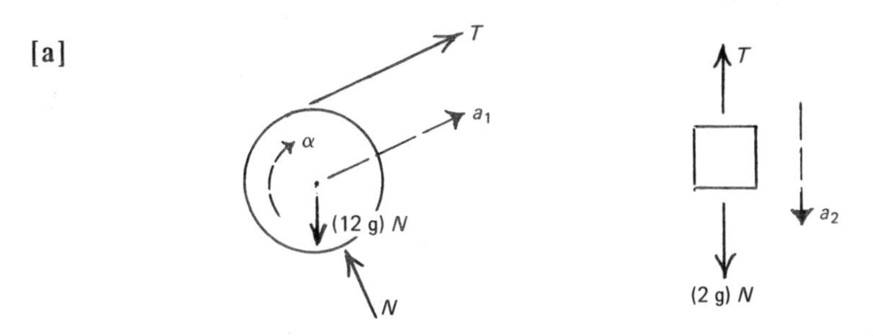

The free-body diagrams of the cylinder and the hanging weight are shown here. We first relate the accelerations α, a_1 and a_2. Because the plane is smooth, the

cylinder slips on it, and there is thus no direct relation between a_1 and α. However, the linear acceleration of the cord will be the peripheral acceleration of the cylinder plus the linear acceleration of the cylinder centre.

$$a_2 = a_1 + \alpha r = a_1 + 0.05\alpha.$$

Equation of translation of cylinder ($\Sigma F = ma$):

$$T - 12g \sin 15° = 12a_1. \qquad (1)$$

Equation of rotation ($\Sigma M = I\alpha$):
Taking moments about G:

$$Tr = I\alpha.$$

[d]

$$T \times 0.05 = (\tfrac{1}{2}mr^2)\alpha = \tfrac{1}{2} \times 12 \times (0.05)^2 \alpha.$$

$$\therefore\ T = 0.3\alpha. \qquad (2)$$

For the weight,

$$2g - T = 2a_2 = 2(a_1 + 0.05\alpha). \qquad (3)$$

Substituting for T in equations 1 and 3,

$$0.3\alpha - 12g \sin 15° = 12a_1. \qquad (4)$$

$$2g - 0.3\alpha = 2a_1 + 0.1\alpha.$$

Equating α from these two equations,

$$\frac{12a_1 + 12g \sin 15°}{0.3} = \frac{2g - 2a_1}{0.4}.$$

$$\therefore\ 4.8a_1 + 4.8g \sin 15° = 0.6g - 0.6a_1.$$

$$\therefore\ a_1 = \frac{0.6 \times 9.81 - 4.8 \times 9.81 \sin 15°}{4.8 + 0.6}.$$

$$= \underline{-1.167 \text{ m s}^{-2}.}$$

Substituting in equation 4,

$$\alpha = \frac{12a_1 + 12g \sin 15°}{0.3} = \underline{54.88 \text{ rad s}^{-2}.}$$

$$a_2 = a_1 + \alpha r = -1.167 + 54.88 \times 0.05 = \underline{1.577 \text{ m s}^{-2}.}$$

From the results it is seen that the weight accelerates downwards, the cylinder accelerates clockwise, but also accelerates down the plane.

Example 2.14

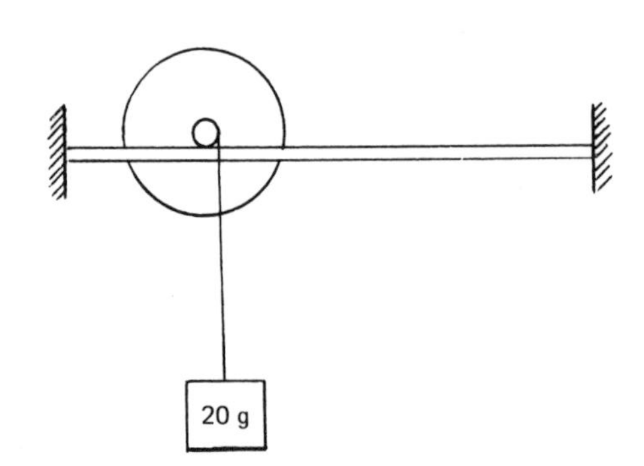

A disc has a mass of 0.4 kg and a moment of inertia about its axis of 0.000 32 kg m^2. It is fixed to an axle of negligible mass and of radius 4 mm, and the axle rests on a pair of horizontal rails, as shown in the diagram. A weight of mass 20 g hangs from a string which is attached to and wound round the axle. If the system is released from rest, calculate the time for the disc to travel 0.1 m along the rail, and determine its angular velocity at this point. The axle rolls along the rails without slipping. It may be assumed that the string hangs vertically at all times.

Solution 2.14

[a]

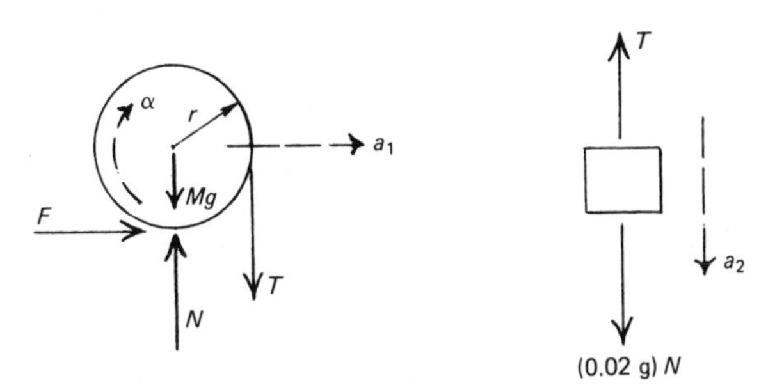

Free-body diagrams of the disc and the weight are shown here. The axle is shown very enlarged, to show clearly the forces acting. These include a normal rail reaction force N, and a tangential friction force F. The string tension will cause a clockwise rotation of the axle, which thus thrusts against the rail to the left; the rail reaction must therefore be equal and opposite to this. The mass of the disc is designated M, and the axle radius r.

We first obtain the kinematic relationship between α, the angular acceleration of the disc, a_1, the linear acceleration, and a_2, the acceleration of the weight.

Because the axle does not slip,

$$a_1 = \alpha r.$$

Because the string hangs vertically, a_2 will be the same as the peripheral acceleration of the axle. Thus

$$a_2 = \alpha r.$$

Note that the problem has been deliberately simplified. In fact, the string cannot hang vertically. When in dynamic equilibrium, the weight will accelerate horizontally with the same acceleration as the disc; it can do this only if the string tension will actually be at a slight angle to the vertical, therefore, but the question has been simplified to exclude this consideration, and furthermore the effect is small enough to be negligible.

Equation of translation ($\Sigma F = ma$):

$$F = Ma_1 = M\alpha r. \tag{1}$$

Equation of rotation ($\Sigma M = I\alpha$):

$$Tr - Fr = I\alpha. \tag{2}$$

Equation of translation for weight:

$$mg - T = ma_2 = m\alpha r. \tag{3}$$

Substitute for F and T in equation 2:

$$r(mg - m\alpha r) - r(M\alpha r) = I\alpha.$$

Rearranging,

$$\alpha(I + Mr^2 + mr^2) = mgr.$$

$$\therefore \ \alpha = \frac{mgr}{I + (M + m)r^2}.$$

Substituting values,

$$\alpha = \frac{0.02 \times 9.81 \times 0.004}{0.000\,32 + (0.4 + 0.02)\,(0.004)^2}$$

$$= 2.402 \text{ rad s}^{-2}.$$

$$a_1 = \alpha r = 2.402 \times 0.004 = 0.009\,61 \text{ m s}^{-2}.$$

For the disc motion,

$$x = ut + \tfrac{1}{2}at^2.$$

$$\therefore \ 0.1 = 0 + \tfrac{1}{2} \times 0.009\,61t^2$$

$$\therefore \ t = \underline{4.562 \text{ s.}}$$

$$\omega_2 = \omega_1 + \alpha t$$

$$= 0 + 2.402 \times 4.562$$

$$= \underline{10.96 \text{ rad s}^{-1}}.$$

Example 2.15

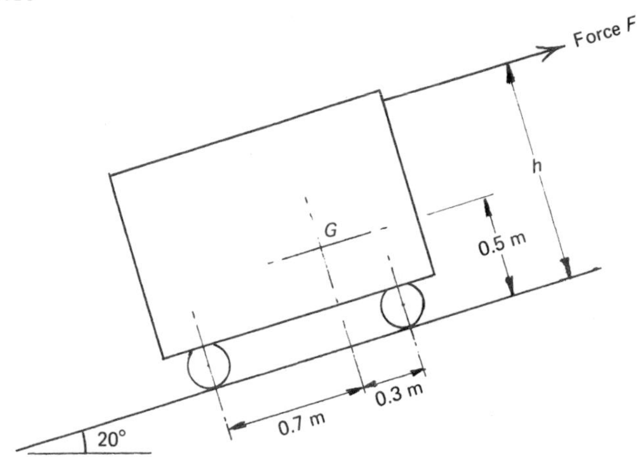

The diagram shows a truck of mass 1550 kg, with some principal dimensions. It is pulled up a slope of 20° by a cable attached at point P which is a height h above the ground. The line of action of the cable is parallel to the ground. The force F in the cable causes the truck to have a forward acceleration of 2.1 m s^{-2}.

(a) Calculate the value of force F.

(b) Given that h is 0.8 m, calculate the magnitudes of the front and rear wheel reaction forces.

(c) Determine what value for h is required in order that the front and rear wheel reaction forces would be equal.

Neglect all friction forces and the inertia effects of the wheels.

PCL

Solution 2.15

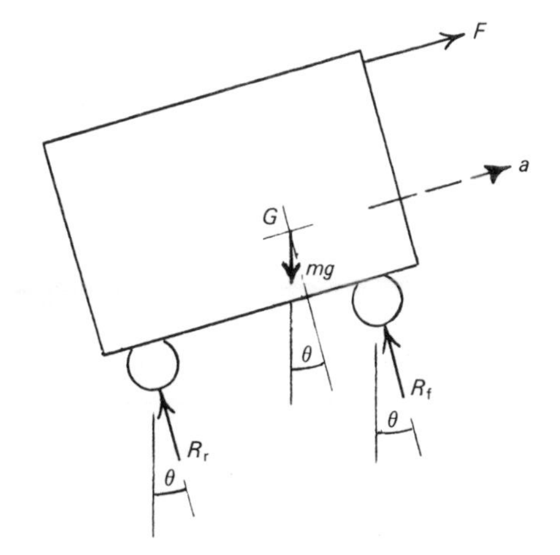

The appropriate free-body diagram shows the forces as consisting of the weight at *G*, the two wheel reaction forces, and the tensile force *F*. Since the truck is pulled along by the cable there is, of course, no traction force between the wheels and the ground. We may write the three equations.

In the direction of the acceleration a ($\Sigma F = ma$),

$$F - mg \sin\theta = ma. \tag{1}$$

Perpendicular to above ($\Sigma F = 0$),

$$R_f + R_r - mg \cos\theta = 0. \tag{2}$$

Taking moments of all forces about G ($\Sigma M = 0$),

$$0.7R_r + F(h - 0.5) - 0.3R_f = 0. \tag{3}$$

Part (a)

Substituting values in equation 1,

$$\begin{aligned} F &= ma + mg \sin\theta \\ &= 1550(2.1 + 9.81 \sin 20°) \\ &= \underline{8456 \text{ N.}} \end{aligned}$$

Part (b)

Rearranging equation 2 and substituting,

$$\begin{aligned} R_f + R_r &= mg \cos\theta \\ &= 1550 \times 9.81 \cos 20° \\ &= 14\,288 \text{ N.} \end{aligned} \tag{4}$$

Substituting in equation 3 with h = 0.8 m and using equation 4 to substitute for R_r,

$$0.7(14\,288 - R_f) + 8456(0.8 - 0.5) - 0.3R_f = 0.$$

$$\therefore\ 14\,288 \times 0.7 + 8456 \times 0.3 = R_f(0.3 + 0.7).$$

$$\therefore\ R_f = \underline{12\,538 \text{ N.}}$$

$$R_r = 14\,288 - 12\,538$$

$$= \underline{1750 \text{ N.}}$$

Part (c)

Calling front and rear wheel reaction forces R, and substituting in equation 2,

$$R + R = 14\,288.$$

$$\therefore\ R = 7144 \text{ N.}$$

And substituting in equation 3,

$$0.7 \times 7144 + 8456\,(h - 0.5) - 0.3 \times 7144 = 0.$$

$$\therefore\ h - 0.5 = -\frac{0.4 \times 7144}{8456}$$

$$= -0.338 \text{ m.}$$

$$\therefore\ h = \underline{0.162 \text{ m.}}$$

Example 2.16

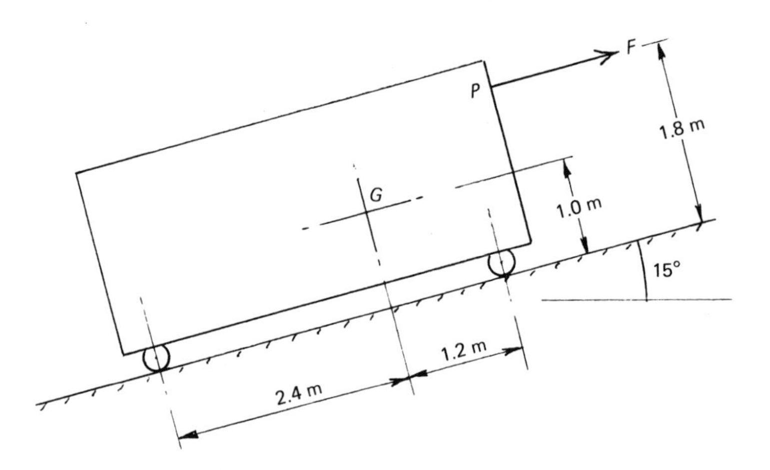

A loaded truck of mass 2000 kg is hauled up a slope of 15° by a wire rope attached at the point P as shown in the diagram. The rope applies a force F, the line of action which is parallel to the surface of the track. Calculate the acceleration of the truck, and the magnitudes of the total front and rear wheel reaction forces when the magnitude of F is 11 kN. Calculate also the distance travelled by the truck in 5 seconds, starting from rest, under these conditions. Neglect all friction and resistance forces. G is the mass centre of the truck.

PCL

Solution 2.16

The question is similar to the previous one. The free-body diagram is shown, with the four forces. The equations are:

In the direction of acceleration, assumed forwards ($\Sigma F = ma$),

$$F - mg \sin\theta = ma. \qquad (1)$$

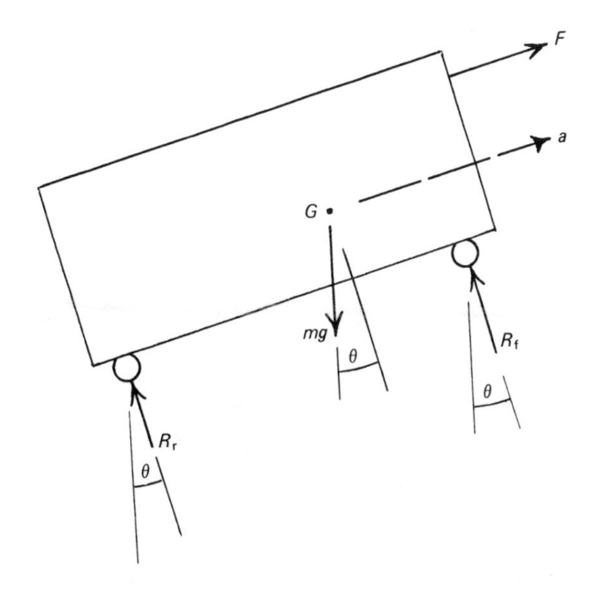

Perpendicular to the direction of acceleration ($\Sigma F = 0$),

$$R_f + R_r = mg \cos \theta . \qquad (2)$$

Taking moments of forces about G ($\Sigma M = 0$),

$$0.8F + 2.4R_r - 1.2R_f = 0. \qquad (3)$$

Substituting values in equation 1,

$$11\,000 - 2000g \sin 15° = 2000a.$$

$$\therefore\ a = \frac{11\,000 - 2000g \sin 15°}{2000}$$

$$= \underline{2.961 \text{ m s}^{-2}}.$$

Substituting values in equation 2,

$$R_f + R_r = 2000g \cos 15°$$
$$= 18\,951 \text{ N}.$$

$$\therefore\ R_r = 18\,951 - R_f.$$

Substituting this in equation 3,

$$0.8 \times 11\,000 + 2.4(18\,951 - R_f) - 1.2R_f = 0.$$

$$\therefore\ 8800 + 2.4 \times 18\,951 = R_f(1.2 + 2.4).$$

$$\therefore\ R_f = \frac{8800 + 45\,482}{3.6}$$

$$= \underline{15\,078 \text{ N}.}$$

$$\therefore\ R_r = 18\,951 - 15\,078$$

$$= \underline{3873 \text{ N}.}$$

Using the equation of kinematics $x = ut + \frac{1}{2}at^2$:

$$x = 0 + \tfrac{1}{2} \times 2.961 \times 5^2$$

$$= \underline{37.01 \text{ m}}.$$

Example 2.17

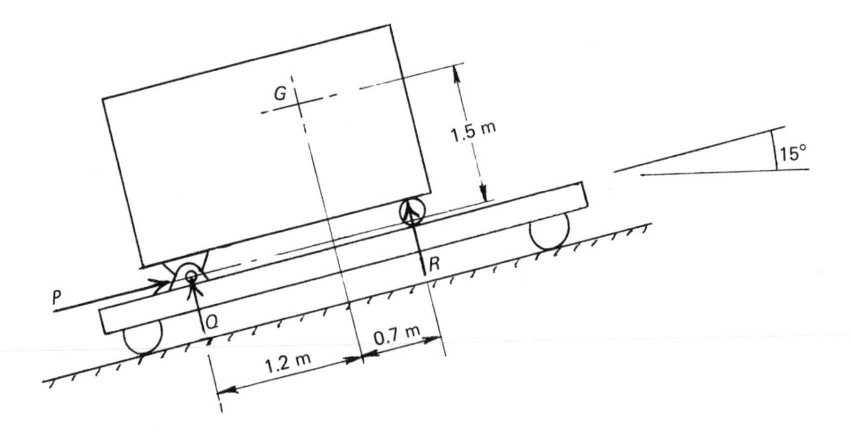

A large crate of mass 450 kg is mounted on the flat floor of a truck by means of a pin-joint at the rear, and a roller support at the front, as shown in the diagram, so that the reaction force at the rear has tangential and normal components P and Q, while that at the front has a normal component R only. The location of the mass centre of the crate, G, is shown.

Calculate the magnitudes of the force components P, Q and R when the truck ascends a slope of 15° with a forward acceleration of 1.6 m s^{-2}.

Assuming that the forward acceleration is then increased so that the value of reaction R is reduced to zero, but the crate is just not tipping backwards, calculate the corresponding values of P and Q.

PCL

Solution 2.17

We may dispense with a free-body diagram as all forces are shown except weight, and the direction of acceleration is given as forwards.

In the forward direction the equation ($\Sigma F = ma$) is

$$P - mg \sin 15° = ma, \tag{1}$$

and perpendicular to this direction the equation ($\Sigma F = 0$) is

$$Q + R = mg \cos 15°. \tag{2}$$

Taking moments about G ($\Sigma M = 0$),

$$1.5P + 0.7R - 1.2Q = 0. \tag{3}$$

For the first part of the question, we substitute values in equation 1:

$$\begin{aligned} P &= ma + mg \sin 15° \\ &= 450(1.6 + 9.81 \sin 15°) \\ &= \underline{1863 \text{ N}}. \end{aligned}$$

and in equations 2 and 3:

$$\begin{aligned} Q + R &= 450g \cos 15° \\ &= 4264 \text{ N}. \end{aligned}$$

$$1.5 \times 1863 + 0.7(4264 - Q) - 1.2Q = 0.$$

$$\therefore\ 1.5 \times 1863 + 0.7 \times 4264 = Q(1.2 + 0.7).$$

$$\therefore\ Q = \frac{5779}{1.9}$$

$$= \underline{3042\ \text{N.}}$$

$$\therefore\ R = 4264 - 3042$$

$$= \underline{1222\ \text{N.}}$$

We use the same equations 1 to 3 to solve the second part, but now $R = 0$ and the value of a will be different.

Substituting values in equation 2,

$$Q = mg \cos 15^\circ$$

$$= \underline{4264\ \text{N.}}$$

and in equation 3,

$$1.5P - 1.2Q = 0.$$

$$\therefore\ P = Q \times \frac{1.2}{1.5}$$

$$= 4264 \times \frac{1.2}{1.5}$$

$$= \underline{3411\ \text{N.}}$$

The new value of a is not required, but could be calculated from equation 1. (The value is 5.041 m s^{-2}.)

Example 2.18

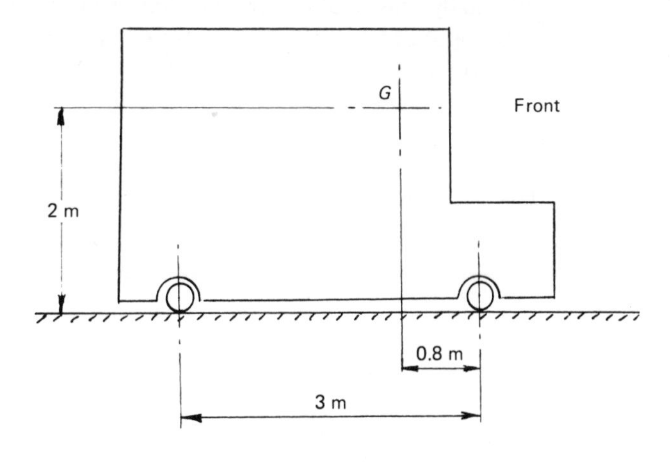

The illustration shows the elevation of a large loaded truck of mass m; G is the mass centre. The truck is descending a slope of 1 in 4 (i.e sin $\theta = \frac{1}{4}$) at a speed of 12 m s^{-1} when a braking force is applied to bring it to rest. Calculate the least distance in which it may be brought to rest so that it would not tip forward over the front wheels.

The same vehicle now ascends the same slope with a forward acceleration of $(\frac{1}{20}g)$. Calculate the values of the front- and rear-wheel reactions in terms of m and g.

PCL

Solution 2.18

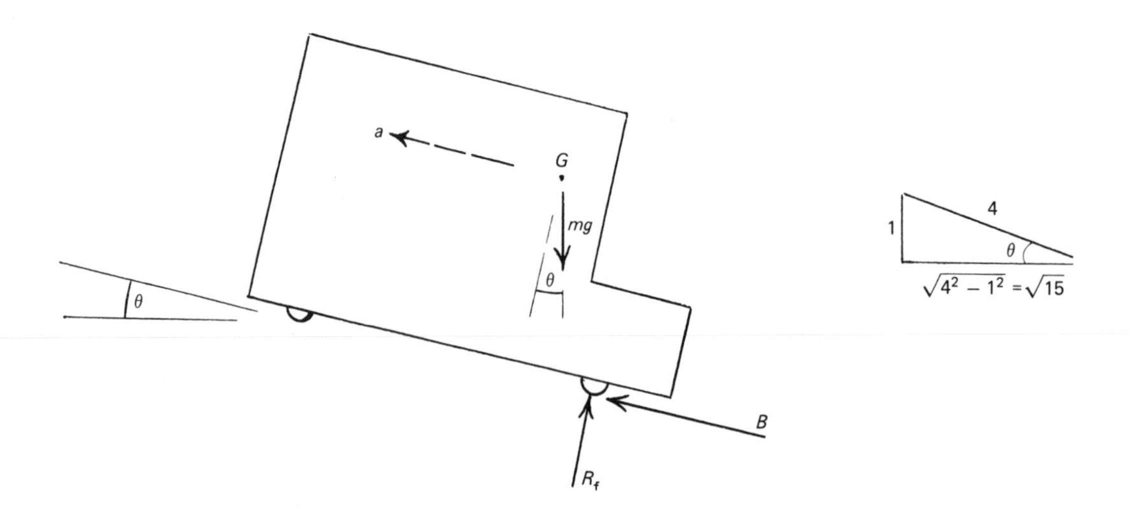

We have encountered braking force before (see Example 2.2). It must be recognised as a force of friction between road and wheels. Note that this question is somewhat academic, in that it is unlikely that any vehicle of this type is capable of applying a braking force sufficient to tip the vehicle forwards; the wheels will almost certainly begin to slip before this condition is achieved. But since no information is supplied concerning friction coefficient, we must assume that the conditions stated in the question are valid.

We have also met 'limiting conditions' before. If the vehicle is assumed to be about to tip over the front wheels, we may assume that the retardation is sufficient to make the rear-wheel reaction zero. This is assumed in the diagram above. (Incidentally, this makes the question even more improbable, because if the rear wheels are about to lift off the ground then they are unable to contribute to the braking force.)

In the diagram the acceleration has for obvious reasons been assumed to be up the slope.

The equation along the slope ($\Sigma F = ma$) is therefore

$$B - mg \sin \theta = ma \tag{1}$$

and across the slope ($\Sigma F = 0$)

$$R_f = mg \cos \theta . \tag{2}$$

Taking moments about G ($\Sigma M = 0$),

$$B \times 2 = R_f \times 0.8 \tag{3}$$

and substituting in equation 3 for B and R_f,

$$2\,(ma + mg \sin \theta) = 0.8mg \cos \theta .$$

Cancelling m,

$$a = 0.4g \cos \theta - g \sin \theta .$$

θ may be calculated but it is easier in such cases to draw the appropriate right-angled triangle. This is shown in the diagram.

It is easily seen that $\cos\theta = \dfrac{\sqrt{15}}{4} = 0.9682.$

$$a = 9.81\,(0.4 \times 0.9682 - \tfrac{1}{4})$$
$$= 1.347 \text{ m s}^{-2}.$$

Using the kinematic equation $v^2 = u^2 + 2ax$,

$$0 = (12)^2 - 2 \times 1.347x.$$

$$\therefore\ x = \frac{144}{2.694}$$
$$= \underline{53.45 \text{ m.}}$$

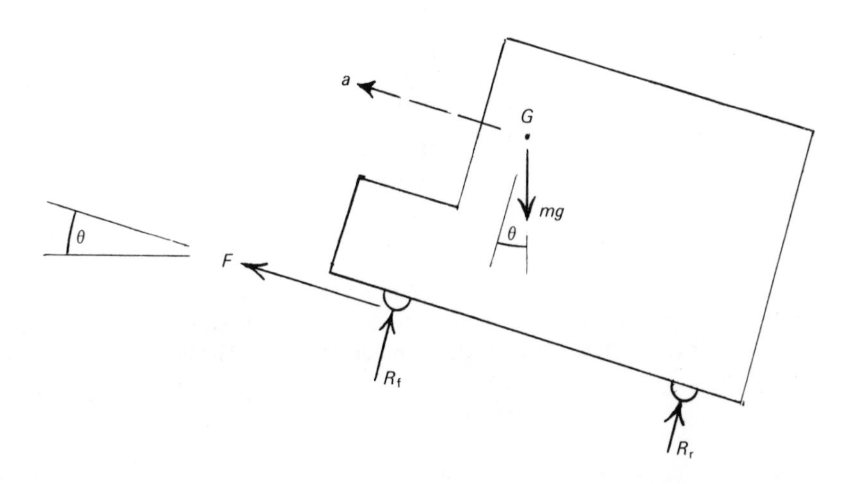

The new free-body diagram for the second part is shown here. The engine must clearly now be exerting a tractive force which we call *F*.

In the direction of acceleration ($\Sigma F = ma$),

$$F - mg\sin\theta = ma. \tag{4}$$

In the perpendicular direction ($\Sigma F = 0$),

$$R_r + R_f = mg\cos\theta. \tag{5}$$

And taking moments about G ($\Sigma M = 0$),

$$F \times 2 + R_f \times 0.8 = R_r(3 - 0.8). \tag{6}$$

From equation 4,

$$F = m\left(\frac{g}{20} + g\sin\theta\right)$$
$$= mg\,(0.05 + \tfrac{1}{4})$$
$$= 0.3mg.$$

Substituting in equation 6 and also for R_r from equation 5:

$$0.6mg + 0.8R_f = 2.2(mg\cos\theta - R_f).$$

$$\therefore\ R_f(0.8 + 2.2) = 2.2mg \times 0.9682 - 0.6mg.$$

$$\therefore\ R_f = mg\left(\frac{2.130 - 0.6}{3.0}\right)$$

$$= \underline{0.51mg.}$$

From equation 5, $R_r = mg(0.9682 - 0.51)$

$$= \underline{0.4582mg.}$$

Example 2.19

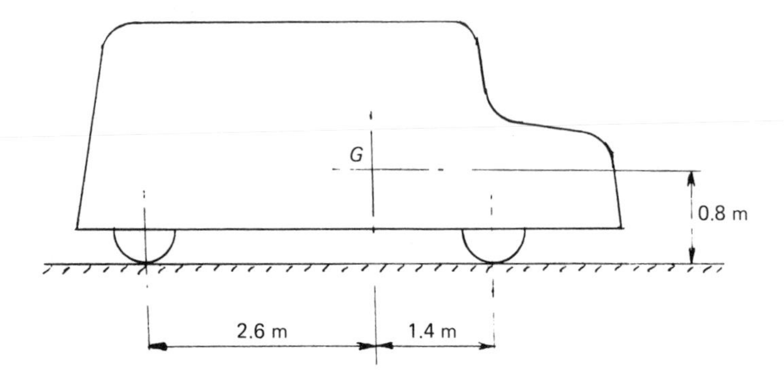

The diagram shows the elevation of a wheeled vehicle and the location of its mass centre G relative to the front and rear wheel centres. The mass of the vehicle is 1200 kg. Calculate the total normal reaction forces at front and rear wheels when the vehicle ascends a slope of 15° with a forward acceleration of 0.8 m s^{-2}. The inertia of the vehicle wheels may be neglected.

PCL

Solution 2.19

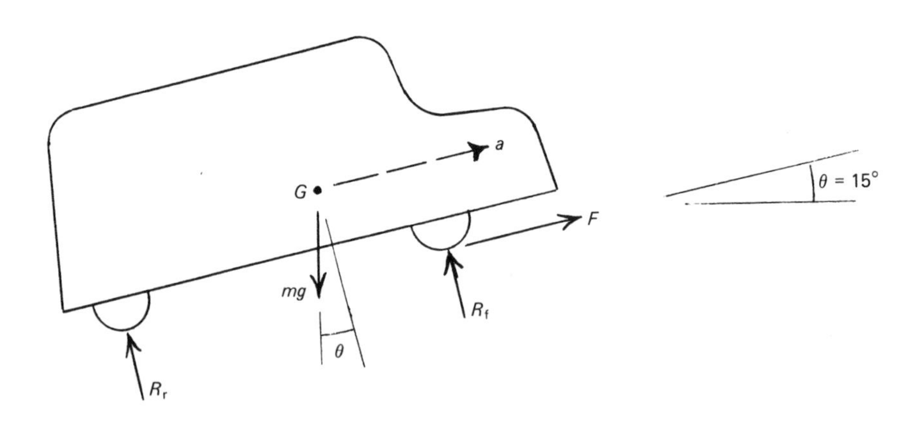

The problem is straightforward, with no complications. The free-body diagram is shown with a tractive force *F*. The reference to wheel inertia will be explored more fully in Chapter 4; it should be realised that if the moment of inertia of the wheels is not negligible then the answers will be different. Here, it is specifically stated that the effect *is* negligible but if it is not so stated, and no information is given about the inertia values, we work on the principle that when data are not stated, the associated effects may be neglected. Notice also that the free-body diagram is not affected by the question of whether it is the front or rear wheels which drive; the tractive force in either case is tangential to the wheels.

The equations are:

Parallel to the track ($\Sigma F = ma$) (upwards is positive):

$$F - 1200g \sin \theta = 1200a.$$

$$\therefore\ F = 1200\,(a + g \sin \theta)$$

$$= 1200\,(0.8 + g \sin 15°)$$

$$= 4007 \text{ N}.$$

Across the track ($\Sigma F = 0$),

$$R_f + R_r = mg \cos \theta$$

$$= 1200g \cos 15°$$

$$= 11\,371 \text{ N}.$$

Taking moments about G ($\Sigma M = 0$),

$$F \times 0.8 + R_f \times 1.4 = R_r \times 2.6.$$

Substituting for F and R_r,

$$4007 \times 0.8 + 1.4R_f = 2.6\,(11\,371 - R_f).$$

$$\therefore\ R_f(1.4 + 2.6) = 2.6 \times 11\,371 - 4007 \times 0.8$$

$$= 26\,359.$$

$$\therefore\ R_f = \underline{6590 \text{ N.}}$$

$$\therefore\ R_r = 11\,371 - 6590$$

$$= \underline{4781 \text{ N.}}$$

Example 2.20

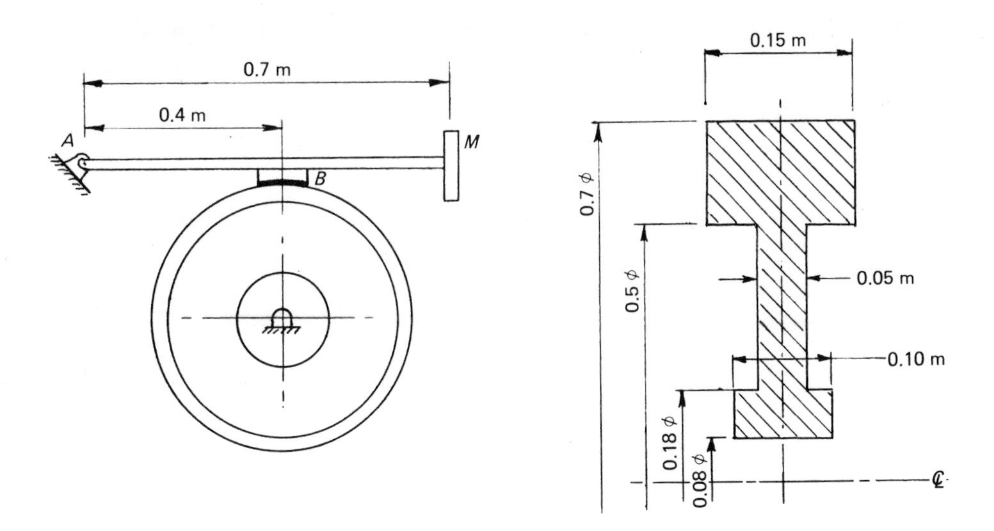

The basic design for a simple brake system on a flywheel is shown, together with the cross-section of the flywheel. A mass M is attached at the end of a lever, pivoted at A, on which a friction pad at B rubs on the edge of the flywheel. The coefficient of friction is 0.15. Find the mass M required to decelerate the flywheel at 0.25 rad s^{-2}. Density of flywheel material 7800 kg m^{-3}.

PCL

Solution 2.20

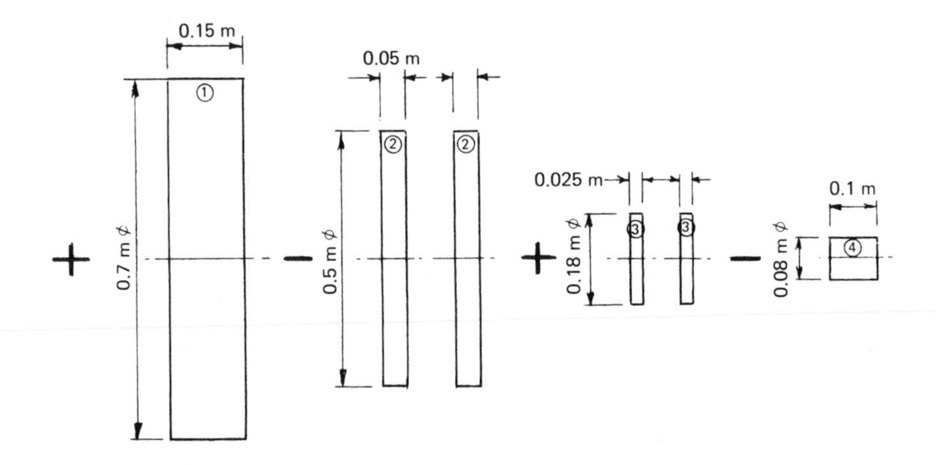

It is first necessary to calculate the moment of inertia of the flywheel. You are clearly expected to know that the moment of inertia of a uniform solid disc of radius r is $\frac{1}{2}mr^2$. The technique required is to 'build up' the wheel as a number of solid discs – in this example, four are required. The diagram shows the four solid-disc 'elements' of the wheel. Notice that elements 1 and 3 are positive, while 2 and 4 are negative. We may write down the complete expression for the moment of inertia I:

$$I = \tfrac{1}{2} \times 7800 \times \pi \times (0.35)^2 \times 0.15 \times (0.35)^2$$
$$- \tfrac{1}{2} \times 7800 \times \pi \times (0.25)^2 \times 0.1 \times (0.25)^2$$
$$+ \tfrac{1}{2} \times 7800 \times \pi \times (0.09)^2 \times 0.05 \times (0.09)^2$$
$$- \tfrac{1}{2} \times 7800 \times \pi \times (0.04)^2 \times 0.1 \times (0.04)^2.$$
$$= 3900\pi\ [(0.35)^4 \times 0.15 - (0.25)^4 \times 0.1 + (0.09)^4 \times 0.05 - (0.04)^4 \times 0.1].$$
$$= 22.83 \text{ kg m}^2.$$

Referring to the diagram of the basic brake design, we may calculate the reaction force R between brake pad and wheel by taking moments of forces on the arm about the pivot A:

$$R \times 0.4 = Mg \times 0.7. \qquad (1)$$

The friction force, acting tangentially to the flywheel, is (μR) and its torque about the wheel centre is $(\mu R)0.35$.

The equation of motion $(\Sigma M = I\alpha)$ is thus

$$0.35\,(\mu R) = I\alpha.$$

Substituting for R and inserting given values:

$$0.35 \times 0.15 \times Mg\,\frac{0.7}{0.4} = 22.83 \times 0.25.$$

$$\therefore\ M = \frac{22.83 \times 0.25 \times 0.4}{0.35 \times 0.15 \times 9.81 \times 0.7}$$

$$= \underline{6.333 \text{ kg.}}$$

Example 2.21

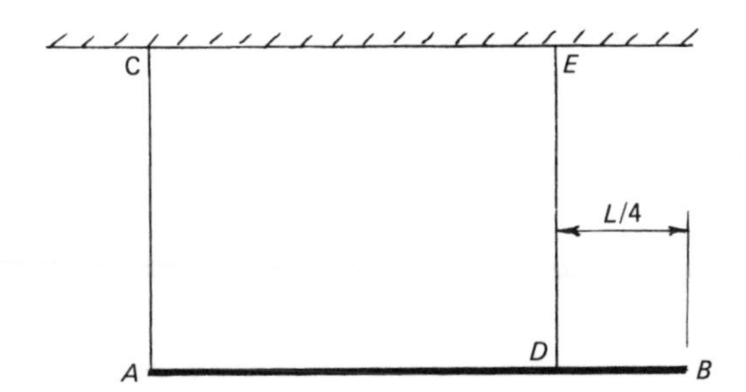

A slender, uniform rod AB, of length L and mass m, is suspended in the horizontal plane by two vertical, light, inextensible wires AC and DE as shown in the diagram. If, when the rod is at rest, wire AC is cut, show that the tension in the wire DE reduces to an instantaneous value of $\frac{6}{7}$ of that which it was before AC was cut.

U. Surrey

Solution 2.21

The free-body diagram here is of the rod *after* the wire is cut.

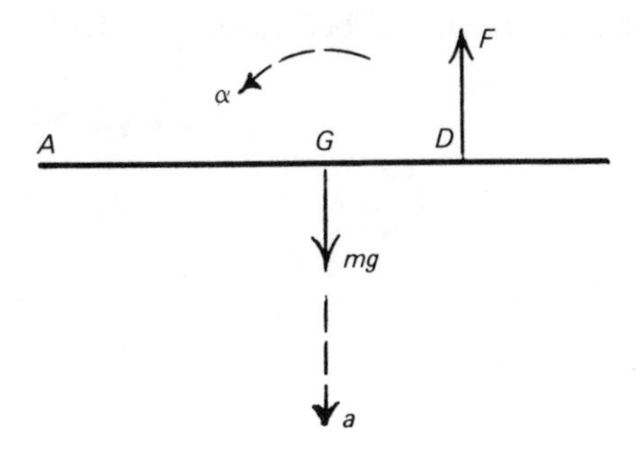

There is, of course, no force associated with AC, but the diagram refers to the instant of cutting, so that the rod is still horizontal.

We may analyse the motion of the rod as comprising the two components:

(a) translation of the mass centre, G (vertically downwards);

(b) rotation of the rod about the mass centre.

The equation of translation ($\Sigma F = ma$) is

$$mg - F = ma \tag{1}$$

and that of rotation about G ($\Sigma M = I\alpha$) is

$$F \times \frac{L}{4} = I_g \alpha \tag{2}$$

(assuming an obvious anticlockwise angular acceleration).

We relate linear acceleration of G and angular acceleration. Since the right-hand wire is not cut, D remains a fixed point and the rod turns about this point. Hence

$$a = \alpha R = \alpha(\tfrac{1}{4}L).$$

This is one of several questions in this book wherein the student is expected to know the expression for the moment of inertia of a uniform rod. In this case, for a rod about a transverse axis through the centre:

[d] $$I = \tfrac{1}{12}mL^2.$$

Substituting in equation 2,

$$F \times \tfrac{1}{4}L = \tfrac{1}{12}mL^2\alpha.$$

$$\therefore\ F = \tfrac{1}{3}mL\alpha.$$

$$\therefore\ \alpha = \frac{3F}{mL}.$$

Substituting this in equation 1,

$$mg - F = ma$$

$$= m(\tfrac{1}{4}L)\alpha$$

$$= m(\tfrac{1}{4}L)\frac{3F}{mL}$$

$$= \tfrac{3}{4}F.$$

$$\therefore\ mg = 1\tfrac{3}{4}F.$$

$$\therefore\ F = \frac{4}{7}mg.$$

Let the force in DE before cutting AC be F_0. Taking moments about A,

$$mg \times \tfrac{1}{2}L = F_0 \times \tfrac{3}{4}L.$$

$$\therefore\ F_0 = \frac{2}{3}mg.$$

$$\therefore\ \frac{F}{F_0} = \frac{4}{7} \times \frac{3}{2} = \underline{\frac{6}{7}}.$$

Example 2.22

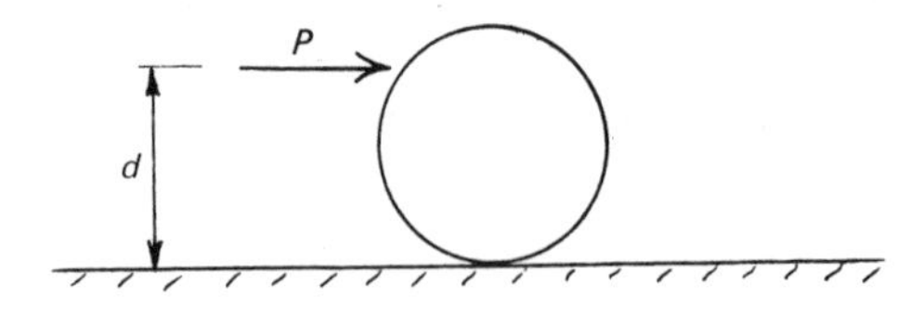

A homogeneous sphere of mass m and radius r lies on a horizontal table and is subject to a constant force of magnitude P which acts in a horizontal plane at a distance d above the surface of the table as shown in the diagram.

If the coefficient of sliding friction between the sphere and the table is μ derive an expression for the distance d such that the sphere translates without rotation. What is the value of this translational acceleration?

U. Surrey

Solution 2.22

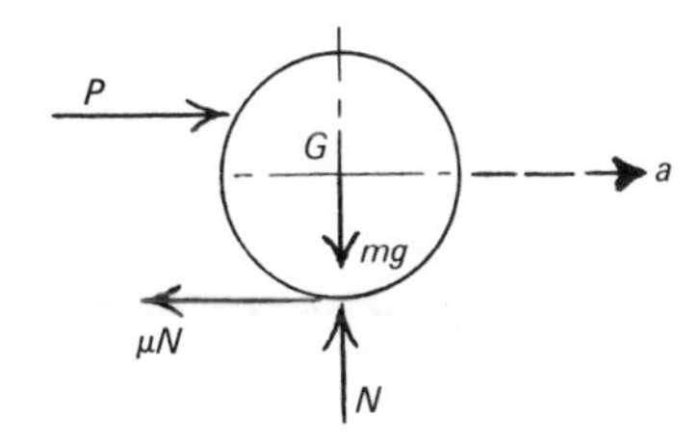

The free-body diagram, shown here, discloses four forces: weight, force *P*, a normal reaction force *N* and a friction force. Since motion must be from left to right, the friction force must be in the opposite direction.

The kinematics are simple because there is no angular acceleration, merely a linear acceleration *a*.

The equation of translation ($\Sigma F = ma$) is

$$P - \mu N = ma.$$

The equation of vertical equilibrium is

$$N = mg.$$

$$\therefore\ P - \mu mg = ma. \tag{1}$$

The equation of rotation ($\Sigma M = I\alpha$) where $\alpha = 0$, taking moments about G, is:

$$P(d - r) + \mu Nr = 0. \tag{2}$$

$$\therefore\ d = -\frac{\mu mgr}{P} + r$$

$$= r\left(1 - \frac{\mu mg}{P}\right). \tag{3}$$

From equation 1,

$$a = \frac{P}{m} - \mu g.$$

And from equation 2,

$$P = -\frac{\mu mgr}{d - r}.$$

$$\therefore\ a = -\frac{\mu gr}{d - r} - \mu g$$

$$= \mu g\left(\frac{r}{r - d} - 1\right)$$

$$= \frac{\mu gd}{r - d}.$$

The changing round of the term ($d - r$) conveniently gets rid of a negative sign in the expression, and in any case it can be seen from equation 3 that *d* must be less than *r*, as the bracketed term must always be less than unity.

Example 2.23

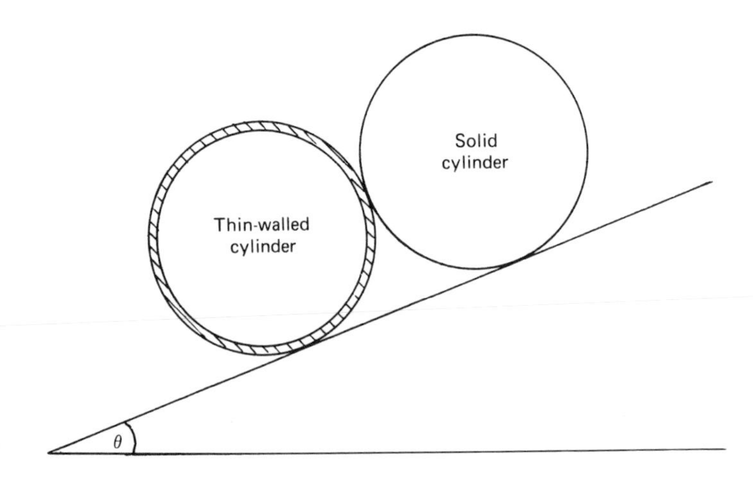

A solid homogeneous cylinder and a thin-walled cylinder lie at rest in contact with one another on a rough plane inclined at an angle θ to the horizontal as shown in the diagram. Both cylinders have mass m and are of radius r. Determine the linear acceleration of the cylinders after release if the coefficient of friction between them is μ.

How would this result be affected if the relative positions of the two cylinders were reversed?

U. Surrey

Solution 2.23

The problem cannot be solved unless it is assumed that both cylinders roll without slipping on the plane. The plane is described as a 'rough plane' and therefore it is legitimate to make this assumption. There are two possibilities in regard to the motion. Either the lower cylinder will accelerate faster than the upper one, and 'run away', or the upper cylinder will try to accelerate faster than the lower one and roll down with it, pushing it all the time. Since the question asks for acceleration and not accelerations, we may assume the latter. (We will see when we come to the last part of the question that this is a correct assumption.)

With this in mind, we can now draw the two free-body diagrams (page 95). A few remarks are necessary. The force between the cylinders is designated as P; it is of course equal and opposite on the two cylinders. The forces between cylinders and plane are called F_1 and F_2. A frequent error is to label such forces as (μN) where N is the corresponding normal force. But friction force is equal to (μN) only when slip is taking place or is just about to. At other times it is less than (μN).

Both cylinders must be assumed to be rolling down the plane without slip; hence they must rub against each other so that the friction force between them acts in a direction to try to prevent this rolling. The equal and opposite friction forces (μP) are in accordance with this requirement.

The cylinders are labelled 'thin-walled' and 'solid'. A full answer cannot be obtained to the question unless we assume the respective moments of inertia to be (mr^2) and $(\frac{1}{2}mr^2)$.

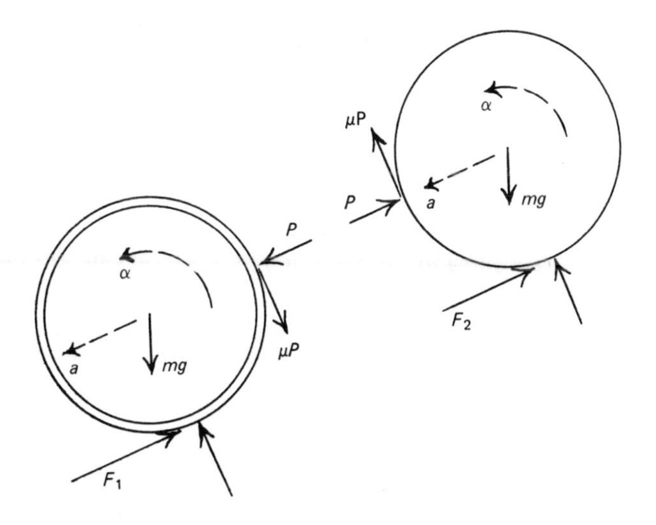

With these important points cleared up, we can now set down some equations. First, we may relate angular and linear acceleration kinematically. As we have encountered in several problems elsewhere, roll without slip on a surface gives

$$a = \alpha r$$

for both cylinders.

For the lower cylinder translation ($\Sigma F = ma$),

$$mg \sin \theta + P - F_1 = ma, \tag{1}$$

and for rotation ($\Sigma M = I\alpha$),

$$F_1 r - \mu P r = I_1 \alpha$$

$$= mr^2 \left(\frac{a}{r}\right).$$

$$\therefore \; F_1 - \mu P = ma. \tag{2}$$

We can reduce equations 1 and 2 to a single equation by substituting for F_1 in equation 1:

$$mg \sin \theta + P - (ma + \mu P) = ma.$$

$$\therefore \; P(1 - \mu) = 2ma - mg \sin \theta. \tag{3}$$

The equations for the upper cylinder are written and manipulated in the same manner.

Translation:

$$mg \sin \theta - P - F_2 = ma. \tag{4}$$

Rotation:

$$F_2 r - \mu P r = I_2 \alpha$$

$$= \tfrac{1}{2} mr^2 \left(\frac{a}{r}\right).$$

$$\therefore \; F_2 - \mu P = \tfrac{1}{2} ma. \tag{5}$$

$$mg \sin \theta - P - (\tfrac{1}{2} ma + \mu P) = ma.$$

$$\therefore \; P(1 + \mu) = mg \sin \theta - 1\tfrac{1}{2} ma. \tag{6}$$

Eliminating P from equations 3 and 6,

$$\frac{2ma - mg \sin \theta}{1 - \mu} = \frac{mg \sin \theta - 1\frac{1}{2}ma}{1 + \mu}.$$

Cross-multiplying,

$$2ma - mg \sin \theta + 2\mu ma - \cancel{\mu mg \sin \theta}$$
$$= mg \sin \theta - 1\tfrac{1}{2}ma - \cancel{\mu mg \sin \theta} + 1\tfrac{1}{2}\mu ma.$$

Cancelling m throughout and collecting all 'a' terms together,

$$a(2 + 2\mu + 1\tfrac{1}{2} - 1\tfrac{1}{2}\mu) = 2g \sin \theta.$$

$$\therefore\ a = \frac{2g \sin \theta}{3\frac{1}{2} + \frac{1}{2}\mu}$$

$$= \frac{4g \sin \theta}{7 + \mu}.$$

We turn to the final part of the question. We may approach this by considering the cylinders rolling, separately, without each constraining the other. The conditions will be as before except that the force P will be absent, or zero. If we look at equation 3, putting $P = 0$, we find that the lower thin-walled cylinder would have an acceleration of magnitude $\frac{1}{2}g \sin \theta$. Equation 6 informs us similarly that the solid cylinder rolling alone would have an acceleration of $\frac{2}{3}g \sin \theta$. Clearly, if they were set on the plane with the solid one in the lower position, it would run away from the other, having the greater acceleration.

Example 2.24

A cylinder of radius r, mass m and radius of gyration $r/\sqrt{2}$ rolls without slipping upon a fixed cylinder of radius $4r$ as shown in the diagram. If the smaller cylinder starts from rest at A (the highest point on the fixed cylinder), derive an expression for the normal reaction at the point of contact at the instant when the position of the smaller cylinder is defined by the angle θ. Hence show that the cylinders will lose contact when $\cos \theta = \frac{4}{7}$.

U. Surrey

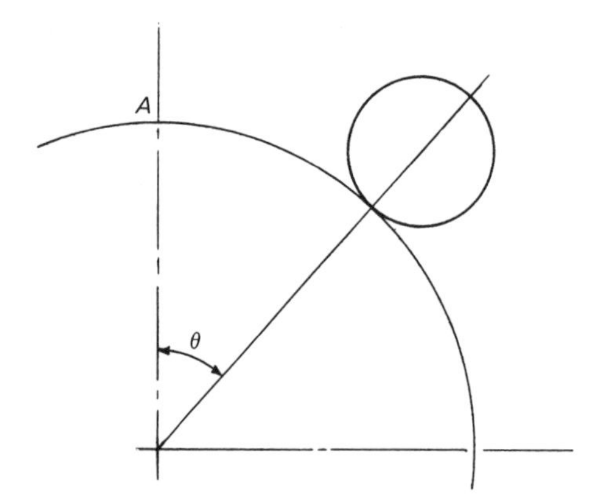

Solution 2.24

It is instructive to consider why the cylinders eventually lose contact with one another. As the small cylinder rolls along the large one, it has a centripetal acceleration which of course is directed towards the centre of the fixed cylinder.

Acceleration requires force to produce it, and the force here is the radially inward component of the weight of the small cylinder. As the velocity of the small cylinder increases, so must the acceleration, and a point is reached when the weight is only just sufficient to produce the acceleration. Beyond this critical point the small cylinder moves away from the large one and the motion will change completely.

The free-body diagram for the rolling cylinder is as shown here.

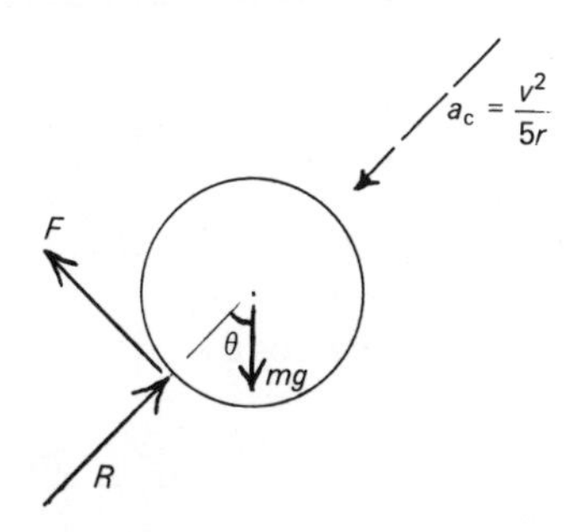

Since the cylinder rolls without slip, there is a friction force between the cylinders; this is shown as *F*. It must be in the direction shown, in order to provide the clockwise rolling consistent with the conditions.

To determine the centripetal acceleration we require an expression for the velocity of the moving cylinder. We may obtain this from an energy equation. (See Chapter 5.) The rolling cylinder loses potential energy and gains, in turn, kinetic energy, by virtue both of the translation around the fixed cylinder, and of rotation about its own axis.

$$\text{Loss of p.e.} = \text{Gain of k.e.}$$

$$mgh = \tfrac{1}{2}mv^2 + \tfrac{1}{2}I\omega^2.$$

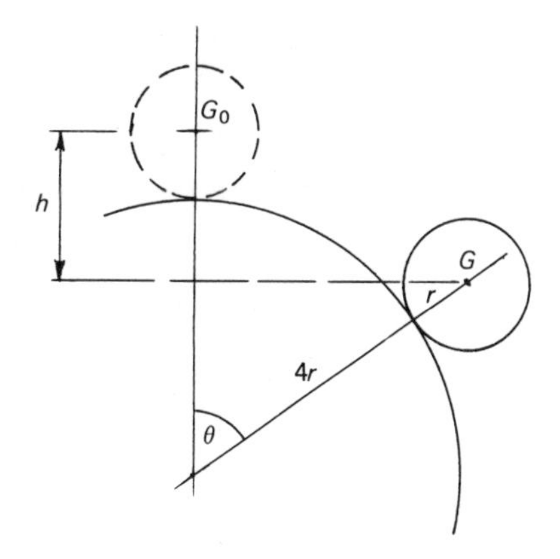

From the diagram it can be seen that the loss of height h is

$$h = (r + 4r) - (r + 4r)\cos\theta$$

$$= 5r(1 - \cos\theta).$$

Kinematically, for roll without slip,

$$v = \omega r.$$

$$mg \times 5r(1 - \cos\theta) = \tfrac{1}{2}mv^2 + \tfrac{1}{2}I\left(\frac{v^2}{r^2}\right).$$

Substituting $I = \frac{1}{2}mr^2$, for a uniform cylinder,

$$5mgr(1 - \cos\theta) = \tfrac{1}{2}mv^2 + \tfrac{1}{2}\left(\tfrac{1}{2}mr^2\right)\left(\frac{v^2}{r^2}\right)$$

$$= \tfrac{3}{4}mv^2.$$

$$\therefore\ v^2 = \frac{20}{3}gr(1 - \cos\theta).$$

The equation of motion ($\Sigma F = ma$) in the radially inwards positive direction is:

$$mg\cos\theta - R = m\left(\frac{v^2}{5r}\right)$$

$$= \frac{m}{5r}\left(\frac{20}{3}gr(1 - \cos\theta)\right).$$

$$\therefore\ R = mg\cos\theta - \tfrac{4}{3}mg + \tfrac{4}{3}mg\cos\theta$$

$$= \underline{\tfrac{1}{3}mg(7\cos\theta - 4)},$$

which is the required expression.

It is easy to see that the bracketed term will reduce to zero when $\cos\theta = \frac{4}{7}$.

Example 2.25

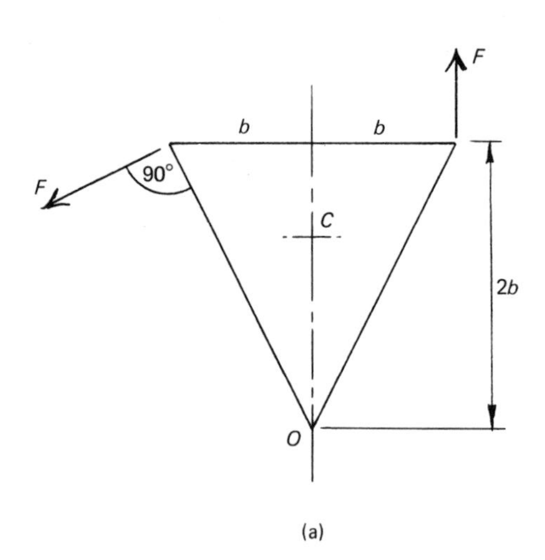

(a)

A triangular plate of uniform thickness and mass m, height $2b$ and base $2b$, rotates in a vertical plane about a fixed point at its apex O. The instant when the mass centre C of the plate is vertically above O is shown in diagram (a), with two forces, each of magnitude F, acting on the plate. Determine in terms of the given quantities:

(a) the moment of inertia of the plate about the fixed point O;
(b) the angular acceleration of the plate for the instant shown in the figure, and
(c) the *horizontal* force acting on the pivot at O for the instant shown in diagram (a).

U. Lond. K.C.

Solution 2.25

Part (a)

The required moment of inertia may be determined by combining the formula for a thin rod with the parallel-axis theorem and the integration process.

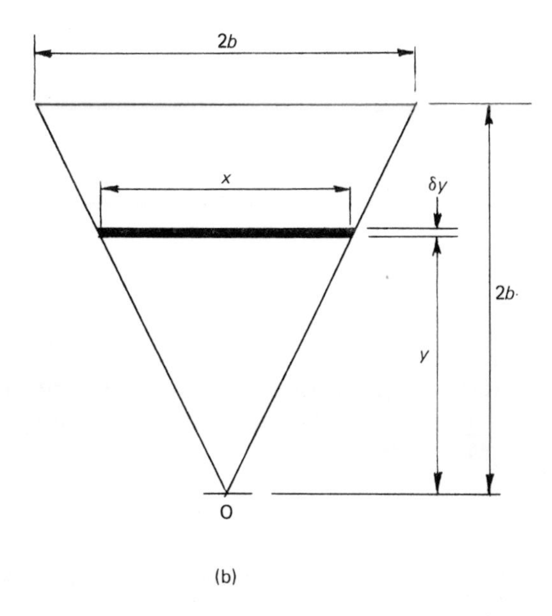

(b)

Consider the element shown in diagram (b), comprising a thin strip of length x, depth δy, and distance y from O. By simple proportion of two similar triangles,

$$\frac{x}{2b} = \frac{y}{2b}.$$

$$\therefore \; x = y.$$

The element area is $x \; \delta y = y \; \delta y$. Hence, by proportion to the whole triangle, the mass δm of the element is:

$$\delta m = \left(\frac{m}{\frac{1}{2}(2b)\,(2b)}\right)(y \; \delta y)$$

$$= \frac{my \; \delta y}{2b^2}.$$

The moment of inertia of the element, δI, using the formula for a thin rod, about its centre is:

$$\delta I = \tfrac{1}{12} \delta m L^2$$

$$= \tfrac{1}{12}\left(\frac{my \; \delta y}{2b^2}\right) y^2$$

and the corresponding moment of inertia about an axis through O, using the parallel-axis theorem, is

$$\delta I_o = \tfrac{1}{12} \frac{my^3 \; \delta y}{2b^2} + \left(\frac{my \; \delta y}{2b^2}\right) y^2$$

$$= \frac{13}{12}\left(\frac{my^3}{2b^2}\right) \delta y \, .$$

For the complete body,

$$I_o = \sum\left\{\frac{13}{12}\left(\frac{my^3}{2b^2}\right)\delta y\right\}$$

$$= \frac{13}{12}\frac{m}{2b^2}\int_0^{2b}(y^3)\,\delta y$$

$$= \frac{13m}{24b^2}\left[\tfrac{1}{4}y^4\right]_0^{2b}$$

$$= \frac{13m}{24b^2}\left(\tfrac{1}{4}(2b)^4\right)$$

$$= \underline{\frac{13}{6}mb^2}.$$

Part (b)

Refer to diagram (a). Both forces F contribute to anticlockwise acceleration. The 'radius-arm' of the inclined force is the length of the side of the triangle which is $\sqrt{b^2 + (2b)^2} = b\sqrt{5}$.

The equation of motion ($\Sigma M = I\alpha$) taking moments about O is

$$F(b\sqrt{5}) + F \times b = \frac{13}{6}mb^2\alpha.$$

$$\therefore\ 3.236\,F = 2.167\,mb\alpha.$$

$$\therefore\ \underline{\alpha = 1.493\,\frac{F}{mb}}.$$

Part (c)

The magnitude of the horizontal component of the force at O is obtained by writing the equation of translation of the body's mass centre, to the left. The linear acceleration of C is $\alpha \times \frac{2}{3}(2b)$. (The mass centre of a triangular lamina is two-thirds of the height from the apex.) Diagram (c) shows a simplified free-body diagram, showing only the horizontal component H of the force at O, since this is all that is required.

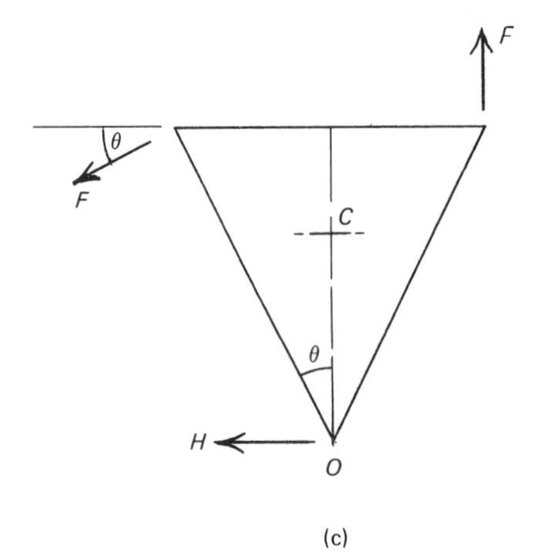

(c)

The equation $(\Sigma F = ma)$ is:

$$F \cos \theta + H = ma$$

$$= m\left(\frac{4}{3}b\right)\left(1.493\ \frac{F}{mb}\right).$$

θ is the half-angle at the triangle apex. Thus $\cos \theta = 2/\sqrt{5}.$

$$\therefore\ H = F\left(1.493 \times \frac{4}{3} - \frac{2}{\sqrt{5}}\right)$$

$$= \underline{1.096F.}$$

Example 2.26

A uniform disc of radius 20 mm and mass 0.1 kg rolls and slips on a horizontal plane. The axis of rotation of the disc is horizontal and it has an angular velocity ω_0 clockwise when its centre moves to the left with velocity v_0.

The disc comes to rest after 0.5 s, just at the instant that it starts to roll without slipping. Given that the coefficient of friction is 0.4, determine the initial velocities ω_0 and v_0. Also determine the distance travelled and the number of revolutions made by the disc before stopping.

U. Lond. U.C.

Solution 2.26

This is an instance of a rigid body having both translational and rotational motion which are kinematically unconnected, as the disc rolls *and* slips. The free-body diagram is shown here.

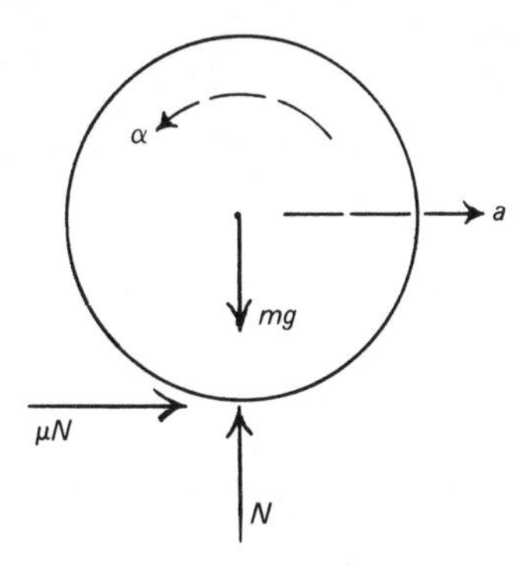

The equation of vertical equilibrium $(\Sigma F = 0)$ is:

$$N = mg. \tag{1}$$

The equation of translation $(\Sigma F = ma)$ taking right as positive is:

$$\mu N = ma.$$

$$\therefore\ \mu mg = ma.$$

Substituting,

$$0.4g = a.$$

Kinematically,

$$v = u + at.$$

$$\therefore\ 0 = -v_0 + at.$$

$$\therefore\ a = \frac{v_0}{t}.$$

$$\therefore\ 0.4g = \frac{v_0}{t}.$$

$$\therefore\ v_0 = 0.4gt$$

$$= 0.4 \times 9.81 \times 0.5$$

$$= \underline{1.962 \text{ m s}^{-1}}.$$

The equation of rotation ($\Sigma M = I\alpha$), taking anticlockwise as positive, is:

$$\mu Nr = I\alpha.$$

$$\therefore\ \mu mgr = \tfrac{1}{2}mr^2\alpha.$$

$$\frac{2\mu g}{r} = \alpha.$$

Kinematically,

$$\omega_2 = \omega_1 + \alpha t.$$

$$0 = -\omega_0 + \alpha t.$$

$$\therefore\ \alpha = \frac{\omega_0}{t}.$$

$$\frac{2\mu g}{r} = \frac{\omega_0}{t}.$$

$$\therefore\ \omega_0 = \frac{2\mu gt}{r}$$

$$= \frac{2 \times 0.4g \times 0.5}{20 \times 10^{-3}}$$

$$= \underline{196.2 \text{ rad s}^{-1}}.$$

For distance travelled,

$$x = \tfrac{1}{2}(u + v)t$$

$$= \tfrac{1}{2}(1.962 + 0) \times 0.5$$

$$= \underline{0.4905 \text{ m}}.$$

And for angular displacement,

$$\theta = \tfrac{1}{2}(\omega_1 + \omega_2)t$$

$$= \tfrac{1}{2}(196.2 + 0) \times 0.5$$

$$= 49.05 \text{ rad}$$

$$= \underline{7.807 \text{ rev}}.$$

Example 2.27

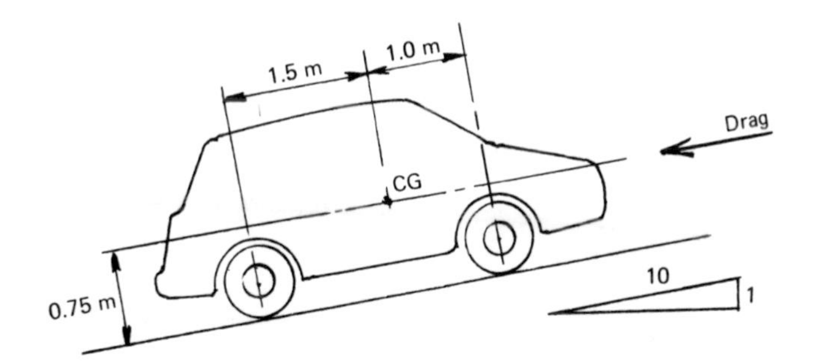

A vehicle of mass 900 kg is driven up an incline of slope $\sin^{-1} \frac{1}{10}$ as shown in the diagram. The drive is through the front wheels, and the drag from all frictional effects is represented by a resultant force of 400 N acting through the mass centre.

(a) Find the driving force at the tyres, and the power required, for a constant speed of 15 m s^{-1}.

(b) Find the maximum possible acceleration at the speed of 15 m s^{-1} if the coefficient of friction between the road and the tyres is 0.75.

Thames Poly.

Solution 2.27

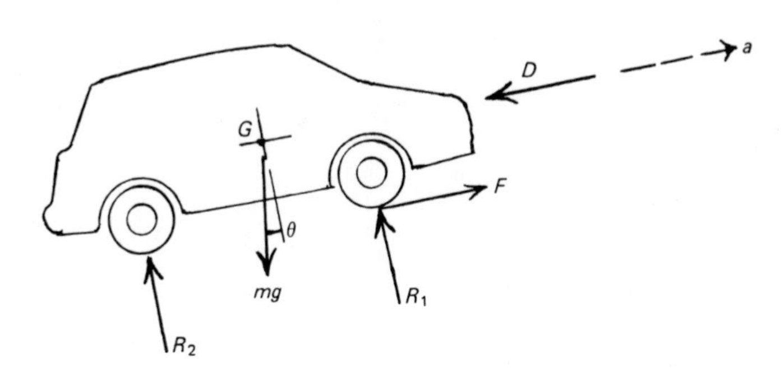

From the free-body diagram shown here the equation of forward motion is:

$$F - D - mg \sin \theta = ma = 0.$$

$$\therefore\ F = D + mg \sin \theta$$

$$= 400 + 900g \times 0.1$$

$$= \underline{1282.9 \text{ N.}}$$

Power:

$$W = Fv$$

$$= 1282.9 \times 15$$

$$= \underline{19.24 \text{ kW}}.$$

Assuming now an upward acceleration a and recognising that the driving force is the tangential force between road and front wheels, and that this force is limited to (μR_1), then

$$\mu R_1 - D - mg \sin \theta = ma. \qquad (1)$$

The equation of equilibrium across the track is

$$R_1 + R_2 = mg \cos \theta$$
$$= 900g \cos \theta$$
$$= 8784.7 \text{ N.} \qquad (2)$$

Taking a moment equilibrium equation about G:

$$\mu R_1 \times 0.75 + R_1 \times 1.0 = R_2 \times 1.5. \qquad (3)$$

Substituting for R_2,

$$0.75R_1 \times 0.75 + R_1 = (8784.7 - R_1) \times 1.5.$$
$$R_1 [(0.75)^2 + 1 + 1.5] = 8784.7 \times 1.5.$$
$$\therefore R_1 = \frac{8784.7 \times 1.5}{3.0625}$$
$$= 4302.7 \text{ N.}$$

Substituting in equation 1,

$$a = \frac{0.75 \times 4302.7 - 400 - 900g \times 0.1}{900} = \underline{2.160 \text{ m s}^{-2}}.$$

Example 2.28

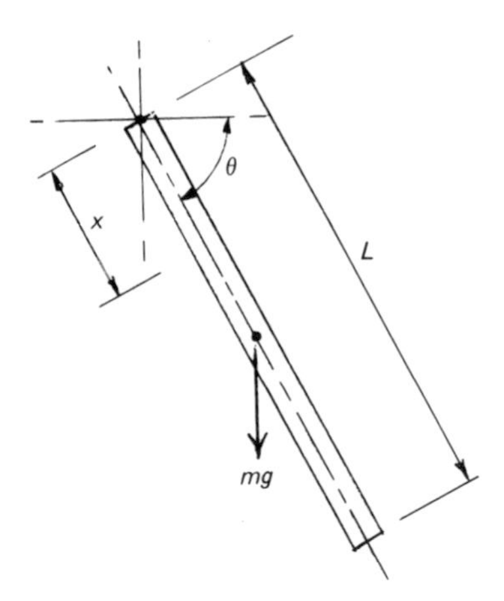

What is d'Alembert's principle?

The diagram shows a uniform rigid bar of mass m, length L, pivoted freely at its upper end about a horizontal axis. Find an expression for the bending moment M in the bar in terms of x and θ, and show that M is a maximum at $x = L/3$ for all θ.

U. Manchester

Solution 2.28

The Fact Sheet (f) refers to d'Alembert's principle. It must be stressed that this is an alternative approach to dynamics problems, and not an essential requisite. You are reminded that the method is to add to a free-body diagram (a) a 'reversed effective force' of magnitude (ma) at the mass centre of a body, in a direction opposite to the linear acceleration of the mass centre, and (b) a 'reversed effective moment' of magnitude ($I\alpha$) in a direction opposite to the angular acceleration α of the body. The augmented free-body diagram may then be treated as if in static

equilibrium, and the equations of dynamic motion become unnecessary. Since the question opens with a demand for a statement of the principle, we will employ the method to solve this problem.

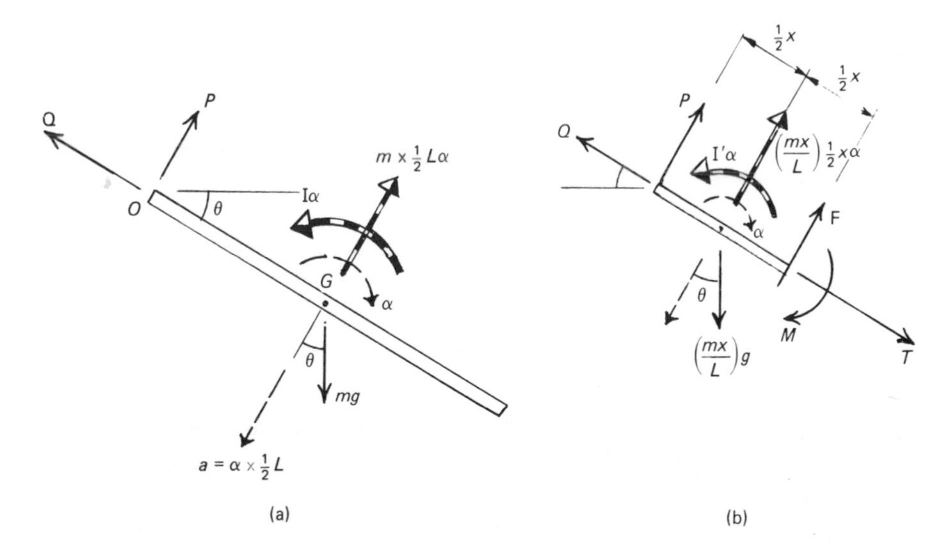

(a) shows the complete rod and (b) the portion of rod, length x, from the pivot, both as free-body diagrams. For the complete rod, only two forces act on the rod: the weight mg acting vertically downwards at G, and the pivot reaction force, shown as two components P and Q, respectively perpendicular to and parallel with the rod axis.

The rod is assumed to have an angular acceleration α in a clockwise direction. Therefore included on the diagram is a 'reversed effective moment' of magnitude ($I\alpha$) in an anticlockwise direction. Although this is shown by a distinguishing arrow around G, it is stressed that a moment does not have a particular point of application; it can be shown anywhere on the diagram.

As a consequence of this angular acceleration, the mass centre G must have a linear acceleration of magnitude ($\alpha \times \frac{1}{2}L$) (since the end O is fixed); furthermore, this acceleration must be in a 'down-left' direction perpendicular to the rod axis. Hence a 'reversed effective force' of magnitude (ma) = ($m \times \frac{1}{2}L\alpha$) is included on the diagram in a direction opposite to that of the acceleration; this reversed effective force acts at G.

Remember that both the reversed effective moment and the reversed effective force are not actual forces or moments; they are imaginary additions to the true free-body diagram, allowing us to treat the system henceforward as if it were in static equilibrium.

(In addition to the two acceleration components stated, the mass centre will also have a centripetal acceleration component, but it will be seen that this will not be needed to solve this particular problem, and it has therefore not been included on the diagram.)

By way of preparation for the solution, you are reminded that the moment of inertia of a thin uniform rod about an axis through the centre is $\frac{1}{12}mL^2$.

We begin by an equilibrium equation of moments about O:

$$I_g\alpha + \tfrac{1}{2}mL\alpha \times \tfrac{1}{2}L - mg\cos\theta \times \tfrac{1}{2}L = 0.$$

$$\therefore\ \alpha\left(I_g + \tfrac{1}{4}mL^2\right) = \tfrac{1}{2}mgL\cos\theta.$$

$$\therefore\ \alpha = \frac{\tfrac{1}{2}mgL\cos\theta}{\tfrac{1}{12}mL^2 + \tfrac{1}{4}mL^2}.$$

$$\therefore\ \alpha = \frac{3}{2}\frac{g}{L}\cos\theta.$$

An equation of force equilibrium in the direction of P gives

$$P + \tfrac{1}{2}mL\alpha = mg\cos\theta.$$

$$\therefore\ P = mg\cos\theta - \tfrac{1}{2}mL \times \frac{3}{2}\frac{g}{L}\cos\theta$$

$$= mg\cos\theta\,(1 - \tfrac{3}{4})$$

$$= \tfrac{1}{4}mg\cos\theta.$$

We now turn our attention to the free-body diagram (b) for the section of rod. Whenever an internal force or moment is required, it is necessary to show a section or part only of a member, as a free-body diagram. The part of the rod not shown will exert forces on the section shown, and these forces are shown as transverse (shear) and longitudinal (tension) components F and T, together with a bending moment, shown as M. The directions of all three components, not being known, are assumed.

The moment of inertia of this portion of the rod, I', using the same formula as before, and recalling that its mass is reduced in proportion to its length, is

$$\tfrac{1}{12}\left(\frac{mx}{L}\right)x^2.$$

The associated reversed effective moment, $I\alpha'$, and the reversed effective force,

$$\left(\frac{mx}{L}\right)\tfrac{1}{2}x\alpha,$$

are shown; it should of course be clear that this portion of rod has the same angular acceleration as the complete rod.

An equation of moment equilibrium about the 'cut' end conveniently eliminates the unknown force components Q, T and F. Writing all clockwise moments first,

$$M + Px + \left(\frac{mx}{L}\right)\alpha\,(\tfrac{1}{2}x)\,(\tfrac{1}{2}x) - I'\alpha - \left(\frac{mx}{L}\right)g\cos\theta\,(\tfrac{1}{2}x) = 0.$$

$$\therefore\ M = -Px - \left(\frac{mx}{L}\right)\frac{x^2}{4}\alpha + I'\alpha + \frac{mx^2}{2L}g\cos\theta$$

$$= -\tfrac{1}{4}mgx\cos\theta - \frac{mx^3}{4L}\left(\frac{3}{2}\frac{g}{L}\cos\theta\right) + \tfrac{1}{12}\left(\frac{mx}{L}\right)x^2\left(\frac{3}{2}\frac{g}{L}\cos\theta\right) + \frac{mx^2}{2L}g\cos\theta.$$

$$\therefore\ M = mg\cos\theta\left(-\frac{x}{4} - \frac{3x^3}{8L^2} + \frac{x^3}{8L^2} + \frac{x^2}{2L}\right)$$

$$= mg\cos\theta\left(-\frac{x}{4} + \frac{x^2}{2L} - \frac{x^3}{4L^2}\right).$$

At this point, we have a useful check on the correctness of this derivation. First, if we insert a value of $x = 0$ then this should give the bending moment at the hinged end of the rod. A frictionless hinge or pin cannot transmit any moment, and the answer should therefore be zero, which it is clearly seen to be. Second, a value of $x = L$ gives us the bending moment at the free end of the rod, and again the end of a bar cannot transmit any moment, so again the value of M should be zero when $x = L$. Again this is seen to be so, so the expression for M is likely to be correct.

In between these two boundary values of zero, there must be a point of maximum M, which is determined by differentiating the above expression with respect to x.

$$\frac{dM}{dx} = mg \cos\theta \left(-\frac{1}{4} + \frac{2x}{2L} - \frac{3x^2}{4L^2}\right) = 0.$$

The contents of the bracket must be 0. Multiplying all terms by $(4L^2)$,

$$-L^2 + 4xL - 3x^2 = 0.$$

Factorising,

$$(-L + x)(L - 3x) = 0$$

giving $x = L$ or $x = \frac{1}{3}L$,

the latter being the answer required. Although the question does not require it, the value of maximum bending moment can be determined by substituting the value of $x = L/3$ in the expression for M; the answer is $-\frac{1}{27}mgL \cos\theta$.

The determination of the bending moment would also have been accomplished by considering the dynamics of the other portion of rod, and you may give yourself some extra practice by solving the problem this way, although it is probable that the resulting algebra may be a little more daunting than in the above solution.

2.3 Problems

Problem 2.1

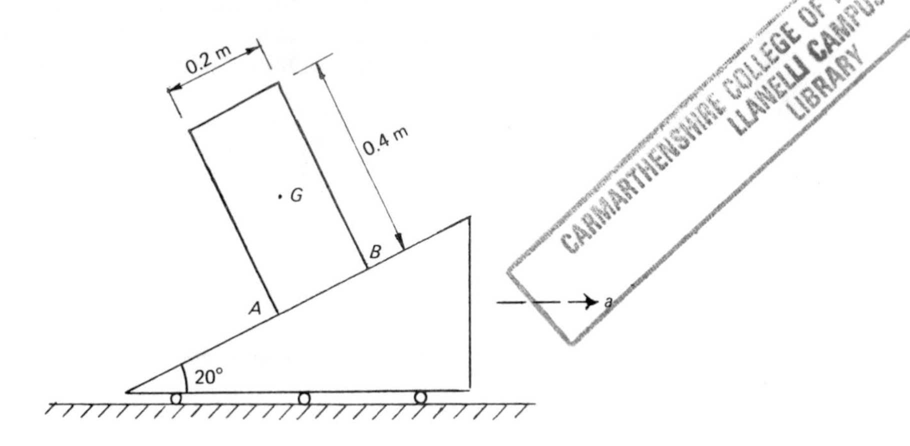

The diagram shows a uniform block of mass 20 kg, 0.2 m wide and 0.4 m high, resting on an inclined plane of slope 20°. The plane itself rests on a horizontal surface along which it can move. Calculate the least horizontal acceleration of the plane sufficient just to cause the block to tip up (a) about the lower edge A, (b) about the upper edge B. Assume in both cases that the block does not slip.

Hint: the direction of acceleration of the block is the same as that of the plane.

Problem 2.2

A uniform rectangular block 0.1 m wide and 0.2 m high and having a mass of 40 kg rests on a perfectly smooth flat horizontal surface. A horizontal string attached to the top passes round a fixed frictionless pulley and supports a hanging weight of mass 6 kg. A second

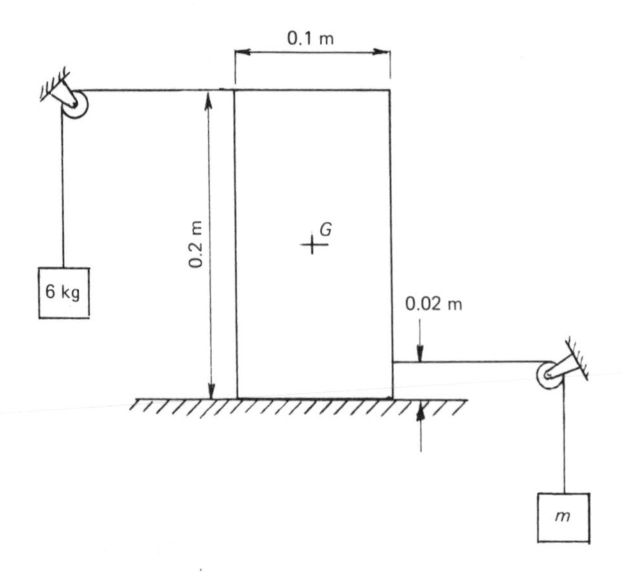

horizontal string attached 0.02 m from the base of the block passes round a second pulley and supports a weight of mass m. The diagram shows the arrangement. Calculate the maximum allowable value of m if the block is not to tip, and determine the corresponding acceleration.

Problem 2.3

A truck of mass 2500 kg has left side and right side wheel centres 1.8 m apart. The mass centre is located centrally between them and 0.8 m above ground level.

(a) Calculate the maximum permissible speed of the truck on a circular curve of mean radius 8 m banked inwards at an angle of 10° if the coefficient of friction between tyres and road is 0.76. State whether the truck is about to slip or tip at this speed.

(b) A crate 1 m wide and 1.8 m high is placed on the flat floor of the truck with the 1 m dimension transverse to the line of motion of the truck. Calculate the maximum permissible speed of the truck round the same curve if the crate is not to slip or tip, given that the coefficient of friction between crate and truck floor is 0.58. State whether the crate is about to slip or tip at this speed. The mass centre of the crate may be assumed to be at its geometric centre. Neglect the effect of the crate on the mass centre of the truck.

Problem 2.4

A vehicle has a total effective mass of 1300 kg. The distance between front- and rear-wheel centres is 2.65 m. The mass centre is located 0.95 m to the rear of the front-wheel centre and is 0.42 m above ground level. Evaluate the magnitudes of the front- and rear-wheel reaction forces, and the tractive (or braking) force when the vehicle is travelling up a slope of $\sin^{-1} 0.1$: (a) with an acceleration of 4.2 m s^{-2}; (b) with a retardation of 0.8 m s^{-2}.

Problem 2.5

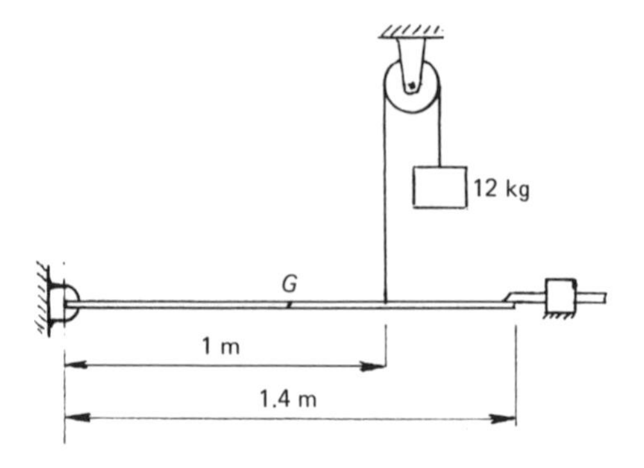

The diagram shows a simple trapdoor in a ceiling. The door may be considered as a uniform flat board of length 1.4 m and mass 15 kg. It is hinged at one end. A wire is attached 1 m from the hinge and passes over a fixed frictionless pulley, a weight of mass 12 kg hanging from the free end. The door is retained in the horizontal position by a bolt at the end. When the door is locked in the horizontal position as shown then the wire is vertical. If the bolt is removed, determine, at the instant of release:

(a) the angular acceleration of the door;
(b) the tension in the wire;
(c) the magnitude of the reaction force at the hinge;
(d) the bending moment at the point of attachment of the wire, assuming this to be uniform across the width of the door.

Problem 2.6

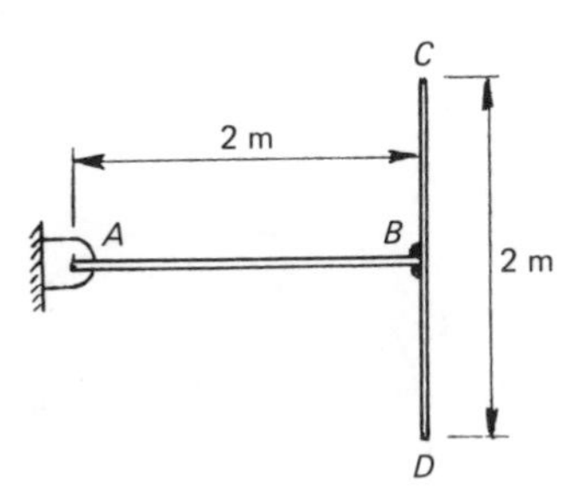

Two uniform rods, AB and CD, each of length 2 m and mass 12 kg, are welded together, the end of one to the mid-point of the other, to form a tee as shown in the diagram. The assembly is attached to a frictionless fixed pivot at A. It is held so that AB is horizontal, and then released from rest so that it swings downwards in a vertical plane. Determine the bending moment at the point B (a) at the instant of release; (b) after the assembly has turned through 45°.

Hint: See Example 2.8. Evaluate I_a. Evaluate α by an equation of rotation about A. Then draw the free-body diagram of CD and write the equation of rotation.

Problem 2.7

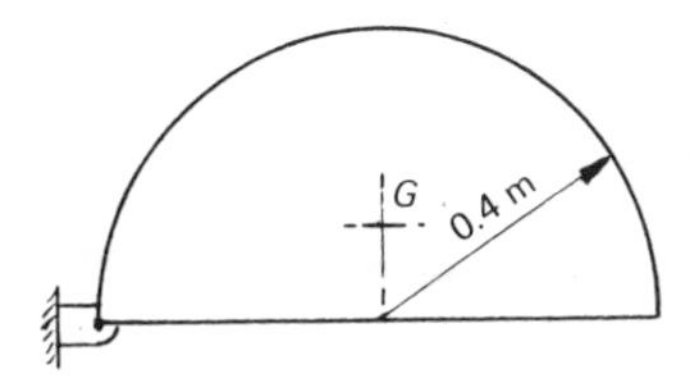

A uniform thin flat semicircular plate of mass 22 kg and radius 0.4 m is pivoted at one end so that it can swing in a vertical plane, as shown in the diagram. It is held so that the diameter is horizontal and released from rest. Calculate the angular velocity and acceleration, and the total reaction force at the pivot, when the plate has turned through 90°. The mass centre of a semicircular plate is distant $4R/3\pi$ from the centre. The moment of inertia of a *circular* disc about a polar axis through the centre is $\frac{1}{2}mR^2$.

Hints: See Example 2.11. Note the catch in the question: I is given for a circular plate, not a semicircular one. This point is alluded to in Example 2.11.

Problem 2.8

A uniform rod of mass 8 kg and length 1 m stands vertically on a horizontal surface. The coefficient of friction between surface and rod is 0.2. The rod is given a small negligible disturbance, causing it to fall. Verify that when the rod has turned through an angle of 15.52° the lower end will begin to slip on the surface. Determine the angular velocity and acceleration and the magnitude of the normal reaction force between rod and surface at this instant.

Hints: Refer to Example 2.10. Determine ω from an energy equation, and α from an equation of rotation about the lower end. Draw the free-body diagram, with the normal and tangential forces R and F at the lower end. Obtain two equations to solve for R and F. Thus show that F is $0.2R$.

Problem 2.9

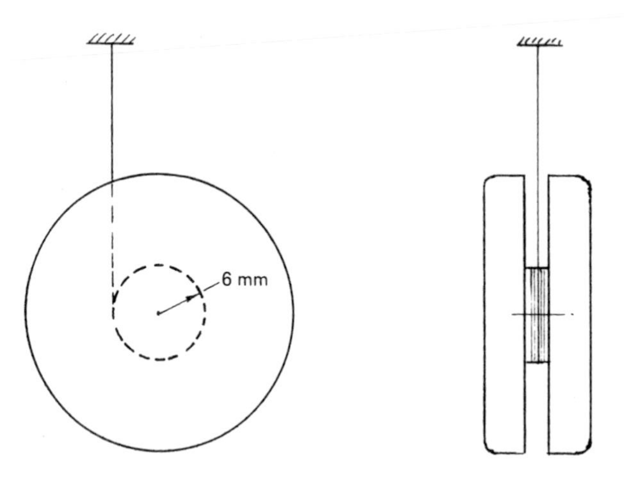

A dynamic toy called a Yo-Yo comprises two wooden discs joined by an axle around which a light string is wound, as shown in the two views in the diagram. The mass of the assembly is 100 g, its moment of inertia about the central axis is 8×10^{-5} kg m^2 and the axle radius is 6 mm. The end of the string is held stationary, and the disc is released and allowed to descend under gravity. Calculate its angular velocity and the tension in the string when the disc has descended 0.5 m, and calculate the time taken to reach this point. Assume the string to be always vertical.

Problem 2.10

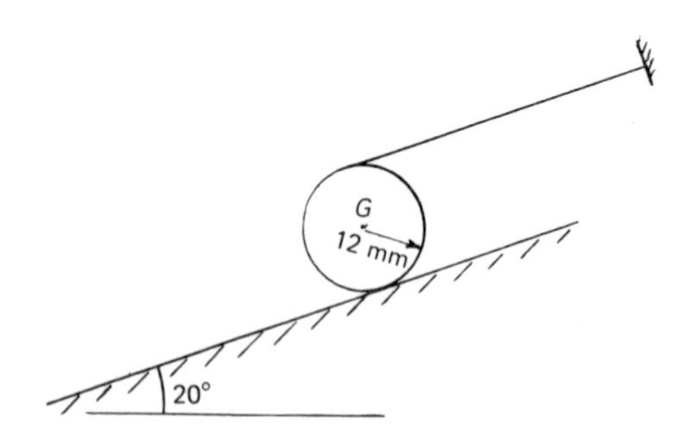

A cylinder of mass 1.8 kg, radius 12 mm and moment of inertia about its axis of 0.000 129 6 kg m^2 has a light cord attached to the periphery and wrapped around it. The cylinder rests transversely across an inclined plane of slope 20° to the horizontal, and the end of the string is secured to a fixed point, as shown in the diagram, so that the string is parallel to the plane. The coefficient of friction between cylinder and plane is 0.1. Calculate the linear acceleration of the cylinder down the plane, and the tension in the string when the cylinder is released from rest.

2.4 Answers to Problems

2.1 (a) 1.129 m s^{-2} in direction shown.
(b) 10.36 m s^{-2} in direction opposite to that shown.

2.2 20.24 kg. 2.109 m s^{-2}.

2.3 (a) 9.211 m s^{-1}; slip. (b) 7.980 m s^{-1}; tip.

2.4 (a) 7072.7 N and 5616 N; 6735.3 N (tractive).
(b) 8102.9 N and 4586.2 N; 235.3 N (tractive).

2.5 (a) 0.675 rad s^{-2}; (b) 109.62 N; (c) 44.618 N; (d) 9.141 N m.

2.6 (a) 20.77 N m. (b) 14.69 N m.

2.7 6.825 rad s^{-1}; 6.939 rad s^{-2}; 644.19 N (components 643.9 N and 17.98 N).

2.8 $\omega = 1.0359$ rad s^{-1}; $\alpha = 3.9374$ rad s^{-2}; $R = 70.12$ N.

2.9 108.32 rad s^{-1}; 0.9387 N; 1.539 s.

2.10 1.007 69 m s^{-2}; 2.566 N.

3 Momentum and Impact

Linear and angular momentum. Conservation of momentum. Collision of two bodies in translation and rotation. Impulse. Restitution. 'Lost' energy. Momentum as a vector quantity. Force of a fluid jet.

3.1 The Fact Sheet

(a) Linear Momentum

Linear momentum of a body = Mass × Velocity.

$$\text{Momentum} = mu.$$

The unit is the kilogram metre per second: kg m s^{-1}.

(b) Angular Momentum

Angular momentum of a body = Moment of inertia × Angular velocity.

$$\text{Angular momentum} = I\omega.$$

The unit is the kilogram metre squared radian per second: kg m^2 rad s^{-1}.

(c) Impulse

The impulse of a force F acting for time t is the product of force and time.

$$\text{Impulse} = Ft.$$

The unit is the newton second: N s.

(d) Conservation of Linear Momentum

When two translating bodies collide, the total momentum of the bodies is unchanged by the collision.

For bodies of masses m_a, m_b having pre-collision and post-collision velocities, u_a, u_b, v_a, v_b:

$$m_a u_a + m_b u_b = m_a v_a + m_b v_b.$$

The total momentum of the system can be changed only by the action of an external force.

(e) Change of Linear Momentum

When an external force acts on a system, the linear momentum of the system is changed in the direction of the force only.

The force is equal to the rate of change of momentum along the line of action of the force.

(f) Restitution

When bodies moving along a straight line collide, the ratio

$$\frac{\text{Relative velocity of separation}}{\text{Relative velocity of approach}}$$

is defined as the coefficient of restitution, e.

$$\frac{v_b - v_a}{u_b - u_a} = -e.$$

The negative sign in front of e recognises that the separation velocity must always be in the opposite direction to the approach velocity. Even if the value of e is given without the negative sign, it must always be given a negative value in calculations.

For 'elastically perfect' bodies $e = 1$. For non-perfect bodies e is less than 1.

(g) 'Lost' Energy

When bodies collide there is always some 'loss' of energy to the system. Energy is used in deforming the bodies, and not all of this is recovered. For this reason, the principle of conservation of energy must not be used for bodies before and after collision, except in the particular case that $e = 1$. This denotes 'elastically perfect' bodies when the energy absorbed in distortion is assumed to be fully restored to the system as the bodies recover after collision.

(h) Energy (Revision):

Potential energy of a mass m at a height h above a datum = mgh.
Kinetic energy of translation = $\frac{1}{2}mu^2$.
Kinetic energy of rotation = $\frac{1}{2}I\omega^2$.

3.2 Worked Examples

Example 3.1

Two bodies, A and B, are arranged to collide on a straight horizontal frictionless track. A has a striking velocity of u and B is initially stationary. The coefficient of restitution for the collision is 1. Determine the velocities of both bodies after collision in terms of u: (a) if they have equal mass; (b) if A has one-third the mass of B.

Solution 3.1

Part (a)

[d] $$mu + m \times 0 = mv_a + mv_b.$$

Simplifying, $$u = v_a + v_b. \qquad (1)$$

[f] $$\frac{v_b - v_a}{0 - u} = -1.$$

Simplifying, $$v_b - v_a = u. \qquad (2)$$

Adding equations 1 and 2,

$$2v_b = 2u.$$

$$\therefore \underline{v_b = u.}$$

Substituting in equation 2,

$$\underline{v_a = 0.}$$

Part (b)

[d] Momentum: $$\tfrac{1}{3}mu + m \times 0 = \tfrac{1}{3}mv_a + mv_b.$$

Simplifying, $$u = v_a + 3v_b. \qquad (3)$$

[f] Restitution: $$\frac{v_b - v_a}{0 - u} = -1.$$

Simplifying, $$v_b - v_a = u. \qquad (4)$$

Adding equations 3 and 4,

$$4v_b = 2u.$$

$$\therefore \underline{v_b = \tfrac{1}{2}u.}$$

Substitute in equation 3:

$$u = v_a + 3(\tfrac{1}{2}u).$$

$$\therefore v_a = u - 1\tfrac{1}{2}u.$$

$$\therefore \underline{v_a = -\tfrac{1}{2}u.}$$

In case (a) the striking body comes to rest and the struck body moves off with the same velocity. In case (b) both bodies rebound with equal velocities, in opposite directions.

Example 3.2

Two ships, A and B, are connected by an initially slack cable. A has a mass of 340 tonnes (1 tonne = 1000 kg) and is initially at rest. B has a mass of 115 tonnes and is moving away from A at a steady speed of 8 m s^{-1} causing the cable eventually to tighten.

(a) Calculate the common speed of the two ships at the instant the cable tightens to its full stretch.

(b) Calculate the reduction of kinetic energy of the two ships at the instant the cable is at full stretch.

(c) Calculate the final speeds of the two ships after the cable has ceased to stretch, assuming that 20 per cent of the energy before the impact is dissipated in the impact. Neglect the resistance of the water, and assume that, during the process, B is not being driven.

Solution 3.2

Part (a)

As the cable tightens, A will start to move, and at the instant the cable is stretched to its maximum, both ships will move at the same speed, v.

[d] Momentum: $$m_a u_a + m_b u_b = m_a v_a + m_b v_b.$$

$$0 + (115 \times 10^3) \times 8 = [(340 + 115) \times 10^3]\, v.$$

$$\therefore\ v = 8 \times \frac{115}{455}$$

$$= \underline{2.022 \text{ m s}^{-1}}.$$

Part (b)

$$\text{Initial k.e.} = \tfrac{1}{2} m_a u_a^2$$

$$= \tfrac{1}{2}(115 \times 10^3) \times 8^2$$

$$= 3680 \text{ kJ}.$$

$$\text{Energy when coupled} = \tfrac{1}{2}(455 \times 10^3)(2.022)^2$$

$$= 930.13 \text{ kJ}.$$

$$\therefore\ \text{reduction} = 3680 - 930.11$$

$$= \underline{2749.9 \text{ kJ}.}$$

Part (c)

[d] Momentum: $$m_a u_a + m_b u_b = m_a v_a + m_b v_b.$$

$$0 + (115 \times 10^3) \times 8 = (340 \times 10^3)\, v_a + (115 \times 10^3)\, v_b.$$

Simplify: $$920 = 340 v_a + 115 v_b. \qquad (1)$$

$$\text{Final k.e.} = 0.8 \text{ initial k.e.}$$

$$= 0.8 \times 3680$$

$$= 2944 \text{ kJ}.$$

$$\therefore\ \tfrac{1}{2} m_a v_a^2 + \tfrac{1}{2} m_b v_b^2 = 2944 \times 10^3.$$

$$\tfrac{1}{2}(340 \times 10^3)\, v_a^2 + \tfrac{1}{2}(115 \times 10^3)\, v_b^2 = 2944 \times 10^3.$$

$$\therefore\ 170 v_a^2 + 57.5 v_b^2 = 2944. \qquad (2)$$

From equation 1, $$v_b = \frac{920 - 340 v_a}{115}$$

$$= 8 - 2.956\, v_a.$$

Substitute in equation 2:

$$170 v_a^2 + 57.5\,(8 - 2.956 v_a)^2 = 2944.$$

$$170 v_a^2 + 57.5\,(64 + 8.738 v_a^2 - 47.3 v_a) = 2944.$$

$$170 v_a^2 + 3680 + 502.4 v_a^2 - 2719.8 v_a = 2944.$$

$$672.6 v_a^2 - 2719.8 v_a + 736 = 0.$$

$$\therefore\ v_a^2 - 4.045 v_a + 1.094 = 0.$$

$$\therefore\ v_a = \frac{4.045 \pm \sqrt{4.045^2 - 4 \times 1.095}}{2}.$$

$$\therefore\ v_a = 0.292 \text{ m s}^{-1} \text{ or } 3.753 \text{ m s}^{-1}.$$

From above: $v_b = 8 - 2.956v_a$,

giving corresponding values:

$$v_b = 7.137 \text{ m s}^{-1} \text{ or } -3.094 \text{ m s}^{-1}.$$

The only possible correct solution is:

$$\underline{v_a = 3.753 \text{ m s}^{-1}} \quad \text{and} \quad \underline{v_b = -3.094 \text{ m s}^{-1}}.$$

The two answers need explaining. The solution to part (a) of the problem tells us that the above answer must be the only possible one, because while the ships are moving together at 2.022 m s^{-1}, the cable is stretched, pulling *B* back and *A* forward. So the final speed of *A* must be more than 2.022 m s^{-1} and that of *B* must be less. The pair of answers rejected are the speeds which would result if the cable joining the ships were in compression (assuming this to be possible). *A* would then reduce in speed from 2.022 m s^{-1} and *B* would increase.

The answer to part (b) is also interesting. Although, ultimately, only 20 per cent of the initial system energy is lost, it is seen that the kinetic energy when both ships move together is very much less than 80 per cent of the original energy. The explanation is that a lot of the energy is stored in the cable itself, momentarily, as strain energy – the work done in stretching the cable. This is released, and partly restored to the ships as kinetic energy when the cable slackens again. Towing cables are required, for this reason, to be of a certain minimum length and special construction, in order to be capable of accepting energy transfers of this order.

Example 3.3

Two steel balls having masses of 2 kg and 3 kg hang vertically by light strings so that they just touch when at rest, their centres then being 0.8 m below the point of suspension. The smaller ball is moved away until the string makes an angle of 60° to the vertical, when it is then released from rest to strike the second ball. The balls collide with a coefficient of restitution of 1. Calculate the angle each string makes to the vertical when each ball has rebounded its maximum distance from the lowest position. By calculating the potential energies, before and after collision, show that no energy is lost to the system by the collision.

Solution 3.3

The ball is raised through a vertical height h. From diagram (a), page 117,

$$h = 0.8 - 0.8 \cos 60^\circ$$
$$= 0.4 \text{ m}.$$

The approach velocity can be calculated from an energy equation:

$$\text{Loss of p.e.} = \text{Gain of k.e.}$$

$$mg\,.\,h = \tfrac{1}{2}mv^2.$$

$$\therefore\ v = \sqrt{2gh} = \sqrt{2 \times 9.81 \times 0.4} = 2.801 \text{ m s}^{-1}.$$

[d] Momentum:

$$2 \times 2.801 + 3 \times 0 = 2v_a + 3v_b.$$

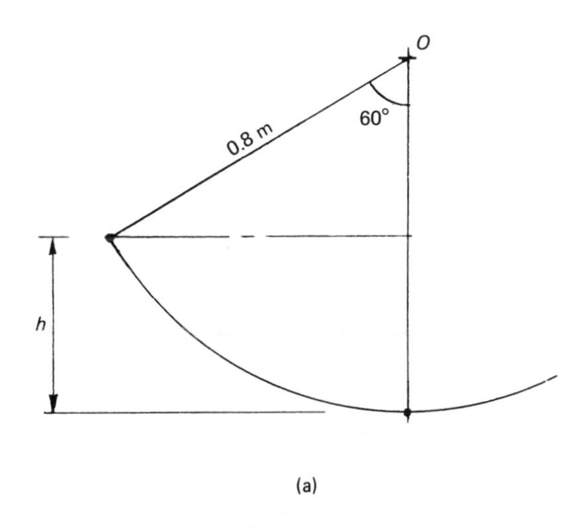

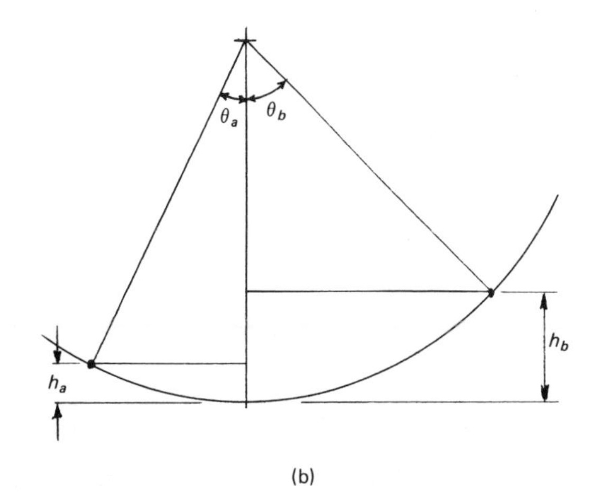

Simplify: $$2v_a + 3v_b = 5.602. \qquad (1)$$

[f] Restitution:

$$\frac{v_b - v_a}{0 - 2.801} = -1 \qquad \text{(taking direction of striking ball as positive).}$$

Simplify: $$v_b - v_a = 2.801. \qquad (2)$$

Substitute in equation 1 for v_b:

$$2v_a + 3(2.801 + v_a) = 5.602.$$
$$\therefore\ 5v_a = 5.602 - 8.403.$$
$$\therefore\ v_a = -0.5602 \text{ m s}^{-1} \text{ (i.e. rebounding).}$$

Substitute in equation 2:

$$v_b = 2.801 + (-0.5602)$$
$$= +2.2408 \text{ m s}^{-1} \text{ (i.e. in opposite direction).}$$

The energy equation used above can be used 'in reverse' to find the heights of rebound.

For 2 kg ball: $$h_A = \frac{v^2}{2g} = \frac{0.5602^2}{2 \times 9.81} = 0.0160 \text{ m.}$$

For 3 kg ball: $$h_B = \frac{2.2408^2}{2 \times 9.81} = 0.2559 \text{ m.}$$

And from the trigonometry of diagram (b):

$$h = 0.8 - 0.8\cos\theta.$$
$$\therefore\ \theta = \cos^{-1}\left(1 - \frac{h}{0.8}\right).$$

For 2 kg ball: $$\theta_a = \cos^{-1}\left(1 - \frac{0.0160}{0.8}\right) = \underline{11.48°}.$$

For 3 kg ball: $$\theta_b = \cos^{-1}\left(1 - \frac{0.2559}{0.8}\right) = \underline{47.14°}.$$

Initial k.e. (2 kg ball only) $= mgh$

$$= 2 \times 9.81 \times 0.4$$
$$= 7.848 \text{ J.}$$

Final k.e. after rebound (using heights already calculated)

$$= 2 \times 9.81 \times 0.0160 + 3 \times 9.81 \times 0.2559$$

$$= 7.845 \text{ J.}$$

The discrepancy is negligible.

Example 3.4

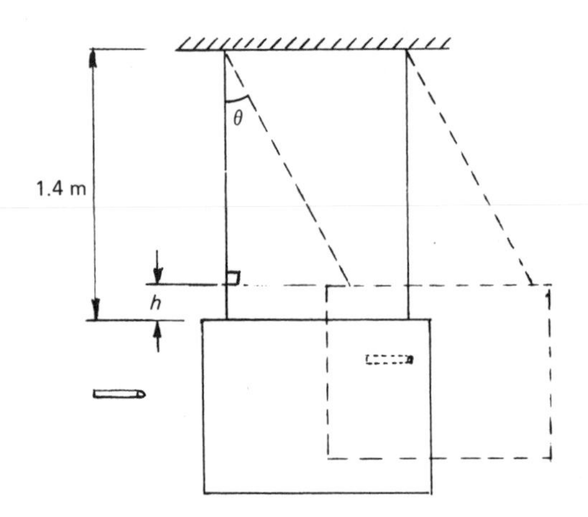

A ballistic pendulum comprises a box full of sand, having a total mass of 25 kg, and suspended by four strings of equal length 1.4 m, so that the box can swing sideways and upwards without turning, as shown in the diagram. A bullet of mass 32 g is fired horizontally into the box. It remains embedded in the sand, and the impact causes the box to swing so that the strings make an angle of 9° to the vertical at the point of maximum swing. Estimate the striking velocity of the bullet, and also calculate the energy dissipated by the impact.

Solution 3.4

In this problem, the coefficient of restitution is zero; the two bodies (box and bullet) do not separate. In such cases, we do not bother to use the restitution formula, but merely state that the velocities after impact are the same, i.e.

$$v_a = v_b.$$

Let the striking velocity of the bullet be u.

[d] Momentum: $0.032u + 25 \times 0 = 0.032v + 25v$

(calling the common velocity after impact v).

Simplify: $$0.032u = 25.032v. \quad (1)$$

v can be found by use of an energy equation, equating loss of k.e. to gain of p.e.

Let the vertical height the box is raised be h.

$$\text{Gain of p.e.} = \text{Loss of k.e.}$$

$$mgh = \tfrac{1}{2}mv^2.$$

$$\therefore\ v = \sqrt{2gh}. \quad (2)$$

From the diagram, $$h = 1.4 - 1.4 \cos\theta$$

$$= 1.4(1 - \cos 9°)$$

$$= 0.0172 \text{ m.}$$

Substitute in equation 2:

$$v = \sqrt{2 \times 9.81 \times 0.0172} = 0.5809 \text{ m s}^{-1}.$$

Substitute in equation 1:

$$u = \frac{25.032 \times 0.5809}{0.032}.$$

$$= \underline{454.4 \text{ m s}^{-1}}.$$

$$\text{Initial k.e.} = \tfrac{1}{2} \times 0.032 \times (454.4)^2 = 3303.8 \text{ J}.$$

$$\text{Final k.e.} = \tfrac{1}{2} \times 25.032 \times (0.5809)^2 = 4.223 \text{ J}.$$

$$\therefore \text{ Loss} = \underline{3299.577 \text{ J}}.$$

It is seen that practically all the initial kinetic energy of the bullet is spent in embedding itself in the sand.

You may recall a warning against using energy equations to solve problems of collision, and may object to the use of an energy equation in this solution. But this energy equation is valid because it concerns the time *after* the impact of the bullet. The only inaccuracy in this equation would be due to neglecting resistance of the air to the swinging box, and this would be very small. The thing you must *not* do is to equate the initial k.e. of the bullet to the final p.e. of (box and bullet) – as the final answer clearly shows.

Example 3.5

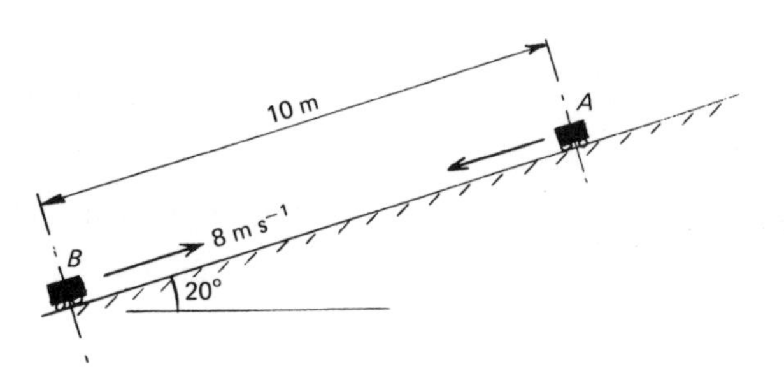

A body A of mass 200 kg is released from rest 10 m from the bottom of a straight frictionless inclined track of slope 20° (see diagram). At the same instant, a body B of mass 300 kg is projected up the track from the bottom with an initial velocity of 8 m s^{-1}. Determine (a) the position of body A on the track when it instantaneously comes to rest after rebounding from body B; (b) the velocity of body B when it reaches the bottom of the track again after rebounding from A. Assume a coefficient of restitution for the collision of $e = 0.8$.

Solution 3.5

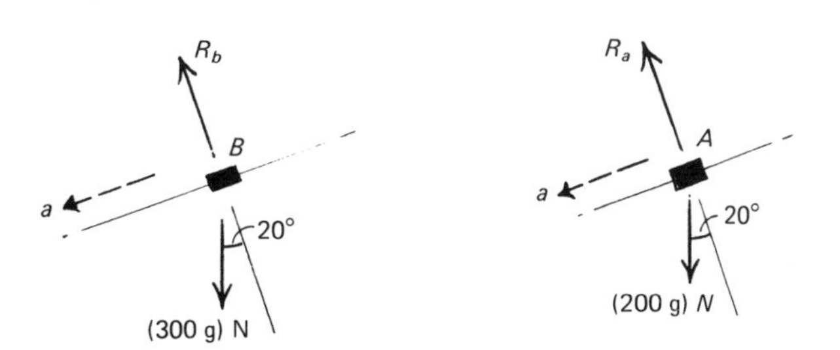

We first require the point on the track at which the bodies collide. The free-body diagrams for the two bodies are on page 119. Each discloses two forces only. Motion is constrained to be along the track.

Resolving along the direction of the track:

$$300g \sin 20° = 300a; \qquad 200g \sin 20° = 200a.$$

Thus, for both A and B, $a = g \sin 20°$ down the track.

The point at which A and B meet is calculated from the fact that both take the same time to get there.

Let A and B meet distance x from the bottom of the track, after time t. Assume a direction positive up the track.

$$s = ut + \tfrac{1}{2}at^2.$$

For A: $$-(10 - x) = 0 + \tfrac{1}{2}(-g \sin 20°)t^2. \qquad (1)$$

For B: $$x = 8t + \tfrac{1}{2}(-g \sin 20°)\, t^2. \qquad (2)$$

Subtract equation 1 from equation 2:

$$10 = 8t.$$

$$\therefore\ t = 1.25 \text{ s}.$$

Substitute in equation 2:

$$x = 8 \times 1.25 - \tfrac{1}{2} \times g \sin 20° \times (1.25)^2 = 7.379 \text{ m}.$$

We can now calculate the velocities at the instant of collision.

$$v = u + at.$$

For A: $$v = 0 - 9.81 \sin 20° \times 1.25$$
$$= -4.194 \text{ m s}^{-1} \text{ (i.e. down track)}.$$

For B: $$v = 8 + (-9.81 \sin 20°) \times 1.25$$
$$= +3.806 \text{ m s}^{-1} \text{ (i.e. up track)}.$$

We now employ the momentum and restitution formulae to calculate velocities immediately after impact. The two velocities just calculated are respectively u_a and u_b. Continuing to assume 'up track' as positive,

[d] Momentum: $200 \times (-4.194) + 300 \times (+3.806) = 200v_a + 300v_b$.

Simplifying, $$200v_a + 300v_b = -838.8 + 1141.8$$
$$= 303.0. \qquad (3)$$

[f] Restitution: $$\frac{v_b - v_a}{3.806 - (-4.194)} = -0.8.$$

Simplifying, $$v_b - v_a = -0.8\,(3.806 + 4.194)$$
$$= -6.4. \qquad (4)$$

Note that although e is stated as having the value 0.8, the minus sign must be included when substituting in the formula.

Determine v_b from equation 4 and substitute in equation 3:

$$200v_a + 300(v_a - 6.4) = 303.0.$$

$$500v_a = 303.0 + 1920.$$

$$v_a = \frac{2223}{500} = 4.446 \text{ m s}^{-1} \text{ (i.e. up track)}.$$

Substitute in equation 4:

$$\begin{aligned} v_b &= -6.4 + v_a \\ &= -6.4 + 4.446 \\ &= -1.954 \text{ m s}^{-1} \text{ (i.e. down track).} \end{aligned}$$

After the collision, both bodies will again be subjected to the same acceleration, down the track, of ($g \sin 20°$) m s^{-2}.

Let A come to rest distant s from the point of collision.

$$v^2 = u^2 + 2as.$$

$$0 = (4.446)^2 + 2(-g \sin 20°)s.$$

$$\therefore \; s = \frac{4.446^2}{2 \times 9.81 \sin 20°} = \underline{2.946 \text{ m.}}$$

(i.e. A overshoots its starting-point by $(7.379 + 2.946 - 10) = 0.325$ m.)

Let the final velocity (at bottom of track) of B be v:

$$\begin{aligned} v^2 &= u^2 + 2as \\ &= 1.954^2 + 2(-g \sin 20°) \times (-7.379) \\ &= 53.33. \end{aligned}$$

$$\therefore \; v = \underline{7.303 \text{ m s}^{-1}.}$$

The position of A and the bottom-of-track velocity of B could have been determined by use of an energy equation instead: Loss of k.e. = Gain of p.e. for A, and Gain of k.e. = Loss of p.e. for B.

Example 3.6

A body A of mass 5 kg rests on a horizontal table. It is connected to a second body B of mass 4 kg by a string of length 5 m, the string passing over a small smooth rigid peg which is 3 m vertically above A. Body B is lifted to a height h above the table and allowed to fall. It can be assumed that the string tightens without rebounding, i.e. that both bodies move together at the same speed after tightening.

(a) If h is 1.2 m, calculate the maximum height that A will be raised above the table.

(b) Calculate what value of h will cause B just to reach the surface of the table.

Calculate the loss of energy to the system in each case.

Solution 3.6

Since this consists of two problems, some time and effort can be saved by deriving a general expression for the height A will be raised, in terms of h.

If B is allowed to hang freely, it will be 1 m above the table. Therefore:
Body B falls freely a distance $(h - 1)$ m before the string tightens.
The velocity u as the string tightens is given by:

$$u = \sqrt{2g(h-1)}.$$

Let common velocity of masses after tightening be v.

[d] Momentum: $\qquad m_a \times 0 + m_b u = (m_a + m_b)v.$

$$\therefore \; v = u\left(\frac{m_b}{m_a + m_b}\right) = \tfrac{4}{9}\sqrt{2g(h-1)}.$$

The coupled system will then be subjected to a retardation of $\frac{g}{9}$.

(This result can be worked out by writing the equations of motion for both masses, but for a simple two-mass coupled system on a pulley like this, a quick answer is obtained by recognising that the total moving mass is (5 + 4) = 9 kg, and the resultant force (i.e. the excess weight of *A* over *B*) is (1*g*) N. Thus: 1*g* = 9*a*, giving *a* = *g*/9.)

B comes to rest in a distance *y*.

$$v^2 = u^2 + 2ax.$$

$$0 = v^2 + 2\left(-\frac{g}{9}\right)y.$$

$$\therefore\ y = \frac{9}{2g}v^2 = \frac{9}{2g} \times \frac{16}{81} \times 2g(h-1)$$

$$= \tfrac{16}{9}(h-1),$$

and this is clearly the height *A* is raised from the table.

Part (a)

Substitute:

$$h = 1.2 \text{ m}$$

$$y = \tfrac{16}{9} \times 0.2$$

$$= \underline{0.356 \text{ m}}.$$

Energy loss is calculated most simply from k.e. just before and just after the string tightens, since both velocities are known.

$$\text{Loss of energy} = \tfrac{1}{2}m_b u^2 - \tfrac{1}{2}(m_a + m_b)v^2$$

$$= \tfrac{1}{2} \times 4 \times 2g \times 0.2 - \tfrac{1}{2} \times 9 \times \tfrac{16}{81} \times 2g \times 0.2$$

$$= 7.848 - 3.488$$

$$= \underline{4.36 \text{ J}}.$$

Part (b)

Substitute:

$$y = 1 \text{ m}.$$

$$1 = \tfrac{16}{9}(h-1).$$

$$\therefore\ h = \underline{1.5625 \text{ m}}.$$

$$\text{Loss of energy} = \tfrac{1}{2} \times 4 \times 2g \times 0.5625 - \tfrac{1}{2} \times 9 \times \tfrac{16}{81} \times 2g \times 0.5625$$

$$= 22.0725 - 9.81$$

$$= \underline{12.2625 \text{ J}}.$$

Example 3.7

A shell of mass 260 kg is fired from the barrel of a gun of mass 3100 kg. The gun rests on a smooth horizontal surface. The muzzle velocity of the shell relative to the gun is 550 m s^{-1}.

(a) Calculate the true velocity of the shell (i.e. relative to the ground) and the recoil velocity of the gun when the barrel is aimed horizontally.

(b) Calculate the true velocity of the shell and its angle of flight when leaving the barrel, and the recoil velocity of the gun, when the barrel is aimed at an elevation of 40°.

Solution 3.7

Part (a)

We may make use of the principle of conservation of momentum along the horizontal, as no external force acts on the system. The only external force is the vertical upward reaction of the surface on the gun. The shell propelling force is an internal force, acting equally on shell and gun.

[d] Momentum before firing = Momentum after firing.

$$0 = m_a v_a + m_b v_b,$$

where suffix a denotes 'gun' and b 'shell'.

$$0 = 3100 v_a + 260 v_b. \quad (1)$$

(Note that v_b is not 550 m s^{-1}; this is the velocity relative to the muzzle. A simple velocity diagram is required. Familiarise yourself with the contents of Chapter 1.)

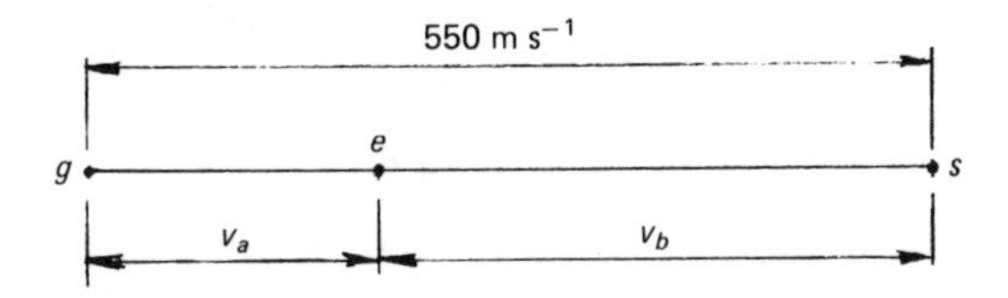

In the velocity diagram here, gun, shell and earth are denoted respectively by g, s and e.

From the diagram: $$v_a + v_b = 550. \quad (2)$$

Substitute from equation 1:

$$v_a + v_a\left(\frac{3100}{260}\right) = 550$$

giving $$v_a = \underline{42.56 \text{ m s}^{-1}}$$

and $$v_b = \underline{507.44 \text{ m s}^{-1}}.$$

Notice that although equation 1 strictly speaking gives v_b as $-\ v_a\left(\frac{3100}{260}\right)$ we do not so substitute in equation 2 as, in drawing the diagram, we have actually shown v_a as negative, i.e. to the left of e.

Part (b)

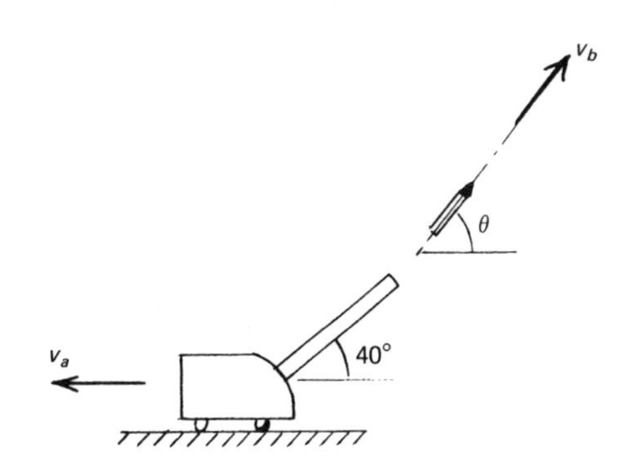

We may use the conservation-of-momentum principle again along the horizontal only, for the same reason as before. This time, we assume what is quite obvious, namely, that the recoil velocity is to the left (see diagram). This will avoid a repetition of the awkward negative sign in the momentum equation.

Horizontal momentum before firing = Horizontal momentum after firing.

$$0 = -m_a v_a + m_b v_b \cos\theta. \qquad (3)$$

A second velocity diagram is required, using the same notation as before:

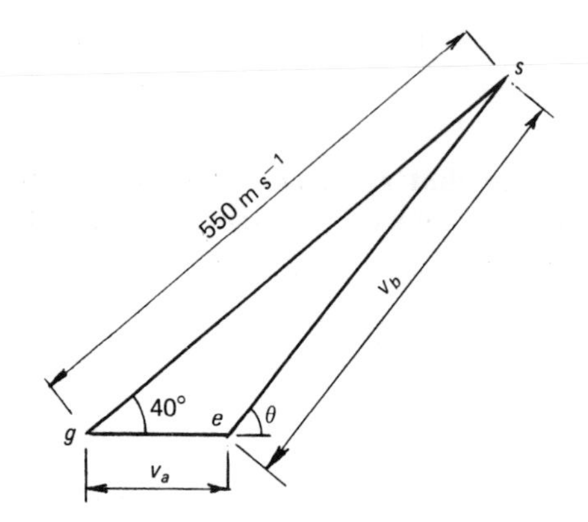

Using the sine rule:

$$\frac{v_a}{\sin(\theta - 40)} = \frac{v_b}{\sin 40}.$$

(Note that the angle at s is $(\theta - 40)°$: external angle = sum of opposite angles.)

$$\frac{v_a}{v_b} = \frac{\sin(\theta - 40)}{\sin 40}. \qquad (4)$$

From equation 3,

$$\frac{v_a}{v_b} = \frac{m_b}{m_a}\cos\theta = \frac{260}{3100}\cos\theta.$$

Equating,

$$\frac{\sin(\theta - 40)}{\sin 40} = 0.083\,87\cos\theta.$$

Expanding and rearranging,

$$\frac{\sin\theta\cos 40 - \cos\theta\sin 40}{\sin 40\cos\theta} = 0.083\,87.$$

$$\frac{\tan\theta}{\tan 40} - 1 = 0.083\,87.$$

$$\therefore\ \theta = \tan^{-1}(1.083\,87\tan 40)$$

$$= 42.286°.$$

The velocities can be calculated by applying the sine rule to the third side of the triangle.

$$\frac{550}{\sin\theta} = \frac{v_b}{\sin 40}.$$

$$\therefore \; v_b = 550 \; \frac{\sin 40}{\sin 42.286}$$

$$= \underline{525.44 \text{ m s}^{-1}}.$$

Substituting in equation 4:

$$v_a = 525.44 \; \frac{\sin 2.286}{\sin 40}$$

$$= \underline{32.606 \text{ m s}^{-1}}.$$

Example 3.8

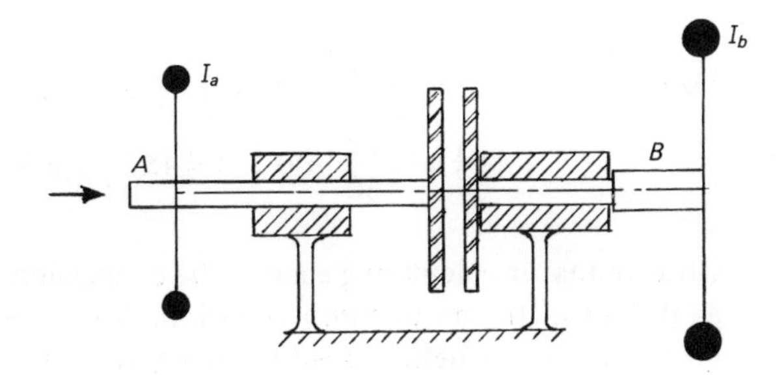

Two shafts, A and B, carrying wheels of moment of inertia $I_a = 24$ kg m^2 and $I_b = 48$ kg m^2, are mounted coaxially. Each shaft carries a friction plate and one of the shafts can move so that the plates are brought into contact. Initially, the plates are separate and shaft A is turning at 900 rev min^{-1}. The plates are then brought into contact and held together until both shafts are revolving at the same speed. Calculate this speed (a) if the initial speed of shaft B is 600 rev min^{-1} in the opposite direction to that of A; (b) if shaft B is initially stationary. Calculate the loss of energy in each case. Calculate also what initial speed of shaft B would cause the final speed of the coupled shafts to be zero.

Solution 3.8

We have no standard formula for this problem and a first-principles approach is recommended. Connecting the plates will result in a friction torque being applied to both shafts, and, in accordance with Newton's third law, these torques will be equal in magnitude and opposite in direction. Let us assume that the torque exerted on shaft A is in the same direction as its initial rotation. Actually, it will not be, but this does not matter. What *does* matter is that we assume an opposite torque on the other shaft. We may call this direction – the initial direction of A – positive. Thus:

For A, $$T = I_a \alpha_a.$$

And for B, $$-T = I_b \alpha_b.$$

$$\therefore \; I_a \alpha_a = -I_b \alpha_b.$$

The speeds will change from ω_{a1} to ω_{a2} and ω_{b1} to ω_{b2}. If the plates are together for time t before they cease slipping then

$$I_a \alpha_a = I_a \left(\frac{\omega_{a2} - \omega_{a1}}{t}\right); \qquad I_b \alpha_b = I_b \left(\frac{\omega_{b2} - \omega_{b1}}{t}\right).$$

$$I_a \left(\frac{\omega_{a2} - \omega_{a1}}{t}\right) = -I_b \left(\frac{\omega_{b2} - \omega_{b1}}{t}\right).$$

Rearranging, $$I_a \omega_{a1} + I_b \omega_{b1} = I_a \omega_{a2} + I_b \omega_{b2}.$$

This is an equation of conservation of angular momentum; the quantity I is defined as the angular momentum of a body. Furthermore, we know that the final speed is the same for both shafts, i.e. $\omega_{a2} = \omega_{b2}$.

Part (a)

Substituting the values,

$$24 \frac{2\pi}{60} 900 + 48 \frac{2\pi}{60} (-600) = (24 + 48) \frac{2\pi}{60} (N)$$

where N is the final common speed, in rev min^{-1}.

$$\therefore\ N = \frac{24 \times 900 - 48 \times 600}{72} = \underline{-100 \text{ rev min}^{-1}} \text{ (i.e. in the opposite direction).}$$

$$\text{Loss of energy} = \tfrac{1}{2}I_a\omega_{a1}^2 + \tfrac{1}{2}I_b\omega_{b1}^2 - \tfrac{1}{2}I_a\omega_{a2}^2 - \tfrac{1}{2}I_b\omega_{b2}^2$$

$$= \tfrac{1}{2}24\left(\frac{2\pi}{60}900\right)^2 + \tfrac{1}{2}48\left(\frac{2\pi}{60}600\right)^2 - \tfrac{1}{2}72\left(\frac{2\pi}{60}100\right)^2 = \underline{197.392 \text{ kJ.}}$$

Observe that in calculating energy, the conversion factor $2\pi/60$ does *not* cancel out, as it does in the momentum equation. Also, note that the directions of rotation do not matter; whether positive or negative, they are in any case squared. Energy is always positive.

Part (b)

Substituting the stated values:

$$24 \frac{2\pi}{60} 900 + 48 \times 0 = (24 + 48) \frac{2\pi}{60} \text{N.}$$

$$\therefore\ N = \frac{24}{72} \times 900 = \underline{+300 \text{ rev min}^{-1}} \text{ (i.e. in the same direction as } A \text{ originally).}$$

$$\text{Loss of energy} = \tfrac{1}{2}24\left(\frac{2\pi}{60}900\right)^2 + 0 - \tfrac{1}{2}72\left(\frac{2\pi}{60}300\right)^2 = 71\,061 \text{ J.}$$

For the final part, $\omega_{a2} = \omega_{b2} = 0$

$$I_a\omega_{a1} + I_b\omega_{b1} = 0$$

$$24 \frac{2\pi}{60} 900 + 48 \frac{2\pi}{60} N_{b1} = 0$$

giving $N_{b1} = \underline{-450 \text{ rev min}^{-1}}$ (i.e in the opposite direction to A).

Example 3.9

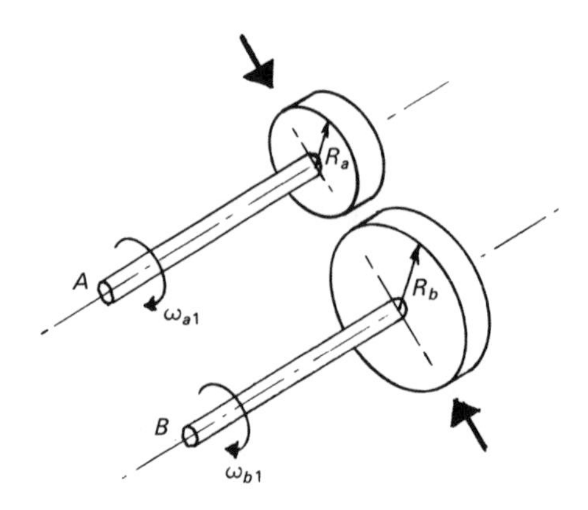

Two rotating wheels are attached to parallel shafts A and B (diagram). Their moments of inertia are $I_a = 16$ kg m^2 and $I_b = 30$ kg m^2. The wheels are arranged so that they can be brought together at their rims, and the contacting surface radii are $R_a = 0.16$ m and $R_b = 0.20$ m. Initially, shaft A is turning at 300 rev min^{-1} in the direction shown. The rims are then brought into contact and held together until they are running without relative slip. Calculate the final speeds of both shafts, given that the initial speed of B is (a) 300 rev min^{-1} in the direction shown; (b) 300 rev min^{-1} in the opposite direction; (c) zero.

What should be the initial speed of shaft B so that there will be no loss of energy when the wheels are brought together?

Solution 3.9

As with Example 3.8, a first-principles approach is called for. Bringing the wheels together means that a tangential friction force acts on each shaft. These forces are equal and opposite, but they do not exert equal torques, because they are at differing radii. If we choose the initial direction of shaft A to be positive, and assume both shafts to turn in this direction, the friction forces are as shown in the diagram.

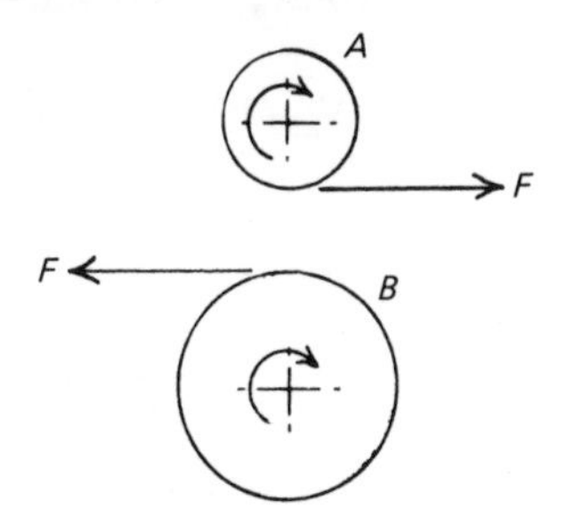

That is, both forces will exert a negative, or retarding, torque on the respective shafts.

Writing the equations of motion,

$$-FR_a = I_a \alpha_a = I_a \left(\frac{\omega_{a2} - \omega_{a1}}{t}\right);$$

$$-FR_b = I_b \alpha_b = I_b \left(\frac{\omega_{b2} - \omega_{b1}}{t}\right).$$

Multiplying by t and dividing by R_a and R_b respectively,

$$-Ft = \frac{I_a}{R_a} (\omega_{a2} - \omega_{a1}) = \frac{I_b}{R_b} (\omega_{b2} - \omega_{b1}). \qquad (1)$$

In Example 3.9 the final speeds of the shafts were the same. This is not so in this problem. But because they run together without slip we may say that the peripheral speed is the same for each wheel, but the direction is opposite. The diagram of the friction forces (above) shows that when the rims are connected the shafts must run in opposite directions.

$$\therefore \ \omega_{a2} R_a = -\omega_{b2} R_b. \qquad (2)$$

Part (a)

Substituting in equations 1 and 2,

$$\frac{16}{0.16} \frac{2\not\pi}{\not60} (N_{a2} - 300) = \frac{30}{0.20} \frac{2\not\pi}{\not60} (N_{b2} - 300)$$

and $$\left(\frac{2\pi}{60}N_{a2}\right) 0.16 = \left(-\frac{2\pi}{60}N_{b2}\right) 0.20.$$

Using the second equation to substitute for N_{b2} in the first,

$$100N_{a2} - 30\,000 = 150\left(-N_{a2}\frac{0.16}{0.20}\right) - 45\,000.$$

$$\therefore\ N_{a2}(100 + 120) = 30\,000 - 45\,000.$$

$$\therefore\ N_{a2} = -\frac{15\,000}{220} = \underline{-68.18 \text{ rev min}^{-1}.}$$

$$\therefore\ N_{b2} = -\quad -68.18 \times \frac{0.16}{0.20} = \underline{+54.55 \text{ rev min}^{-1}.}$$

Part (b)

The calculation is similar, except that N_{b1} is -300 rev min^{-1}.

$$100N_{a2} - 30\,000 = 150\left(-N_{a2}\frac{0.16}{0.20}\right) + 45\,000.$$

$$\therefore\ N_{a2}(100 + 120) = 30\,000 + 45\,000.$$

$$\therefore\ N_{a2} = \frac{75\,000}{220} = \underline{+340.91 \text{ rev min}^{-1}.}$$

$$\therefore\ N_{b2} = -\quad 340.91 \times \frac{0.16}{0.20} = \underline{-272.7 \text{ rev min}^{-1}.}$$

Part (c)

$$100N_{a2} - 30\,000 = 150\left(-N_{a2}\frac{0.16}{0.20}\right) + 0.$$

$$N_{a2}(100 + 120) = 30\,000 + 0.$$

$$N_{a2} = \frac{30\,000}{220} = \underline{+136.4 \text{ rev min}^{-1}.}$$

$$N_{b2} = -\quad 136.4 \times \frac{0.16}{0.20} = \underline{-109.12 \text{ rev min}^{-1}.}$$

For the final part of the question, a detailed calculation is unnecessary if it is clearly understood why there is a loss of energy. When the friction surfaces are brought together, resulting in changes of speed of the two wheels, work is being done at the surfaces which is 'lost' to the system; this is the outlet for the 'lost' energy. For no energy to be lost when the wheels are brought together, the wheels must not slip relatively when brought together, i.e. their peripheral speeds before touching must be the same. We may merely use equation 2, substituting ω_{a1} for ω_{a2} and ω_{b1} for ω_{b2}:

$$\therefore\ \frac{2\pi}{60}N_{a1}R_a = -\frac{2\pi}{60}N_{b1}R_b$$

$$\therefore\ N_{b1} = -N_{a1}\frac{R_a}{R_b} = -300 \times \frac{0.16}{0.20} = \underline{-240 \text{ rev min}^{-1}.}$$

And of course, when the wheels are brought together, they will continue running at these speeds.

Example 3.10

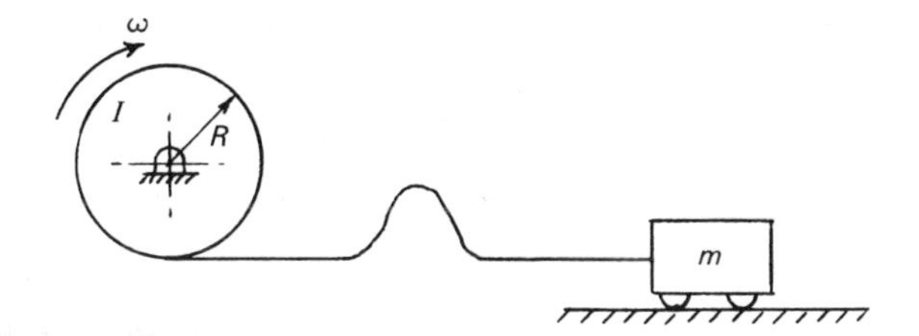

A winding drum of moment of inertia $I = 840$ kg m^2 and winding radius $R = 1.9$ m winds a cable which is connected to a vehicle of mass $m = 184$ kg which is initially at rest on a smooth horizontal track (see diagram). The drum is turning at a steady speed of 12 rev min^{-1} when the rope, which is initially slack, suddenly tightens. Calculate the final speeds of the drum and the vehicle, the energy lost to the system, and the magnitude of the impulse in the cable.

Solution 3.10

This is another problem where a first-principles approach is indicated. Let the force in the cable during the change of speeds be F, acting for a time t.

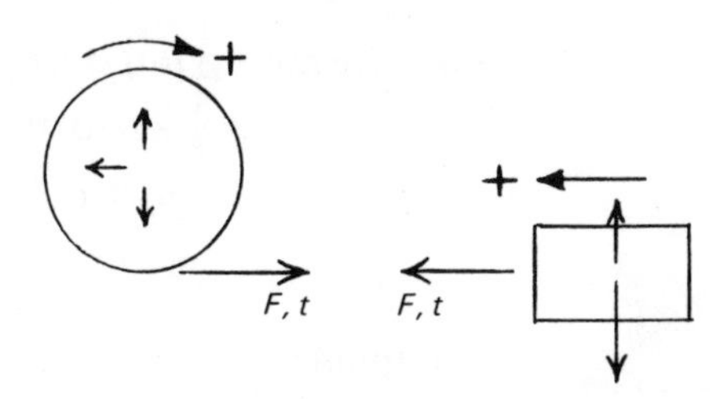

In the free-body diagrams of drum and vehicle shown here, assume clockwise rotation is positive for the drum. The corresponding positive motion of the vehicle (the 'compatible' motion) is right-to-left.

For the drum,
$$-FR = I\alpha = I\left(\frac{\omega_2 - \omega_1}{t}\right).$$
$$-Ft = \frac{I}{R}(\omega_2 - \omega_1).$$

For the vehicle,
$$+F = ma = m\left(\frac{v-u}{t}\right).$$
$$+Ft = m(v-u).$$

Equating (Ft),
$$-\frac{I}{R}(\omega_2 - \omega_1) = m(v-u). \qquad (1)$$

In fact, F is almost certainly not of constant magnitude, and to be strictly accurate (Ft) should be replaced by $\int(F\,dt)$. But since we eliminate this, the simplification does not affect the calculation.

Equation 1 contains two unknown quantities ω_2 and v. A second equation – the compatibility equation – is required. The vehicle, when connected by a tight cable to the drum, must move at the peripheral speed of the drum.

Compatibility: $v = \omega_2 R.$ (2)

Use this to substitute for v in equation 1 and insert the given values:

$$\frac{840}{1.9}\left(\omega_2 - \frac{2\pi}{60}12\right) = -184(\omega_2\ 1.9 - 0).$$

$$\therefore\ 442.1\omega_2 - 555.6 = -349.6\omega_2.$$

$$\therefore\ \omega_2 = \frac{555.6}{442.1 + 349.6}$$

$$= 0.7018 \text{ rad s}^{-1}$$

$$= \left(0.7018 \times \frac{60}{2\pi}\right) \text{rev min}^{-1}$$

$$= 6.702 \text{ rev min}^{-1}.$$

Substitute in equation 2: $v = 0.7018 \times 1.9$

$$= 1.333 \text{ m s}^{-1}.$$

$$\text{Initial energy} = \tfrac{1}{2}I\omega_1^2$$

$$= \tfrac{1}{2} \times 840\left(\frac{2\pi}{60}12\right)^2$$

$$= 663.2 \text{ J.}$$

$$\text{Final energy} = \tfrac{1}{2}I\omega_2^2 + \tfrac{1}{2}mv^2$$

$$= \tfrac{1}{2} \times 840(0.7018)^2 + \tfrac{1}{2} \times 184(1.333)^2$$

$$= 370.3 \text{ J.}$$

$$\therefore\ \text{Loss} = \underline{292.9 \text{ J.}}$$

$$\text{Impulse} = \text{change of linear momentum of vehicle}$$

$$= m(v - u)$$

$$= 184(1.333 - 0)$$

$$= \underline{245.3 \text{ N s.}}$$

Example 3.11

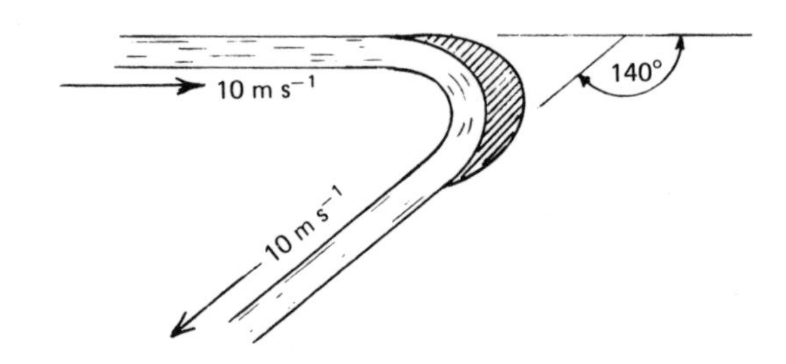

A jet of water has a cross-sectional area of 1600 mm^2 and a velocity of 10 m s^{-1}. It strikes a fixed curved deflector-plate tangentially, as shown in the diagram. This causes the jet to change direction through an angle of 140° without loss of energy. Calculate the force exerted on the plate by the jet. The density of the water is 1000 kg m^3.

Solution 3.11

With this type of problem, we make use of the statement

$$\text{Force} = \text{Rate of change of momentum}$$
$$= \text{Rate of change of } (mv)$$
$$= \frac{d}{dt}(mv).$$

It is helpful to consider what happens in one second of time. In this time, a certain mass of water (we may call it $\dot{m}$, or 'm dot') changes velocity, in direction only, not in magnitude. We can therefore say:

$$\text{Force} = \dot{m} \times \text{Change of velocity}$$

where $\dot{m}$ is the mass-flow rate, which we can determine from the data. We need to recall that momentum is a vector, since it involves velocity. We therefore choose two directions at right angles and consider change of momentum along each; in other words, we resolve the momentum into components. It is convenient to choose the initial direction of the jet as one axis. Positive and negative directions are arbitrary, but once they are chosen you must of course be consistent. We shall also find it easier to solve this problem algebraically, considering an angle θ which is less than 90° for the analysis.

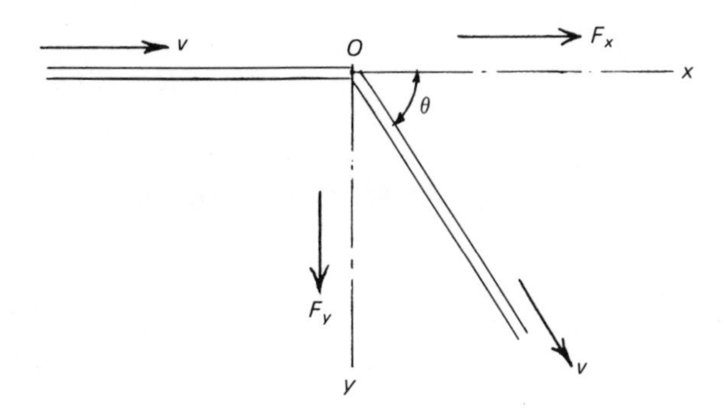

The diagram here shows a jet diverted through angle θ. An x-y coordinate axis system is drawn, with x +ve to right, and y +ve downwards. The force components F_x and F_y are the force components exerted by the plate on the jet, and are assumed +ve along the respective axes.

Considering the x-direction,

$$F_x = \dot{m} \times \text{change of velocity in } x\text{-direction}$$
$$= \dot{m} \times (\text{final } x\text{-velocity} - \text{initial } x\text{-velocity})$$
$$= \dot{m}(v \cos\theta - v)$$
$$= \dot{m}v(\cos\theta - 1). \qquad (1)$$

Considering the y-direction,

$$F_y = \dot{m} \times \text{change of velocity in } y\text{-direction}$$
$$= \dot{m} \times (\text{final } y\text{-velocity} - \text{initial } y\text{-velocity})$$
$$= \dot{m}(v \sin\theta - 0)$$
$$= \dot{m}v \sin\theta. \qquad (2)$$

$$\dot{m} = \text{area of jet} \times \text{velocity} \times \text{density}$$
$$= (1600 \times 10^{-6}) \times 10 \times 1000$$
$$= 16 \text{ kg s}^{-1}.$$
$$\therefore\ F_x = 16 \times 10(\cos 140 - 1)$$
$$= -282.57 \text{ N (i.e. from right to left).}$$
$$F_y = 16 \times 10 \sin 140$$
$$= +102.85 \text{ N (i.e. downwards).}$$

The components, now shown correctly as to direction, and the resultant force R are thus as shown in the diagram.

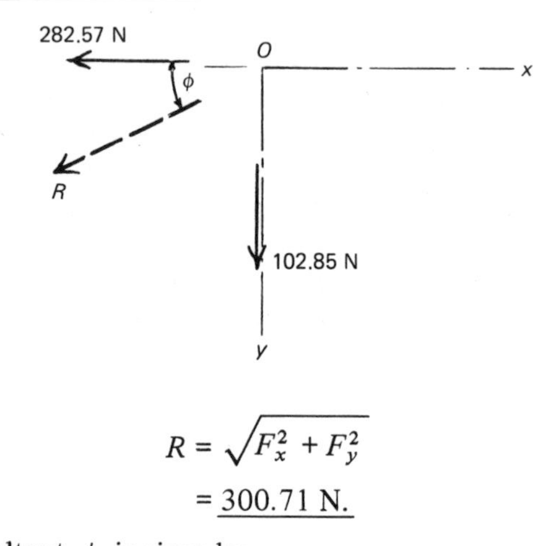

$$R = \sqrt{F_x^2 + F_y^2}$$
$$= \underline{300.71 \text{ N.}}$$

The angle of the resultant, ϕ, is given by

$$\phi = \tan^{-1}(F_y/F_x)$$
$$= \tan^{-1}\left(\frac{102.85}{282.57}\right)$$
$$= \underline{20^\circ.}$$

The line of action of the resultant is thus seen to lie exactly between the paths of the original and the deflected jet. The force exerted on the plate by the jet will be equal and opposite to R, i.e. 300.71 N at 20° anticlockwise to line of original jet. This problem illustrates the importance of choosing arbitrary conventions and subsequently adhering rigidly to them. Notice also how the trigonometry takes care of itself in the calculation for F_x; although the angle is greater than 90° the formula derived still works for us, and indeed works for any angle up to 360°. Avoid the mistake of treating different quadrants as different problems, and learn to work 'round the clock' with your trigonometry.

As an exercise, you could solve this problem using a different set of x-y axes. For instance, the x-direction could be along the deflected jet line. The value of the final force, and its direction, should of course be the same.

Example 3.12

A bullet of mass $m_1 = 0.12$ kg is fired horizontally into a sand box of mass $m_2 = 50$ kg which hangs vertically from four strings of length 2.1 m as shown in the diagram. The striking velocity of the bullet is 320 m s^{-1}. After impact, the bullet remains within the box. Calculate (a) the common velocity of the box and bullet immediately after the impact; (b)

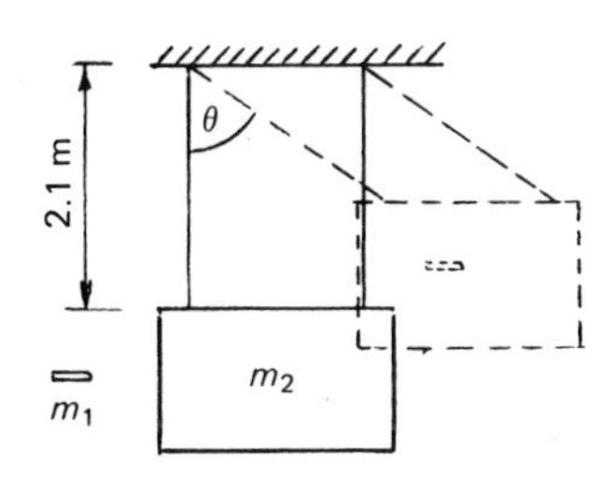

the maximum angle through which the box swings after impact; (c) the amount of energy lost to the system as a result of the impact.

PCL

Solution 3.12

The details are almost exactly the same as those of Example 3.4 and the calculations follow the same lines.

$$m_1u_1 + m_2u_2 = (m_1 + m_2)v.$$

$$\therefore\ v = \frac{0.12 \times 320 + 0}{(0.12 + 50)}$$

$$= \underline{0.7662 \text{ m s}^{-1}}.$$

Energy:

$$\tfrac{1}{2}(m_1 + m_2)v^2 = (m_1 + m_2)gh.$$

$$\therefore\ h = \frac{v^2}{2g}$$

$$= \frac{(0.7662)^2}{2g}$$

$$= \underline{0.0299 \text{ m}}.$$

$$h = 2.1(1 - \cos\theta).$$

$$\therefore\ \cos\theta = 1 - \frac{0.0299}{2.1}$$

$$= 0.9858.$$

$$\therefore\ \theta = \underline{9.68^\circ}.$$

$$\text{Loss of k.e.} = \tfrac{1}{2} \times 0.12 \times (320)^2 - \tfrac{1}{2} \times 50.12 \times (0.7662)^2$$

$$= 6144 - 14.71$$

$$= \underline{6129.29 \text{ J}}$$

and we see that, as with all cases of ballistics, practically all the initial energy is dissipated in the embedding of the missile in the box.

Example 3.13

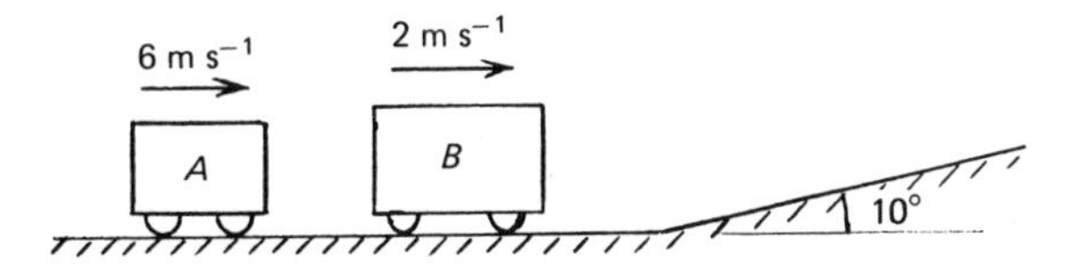

A truck A of mass 800 kg runs from left to right along a straight horizontal track at an initial speed of 6 m s^{-1}. It collides with a second truck B of mass 2000 kg which moves also from left to right along the same track with an initial speed of 2 m s^{-1}. The arrangement is shown in the diagram. After collision, the two trucks remain locked together. They then begin to ascend a slope of 10°.

Calculate the distance along the sloping track that the coupled trucks will travel. Neglect any energy losses due to friction.

PCL

Solution 3.13

The first part of the solution is a simple application of the momentum equation, with the final velocities being the same.

$$m_a u_a + m_b u_b = (m_a + m_b)\,v.$$

$$\therefore\ v = \frac{800 \times 6 + 2000 \times 2}{800 + 2000}$$

$$= 3.143 \text{ m s}^{-1}.$$

At the highest point up the slope the trucks will come to rest.

$$\text{Loss of k.e.} = \text{Gain of p.e.}$$

$$\therefore\ \tfrac{1}{2}(m_a + m_b)\,v^2 = (m_a + m_b)\,gh.$$

$$\therefore\ h = \frac{v^2}{2g}$$

$$= \frac{(3.143)^2}{2g}$$

$$= 0.5035 \text{ m}.$$

This is the vertical height reached. The corresponding distance along the track, L, is given by

$$L = \frac{h}{\sin\theta}$$

$$= \frac{0.5035}{\sin 10°}$$

$$= \underline{2.900 \text{ m}}.$$

Example 3.14

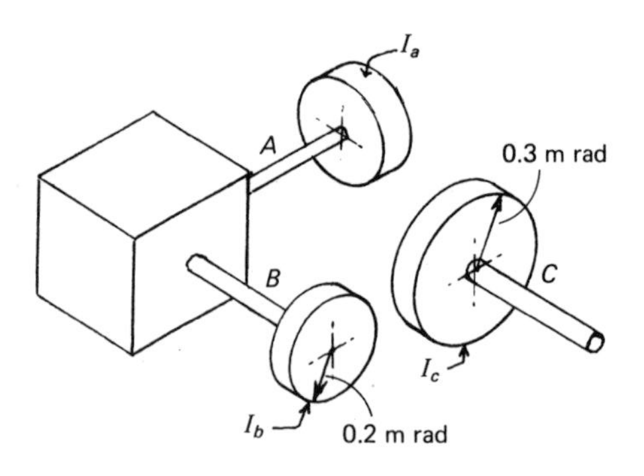

A shaft A carries a wheel of moment of inertia 2.4 kg m^2. This shaft is connected through a gearbox to a shaft B; the speed of B is four times that of A. B carries a wheel of moment of inertia 0.42 kg m^2 and its peripheral radius is 0.2 m. Calculate the torque required on shaft

A to accelerate shaft B to a speed of 600 rev min^{-1} from rest in 12 seconds. Neglect any inertial effects of the gearbox itself.

When shaft B is running at 600 rev min^{-1} the torque is removed from A and a third stationary wheel C is brought into rim contact with B. The moment of inertia of C is 1.6 kg m^2 and its peripheral radius is 0.3 m. The wheels are held together until all slip has ceased. Calculate the final speed of wheel C.

PCL

Solution 3.14

This is an example of two 'topics' covered in one question. The first part of the problem requires the methods of Chapter 4. The working is given here without comment, and you should look at Chapter 4 if you need explanation.

Let equivalent inertia of the system referred to A be I_{ea}.

$$\begin{aligned} I_{ea} &= I_a + G^2 I_b \\ &= 2.4 + (4)^2 \times 0.42 \\ &= 9.12 \text{ kg m}^2. \\ \omega_2 &= \omega_1 + \alpha_t. \\ 2\pi \times \frac{600}{60} &= 0 + \alpha_b \times 12. \\ \therefore\ \alpha_b &= 5.2360 \text{ rad s}^{-2}. \\ \therefore\ \alpha_a &= \tfrac{1}{4} \times 5.2360 \\ &= 1.3090 \text{ rad s}^{-2}. \\ T_a &= I_{ea}\alpha_a \\ &= 9.12 \times 1.3090 \\ &= \underline{11.94 \text{ N m}}. \end{aligned}$$

Again using Chapter 4, we reduce the wheels I_a and I_b and the gearbox to a single inertia I_{eb} referred to B.

$$\begin{aligned} I_{eb} &= 0.42 + \frac{2.4}{(4)^2} \\ &= 0.57 \text{ kg m}^2. \end{aligned}$$

The problem of two wheels in rim contact was examined in Example 3.9.

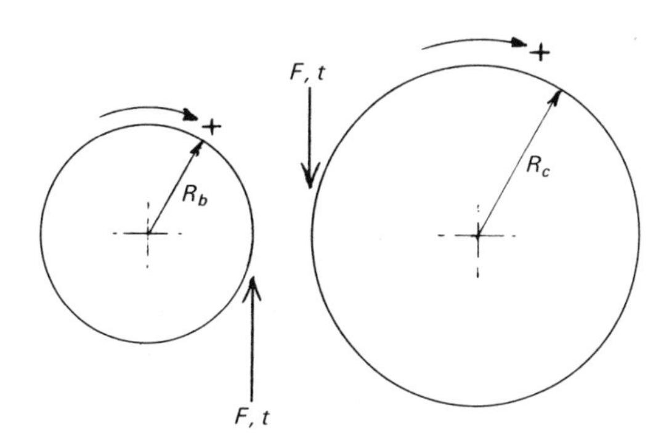

Contact results in a tangential force F acting for a time t. The two equations are:

for B $$-FR_b = I_{eb}\left(\frac{\omega_{b2} - \omega_{b1}}{t}\right)$$

and for C $$-FR_c = I_c\left(\frac{\omega_{c2} - \omega_{c1}}{t}\right).$$

(Notice, as before (Example 3.9) that the torque exerted by force F is negative in both cases.)

Eliminating (Ft) from the equations gives

$$\frac{I_{eb}}{R_b}(\omega_{b2} - \omega_{b1}) = \frac{I_c}{R_c}(\omega_{c2} - \omega_{c1}).$$

For compatibility,

$$\omega_{b2}R_b = -\omega_{c2}R_c.$$

$$\frac{I_{eb}}{R_b}(\omega_{b2} - \omega_{b1}) = \frac{I_c}{R_c}\left(-\omega_{b2}\frac{R_b}{R_c} - \omega_{c1}\right).$$

If you check with Example 3.9 you will see that we may replace the ω terms by N (rev min^{-1}) because the conversion factor $2\pi/60$ will cancel throughout.

Substituting numbers,

$$\frac{0.57}{0.2}(N_{b2} - 600) = \frac{1.6}{0.3}\left(-N_{b2} \times \frac{0.2}{0.3} - 0\right).$$

$$\therefore\ N_{b2} - 600 = -\frac{1.6 \times 0.2 \times 0.2}{0.57 \times 0.3 \times 0.3}N_{b2}$$

$$= -1.248N_{b2}.$$

$$\therefore\ N_{b2} = \frac{600}{2.248}$$

$$= +266.9 \text{ rev min}^{-1}.$$

$$\therefore\ N_c = -\tfrac{2}{3} \times 266.9$$

$$= \underline{-177.9 \text{ rev min}^{-1}}.$$

Example 3.15

A block of mass $5m$ is suspended from a rigid ceiling by a light inextensible string of length L. A pellet of mass m travelling in a downward direction at an angle θ to the horizontal strikes the block and becomes embedded. Derive an expression for the impulse in the string at the instant of impact.

If after the impact the block and pellet are observed to swing through an angle of 90°, determine the velocity of the pellet immediately prior to hitting the block.

U. Surrey

Solution 3.15

In this problem, we have to consider change of momentum in two separate directions. Considering the horizontal (x) direction in the diagram, it can be seen that no *external* force acts on the system comprising pellet and block; the only force is the string force, which is vertical.

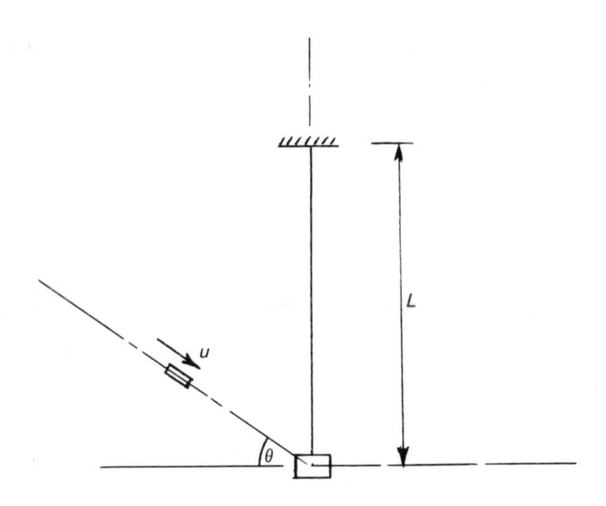

Hence, horizontally:

$$\text{Momentum before impact} = \text{Momentum after impact:}$$

$$m \times u \cos\theta = 6m \times v.$$

$$\therefore\ v = \frac{u\cos\theta}{6}.$$

In the vertical direction,

$$\text{Impulse} = \text{Change of momentum}$$

$$= m \times u \sin\theta.$$

$$\text{Required impulse} = \underline{mu \sin\theta.}$$

After impact,

$$\text{Loss of k.e.} = \text{Gain of p.e.},$$

$$\therefore\ 6mgL = \tfrac{1}{2}(6m)\,v^2.$$

$$\therefore\ 2gL = v^2$$

$$= \frac{u^2\cos^2\theta}{36}.$$

$$\therefore\ u = \frac{\sqrt{72Lg}}{\cos\theta}$$

$$= \underline{\frac{6\sqrt{2Lg}}{\cos\theta}}.$$

Example 3.16

Two bodies of mass m and $3m$ respectively lie on a smooth horizontal table and are joined together by a light spring of stiffness k which itself lies parallel to the table. The bodies are pushed together, compressing the spring by an amount x from its free length, and are then released from rest. Find the velocities of the two bodies when the spring has become extended by an amount $\frac{1}{2}x$.

U. Surrey

Solution 3.16

Before release, the system is subjected to no external force. Although the bodies are pushed together, they are subjected to equal and opposite forces; the pair of

bodies remain at rest. Thus, when released, the total momentum of the system will still be zero.

Let the two velocities be v_1, v_2.

$$mv_1 + 3mv_2 = 0.$$

$$v_1 = -3v_2. \qquad (1)$$

Strain energy released by the spring is converted to kinetic energy of the two bodies. The spring expands from compression x to $\frac{1}{2}x$.

$$\tfrac{1}{2}k\left[x^2 - (\tfrac{1}{2}x)^2\right] = \tfrac{1}{2}mv_1^2 + \tfrac{1}{2}(3m)\,v_2^2. \qquad \text{(See Chapter 5.)}$$

Using equation 1 to substitute for v_1,

$$\tfrac{1}{2}k \times \tfrac{3}{4}x^2 = \tfrac{1}{2}m(9v_2^2) + 1\tfrac{1}{2}mv_2^2.$$

$$\therefore\ \tfrac{3}{4}kx^2 = 12mv_2^2.$$

$$\therefore\ v_2 = \tfrac{1}{4}x\sqrt{\frac{k}{m}},$$

$$\therefore\ v_1 = \tfrac{3}{4}x\sqrt{\frac{k}{m}},$$

and of course, the velocities will be in opposite directions as indicated by equation 1.

Example 3.17

An inertial starter consists of a small flywheel which can be connected to an engine through a clutch and a 20 : 1 speed reduction gear. The flywheel is run up to a speed of 10 000 rev min^{-1} and the clutch is engaged. The speed of the engine is to be 300 rev min^{-1} immediately after engagement and the moment of inertia of the rotating parts of the engine is 50 kg m^2. Determine the moment of inertia of the starter flywheel and the energy lost during engagement.

U. Lond. U.C.

Solution 3.17

The problem is similar to Example 3.8. Again, we require to make use of the principle of equivalent inertias covered in Chapter 4. Let the equivalent inertia of the engine parts, referred to the high-speed shaft of the gearbox, be I_{eH}:

$$I_{eH} = \frac{50}{(20)^2}$$

$$= 0.125 \text{ kg m}^2.$$

'Borrowing' from Example 3.8 and writing an equation of angular momentum before and after connection, called the required flywheel inertia I:

$$I_{eH}\,\omega_1 + I\omega_2 = (I_{eH} + I)\,\omega.$$

Substituting values,

$$0.125 \times 0 + I \times 10\,000 = (0.125 + I)\,(300 \times 20)$$

(remembering that the required final speed of the high-speed shaft is 20 times the required speed of 300 rev min^{-1}).

$$\therefore I(10\,000 - 6000) = 6000 \times 0.125$$

$$\therefore I = \underline{0.1875 \text{ kg m}^2}.$$

$$\text{Energy loss} = \tfrac{1}{2} \times 0.1875 \left(2\pi \times \frac{10\,000}{60}\right)^2 - \tfrac{1}{2}(0.125 + 0.1875)\left(2\pi \times \frac{6000}{60}\right)$$

$$= \underline{41.12 \text{ kJ.}}$$

Example 3.18

In the diagram a trolley of mass 100 kg is to be stopped by colliding with a stationary mass of 400 kg backed up by springs of total stiffness $k = 50$ kN m^{-1}.

(a) Assuming perfectly plastic impact, and an initial trolley speed of 5 m s^{-1}, find the maximum deflection of the springs.
(b) If the collision is now assumed perfectly elastic, what is the maximum deflection of the springs?

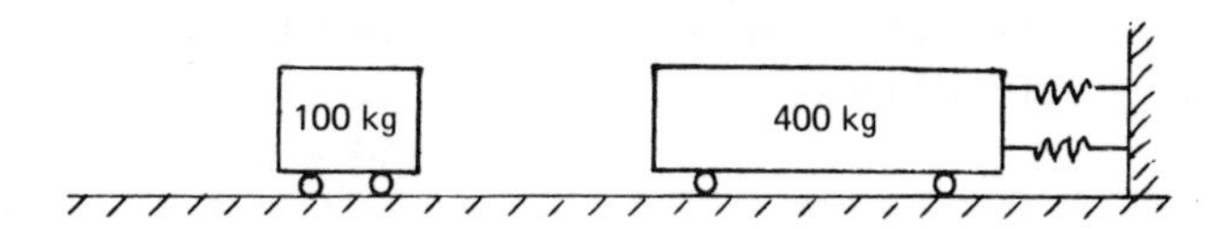

Thames Poly.

Solution 3.18

Part (a)

Perfectly plastic impact means that the relative velocity of separation is zero, i.e. the bodies remain together after collision.

[d]
$$m_a u_a + m_b u_b = (m_a + m_b)\,v.$$
$$100 \times 5 + 0 = 500v.$$
$$\therefore\ v = 1 \text{ m s}^{-1}.$$

The k.e. after collision is therefore

$$E = \tfrac{1}{2}(m_a + m_b)\,v^2$$
$$= \tfrac{1}{2} \times 500 \times (1)^2$$
$$= 250 \text{ J}.$$

When the moving masses are brought to rest, all this energy is assumed to be converted to spring strain energy.

$$\tfrac{1}{2}kx^2 = 250. \qquad \text{(See Chapter 5.)}$$

$$\therefore\ x = \sqrt{\frac{2 \times 250}{50\,000}}$$
$$= \underline{0.1 \text{ m.}}$$

Part (b)

For perfectly elastic impact, the coefficient of restitution is 1.

$$m_a u_a + m_b u_b = m_a v_a + m_b v_b.$$

$$100 \times 5 + 0 = 100v_a + 400v_b.$$

Simplifying,

$$5 = v_a + 4v_b. \quad (1)$$

[f] Restitution:

$$-1 = \frac{v_b - v_a}{u_b - u_a}$$

$$= \frac{v_b - v_a}{0 - 5}.$$

$$\therefore\ 5 = v_b - v_a. \quad (2)$$

Adding equations 1 and 2,

$$10 = 5v_b.$$

$$\therefore\ v_b = +2 \text{ m s}^{-1}.$$

The trolley will recoil with its own k.e. The springs have to absorb only the k.e. of the block.

$$\tfrac{1}{2}kx^2 = \tfrac{1}{2} \times 400 \times (2)^2.$$

$$\therefore\ x = \sqrt{\frac{1600}{50\,000}}$$

$$= \underline{0.1789 \text{ m.}}$$

3.3 Problems

Problem 3.1

A wagon of mass 825 kg travelling at 4 m s^{-1} along a straight level track collides with another wagon of mass 542 kg travelling in the opposite direction at 6.25 m s^{-1}. Calculate the speeds of the two wagons after the collision (a) if they lock together on colliding; (b) if the coefficient of restitution of the collision is 1.0; (c) if the coefficient of restitution of the collision is 0.9. In each case, calculate the loss of kinetic energy to the system.

Problem 3.2

Two bodies, having masses m_a and m_b, collide on a smooth horizontal surface with a coefficient of restitution of 1.

(a) If the bodies approach each other with the same velocity, u, calculate the ratio of m_a to m_b such that m_a is brought to rest by the collision, and evaluate the final velocity of m_b in terms of u.

(b) If m_b is initially stationary and the initial velocity of m_a is u, calculate the ratio of m_a to m_b such that the bodies separate with equal velocities v, and calculate this velocity in terms of u.

(c) If m_b is four times m_a and the bodies approach each other with equal velocities u, calculate the separating velocities.

Problem 3.3

A hammer has a mass of 0.7 kg and when it strikes a nail in wood it is moving at 25 m s^{-1}. The mass of the nail is 28 g. Assuming the hammer head moves horizontally, and does not rebound on collision, calculate the common velocity of hammer and nail immediately after impact. If the nail is driven an additional 4 mm into the wood by the blow, calculate the approximate resisting force of the wood to the nail.

Problem 3.4

Three wagons, A, B and C, are on a horizontal frictionless track, in that order. Initially, B and C are stationary and A moves towards B with a velocity of 4 m s^{-1}. The masses are 800 kg, 1200 kg and 400 kg respectively. Calculate the final velocity of each wagon after all collisions. Assume a coefficient of restitution of 1.

Problem 3.5

Two steel balls hang from fixed points on light strings 0.4 m long. Ball A is pulled to one side until the string makes an angle of 60° to the vertical and is released from this position. It strikes ball B which has a mass of 2 kg. If B swings through an angle of 30° as a result of the impact, calculate the mass of A. Assume a coefficient of restitution of 1.

Problem 3.6

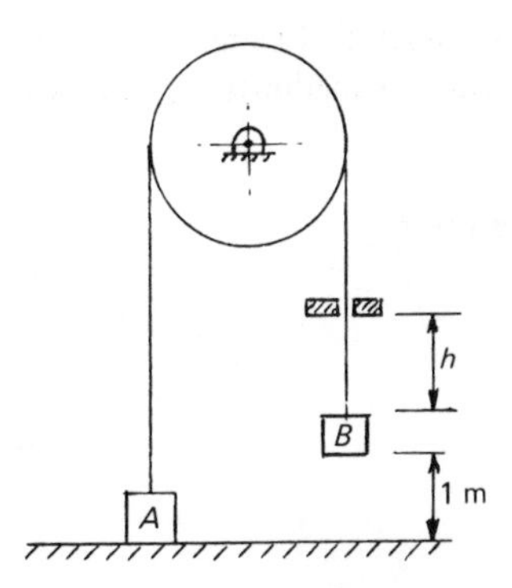

Two weights, A and B, having masses m_a = 2 kg and m_b = 1 kg are connected by a light extensible string passing round a light frictionless pulley. A rests on a horizontal surface and B hangs 1 m above the surface, as shown in the diagram. A third weight C of mass m_c = 0.8 kg is dropped on B from a height h. Calculate the value of h to cause B just to reach the surface with zero velocity.

Hints: Use an energy equation to find the velocity immediately after collision: Loss of k.e. (A, B and C) + Loss of p.e. (B, C) = Gain of p.e. (A). Answer: 1.016 m s^{-1}. For the momentum equation, treat A and B as a single body.

Problem 3.7

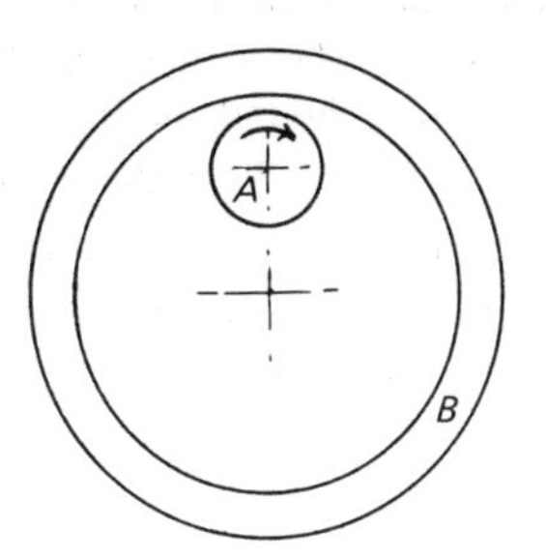

In a device for starting an engine, a small wheel A spinning at high speed is brought into contact with the inside rim of a flywheel, as shown in the diagram. The contacting rims have friction surfaces. The small wheel radius is 0.05 m and its moment of inertia is 0.014 kg m^2.

The flywheel rim radius is 0.24 m and its moment of inertia is 1.1 kg m^2. If the small wheel is spinning at 150 rev min^{-1} when the rims make contact, calculate the final speeds when relative slip has stopped. If the coefficient of friction between the rim surfaces is 0.4, and the rims are held together with a force of 25 N, calculate the peripheral distance over which the wheels slip before finally running together.

Hints: Refer to Example 3.9 and work from first principles. Unlike the case of Example 3.9, the impulse force exerts a negative torque on one wheel but a positive on the other. Also, final 'compatible' speeds are both in the same direction. For the final part, determine loss of k.e. This will equal work done against friction.

Problem 3.8

A drum of radius 0.4 m and moment of inertia 1.8 kg m^2 has a rope round it which carries a load of mass 2 kg. The load is held a height of 2 m above the ground and released from rest. It falls a distance of 1 m before the rope tightens. Calculate the time for the load to reach the ground. Assume that when the rope tightens, load and drum change velocity instantaneously.

Hints: Refer to Example 3.10. Do not include the weight of the load when writing the momentum equation; if the change of velocity is instantaneous, the weight force is negligible.

Problem 3.9

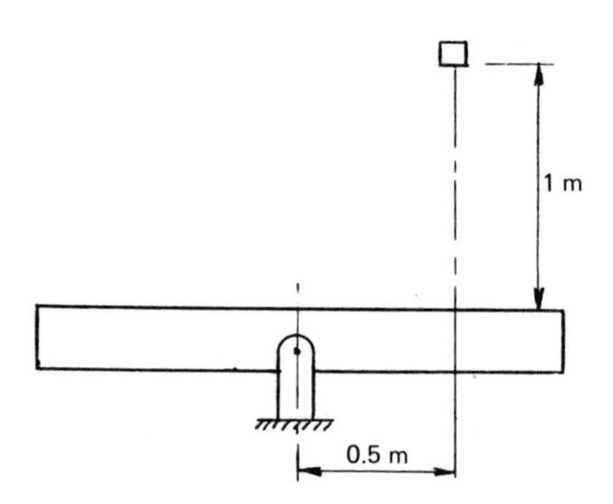

A beam is mounted on a central frictionless pivot so that it may swing in a vertical plane. Initially, it is horizontal and at rest. The moment of inertia of the beam about an axis through the pivot is 12.8 kg m^2. A body having a mass of 2 kg is dropped on to the beam from a height h = 1 m and strikes the beam 0.5 m from the pivot.

(a) If the body is automatically secured to the beam on impact, calculate the angular velocity of the beam immediately after impact.

(b) If the body rebounds on impact and there is no loss of energy to the system due to the impact, calculate the height of rebound of the weight.

Hints: For (a) $v = 0.5\omega$ since both move together. For (b) assume force F between bodies for time t, and write equations; eliminate (Ft): see Ex. 3.10. Write energy conservation equation. Eliminate ω and thus find v by solving the quadratic. $h = v^2/2g$.

Problem 3.10

A steel water channel is 1.2 m wide and carries water with a constant depth of 0.8 m. The average velocity of flow is 2.4 m s^{-1}. The channel curves, to divert the water through an angle of 50°. Calculate the magnitude and direction of the force exerted on the sides of the channel due to this change of direction. The density of water is 1000 kg m^3.

3.4 Answers to Problems

3.1 (a) 0.064 m s^{-1} in direction of smaller wagon; 17 183 J.
(b) −4.128 m s^{-1}, +6.122 m s^{-1}; 0.
(c) −3.722 m s^{-1}, +5.503 m s^{-1}; 3264.8 J.

3.2 (a) $m_a = 3m_b$; $v = 2u$. (b) $m_b = 3m_a$; $v = \frac{1}{2}u$. (c) $v_a = -2.2u$, $v_b = -0.2u$.

3.3 24.038 m s^{-1}; 52.58 kN.

3.4 −0.8 m s^{-1}; +1.6 m s^{-1}; +4.8 m s^{-1}. (Intermediate velocity of *B* is +3.2 m s^{-1}.)

3.5 0.698 kg.

3.6 1.187 m.

3.7 34.01 rev min^{-1} and 7.085 rev min^{-1}. 133.6 mm.

3.8 0.452 + 0.795 = 1.247 s.

3.9 (a) 0.333 rad s^{-1}. (b) 0.855 m.

3.10 4.674 kN at 65° to original direction of flow.

4 Coupled and Geared Systems

The free-body diagram; compatibility. The gearbox; velocity ratio; efficiency. Equivalent mass. Equivalent inertia. Hoists. Vehicles. Optimum gear ratio for maximum acceleration.

4.1 The Fact Sheet

(a) Gearbox Ratio

A gearbox is a rotational machine for changing rotational speed and torque. The gearbox ratio, G, is defined as

$$G = \frac{\text{Angular speed of high-speed shaft}}{\text{Angular speed of low-speed shaft}} = \frac{\omega_H}{\omega_L}.$$

By this definition, G is always greater than 1.
G is sometimes given the name *velocity ratio.*

Angular displacement and angular acceleration of high-speed and low-speed shafts are related by the same ratio, i.e.

$$G = \frac{\omega_H}{\omega_L} \left(= \frac{\theta_H}{\theta_L} = \frac{\alpha_H}{\alpha_L} \right).$$

(b) Input and Output Torques

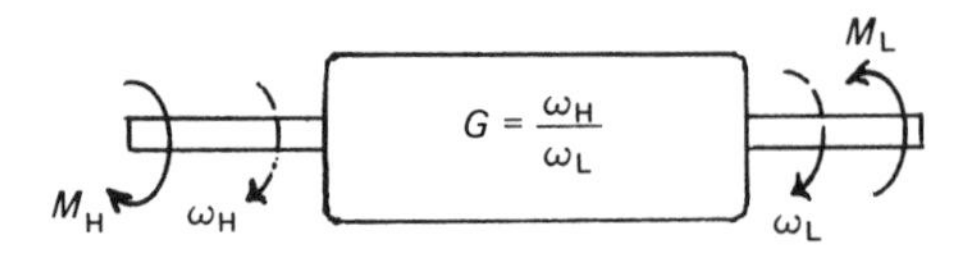

Fig. 4.1

Assuming no energy loss in a gearbox, the input and output torques, M_H and M_L, are related inversely to G, i.e.

$$G = \frac{M_L}{M_H}.$$

(c) Efficiency

When energy is absorbed by the gearbox (due to internal friction), the ratio

$$\frac{\text{Work given out per second}}{\text{Work put in per second}}$$

is called the efficiency, η ($\eta < 1$).

Assuming a 'reduction' gearbox, i.e. M_H is the input, or driving torque, then

$$\eta = \frac{M_L \omega_L}{M_H \omega_H} = \frac{1}{G} \frac{M_L}{M_H}.$$

$$\therefore M_L = M_H G\eta.$$

(d) Efficiency Measurement

If efficiency is stated as a percentage then

$$\eta = \frac{\text{Percentage efficiency}}{100}.$$

(e) Equivalent inertia

When a connected system of masses and rotors is subjected at some point to a torque M producing at that point an angular acceleration α, the equation of motion for the system may be written

$$M = I_e \alpha.$$

I_e is defined as the equivalent inertia of the system, referred to that point.

(f) Equivalent Mass

When a connected system of masses and rotors is subjected at some point to a force F, producing at that point a linear acceleration a, the equation of motion of the system may be written

$$F = am_e.$$

m_e is defined as the equivalent mass of the system, referred to that point.

4.2 Worked Examples

Example 4.1

A simple gearbox of ratio G has rotors of moment of inertia I_H and I_L attached to the high-speed and low-speed shafts respectively. The gearbox has a transmission efficiency E. Derive expressions for the effective inertia of the system (a) referred to the high-speed shaft, (b) ref-ferred to the low-speed shaft.

If $I_H = 2$ kg m^2, $I_L = 24$ kg m^2, $E = 0.8$ and a torque of 0.5 kN m is applied to the high-speed shaft, calculate the value of angular acceleration of the *low*-speed shaft for values of G of 2, 3, 4 and 5.

Solution 4.1

Part (a)

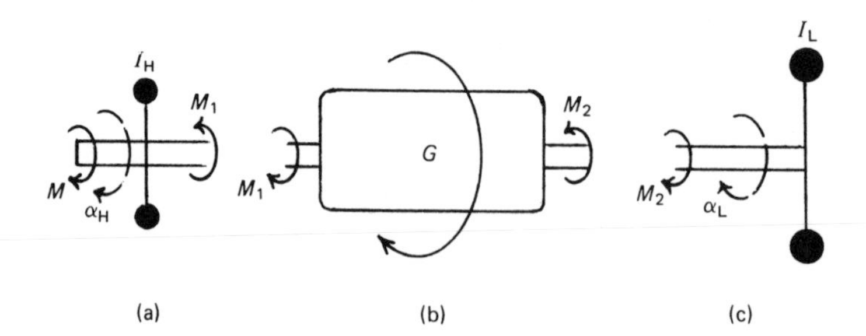

Of the free-body diagrams for the three 'elements' of the system given here, (a) shows the high-speed rotor, (b) the gearbox, and (c) the low-speed rotor. Because we first require the equivalent inertia referred to the high-speed shaft, we assume a torque M acting here. Study diagram (a) carefully. In 'isolating' this rotor, we recognise that the shaft to the right of the rotor transmits a torque (shown as M_1) to the gearbox. But this means that a corresponding reverse torque must act on the rotor. The situation is exactly analogous to a rope connecting two masses; the rope pulls one mass one way, and the second the other way.

Thus, from diagram (a) the equation of motion for I_H is

$$M - M_1 = I_H \alpha_H . \qquad (1)$$

For the gearbox (diagram (b))

[c] $$M_2 = M_1 G\eta. \qquad (2)$$

(By reason of its function, the torques on the two ends of a gearbox are almost always different. Because of this, the gearbox must be held down to some fixed point, in order that this difference of torques does not cause the box to rotate. For this reason, a 'holding torque' is also shown on the diagram. We do not require this for calculation, but it is shown to remind you that a free-body diagram should show all forces and torques acting on a body.)

From diagram (c) the equation of motion for I_L is

$$M_2 = I_L \alpha_L . \qquad (3)$$

[a] Also $$G\alpha_L = \alpha_H . \qquad (4)$$

Eliminate M_2 from equations 2 and 3:

$$M_1 G\eta = I_L \alpha_L .$$

Use this to substitute for M_1 in equation 1. At the same time, substitute for α_L from equation 4:

$$M - \frac{I_L}{G\eta}\left(\frac{\alpha_H}{G}\right) = I_H \alpha_H .$$

$$\therefore\ M = \underline{\alpha_H \left(I_H + \frac{I_L}{G^2 \eta}\right)} . \qquad (5)$$

The term in brackets is the required equivalent inertia.

Part (b)

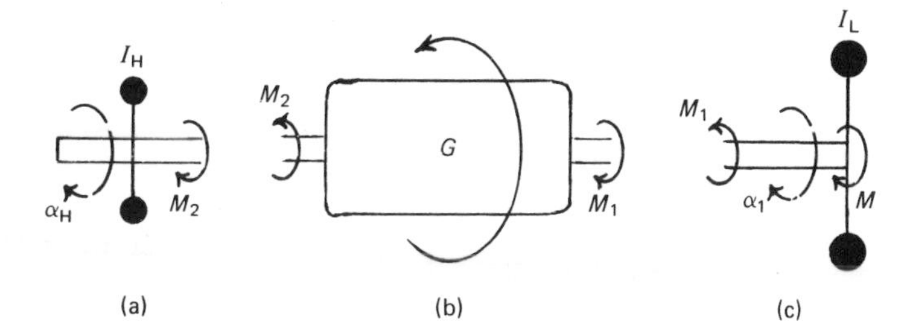

The corresponding free-body diagrams are shown, with the torque M applied to the low-speed shaft. The equations of motion are

from diagram (c) $$M - M_1 = I_L \alpha_L ; \qquad (6)$$

from diagram (b) $$M_2 = \frac{1}{G} M_1 \eta; \qquad (7)$$

from diagram (a) $$M_2 = I_H \alpha_H ; \qquad (8)$$

[a] $$G\alpha_L = \alpha_H . \qquad (9)$$

Algebraic manipulation similar to part (a) will yield the system equation

$$\underline{M = \alpha_L \left(I_L + \frac{G^2}{\eta} I_H \right)} . \qquad (10)$$

For solution of the numerical part we modify equation 5:

$$M = G\alpha_L \left(I_H + \frac{I_L}{G^2 \eta} \right) .$$

Substituting the stated values, with $G = 2$,

$$0.5 = 2\alpha_L \left(2 + \frac{24}{4 \times 0.8} \right) = 19\alpha_L .$$

$$\therefore \ \alpha_L = \underline{0.0263 \text{ rad s}^{-2}} .$$

Similar calculations give the following results

$G = 3$; $\alpha_L = \underline{0.031\,25 \text{ rad s}^{-2}}$.

$G = 4$; $\alpha_L = \underline{0.032\,26 \text{ rad s}^{-2}}$.

$G = 5$; $\alpha_L = \underline{0.031\,25 \text{ rad s}^{-2}}$.

These four answers suggest that it might be possible to find a value of the gearbox ratio G which, for a given input torque, would give a maximum value of acceleration of the output shaft. Example 4.13 explores this in more detail.

Example 4.2

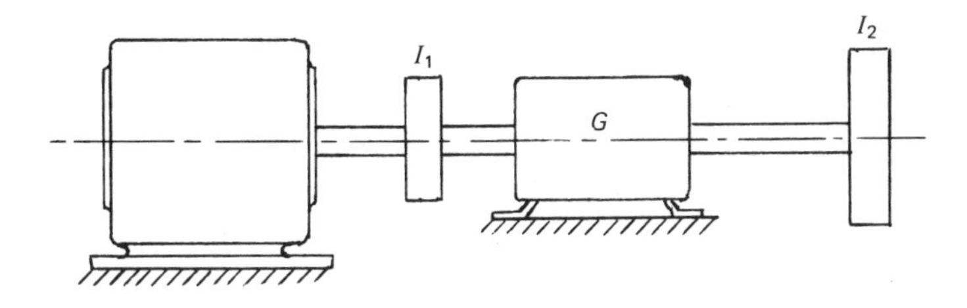

The diagram shows a gearbox with flywheels of moments of inertia $I_1 = 2.1$ kg m^2 and $I_2 = 48.0$ kg m^2 mounted respectively on the high-speed and low-speed shafts. The gearbox ratio G is 5.8.

(a) An electric motor having an armature (the rotating part) of moment of inertia 4.2 kg m^2 is coupled to the high-speed shaft, as shown. The motor is capable of a constant output shaft torque of 15 N m. Calculate for how long the motor must be switched on for the low-speed shaft of the gearbox to reach 500 rev min^{-1} from rest.

(b) If the motor is then switched off and a braking torque applied to the low-speed shaft, calculate the magnitude of this torque if it is required to bring the shaft to rest in not more than 30 revolutions. Neglect any energy loss in the gearbox.

Solution 4.2

Part (a)

Since the question refers to a torque on the high-speed shaft, we 'refer' all inertias to that shaft. Using the result of Example 4.1, part (a),

$$\text{Equivalent inertia (high-speed)} = I_H + \left(\frac{1}{G^2}\right) I_L .$$

$$I_{eH} = 2.1 + \frac{48}{(5.8)^2}$$

$$= 3.527 \text{ kg m}^2 .$$

(We do *not* include the inertia of the motor armature; the motor torque given is the *shaft output* torque.)

$$M = I_{eH}\,\alpha_H .$$

$$\therefore\ 15 = 3.527\alpha_H .$$

$$\therefore\ \alpha_H = 4.253 \text{ rad s}^{-2} .$$

[a] $$\therefore\ \alpha_L = \frac{4.253}{5.8} = 0.7333 \text{ rad s}^{-2} .$$

The required time is calculated from

$$\omega_2 = \omega_1 + \alpha t = 0 + \alpha t \text{ (from rest)}.$$

$$\therefore\ t = \frac{\omega_2}{\alpha} = \left(\frac{2\pi \times 500}{60}\right)\left(\frac{1}{0.7333}\right) .$$

$$\therefore\ t = \underline{71.4 \text{ s.}}$$

Part (b)

Data now refer to the low-speed shaft. We calculate the equivalent inertia I_{eL} from the result of Example 4.1 part (b)

$$I_{eL} = 48.0 + (2.1 + 4.2)\,(5.8)^2$$

$$= 259.9 \text{ kg m}^2 .$$

(Notice that in this case we *do* include the motor armature, as this is coupled to the high-speed shaft and has to be brought to rest.)

We calculate the required retardation from

$$\omega_2^2 = \omega_1^2 + 2\alpha\theta .$$

$$\therefore\ 0 = \left(2\pi\ \frac{500}{60}\right)^2 + 2\,(\alpha_L)\,(30 \times 2\pi).$$

$$\therefore\ \alpha_L = -7.272 \text{ rad s}^{-2}.$$

$$M = I_{eL}\,\alpha_L$$

$$= 259.9 \times 7.272.$$

$$\therefore\ M = \underline{1.89 \text{ kN m}}.$$

Example 4.3

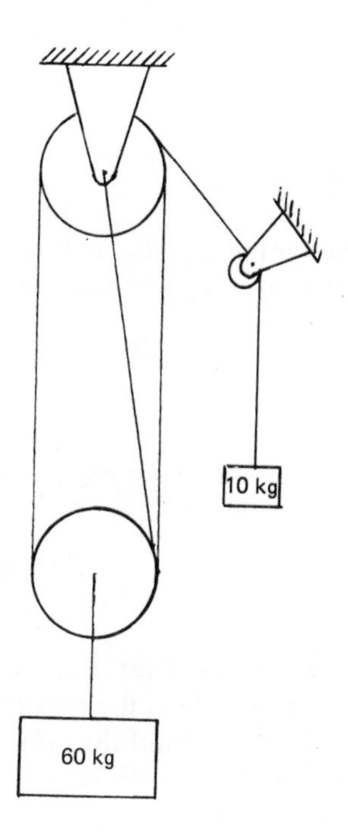

The diagram shows a pulley system of velocity ratio 6 suspended from a fixed point. Bodies of mass 10 kg and 60 kg hang from the high-speed and low-speed ropes respectively. Friction may be neglected.

(a) If a body of mass 1 kg is added to the smaller body when the system is at rest, calculate how long it will take for the larger body to rise a distance of 1 m.

(b) If a body of mass 1 kg is added to the larger body while the system is at rest, calculate how long it will take for the larger body to fall a distance of 1 m.

Solution 4.3

Part (a)

A pulley system has the same function as a gearbox, except that the input and output motions are translations, not rotations. The diagram shows the free-body diagrams of the three 'elements' of the system. Forces T_1 and T_2 are the intermediate rope tension forces between the pulley system and the two loads. Note how a rope can exert a tension force only: the ropes pull upwards on the loads and downwards on the pulley.

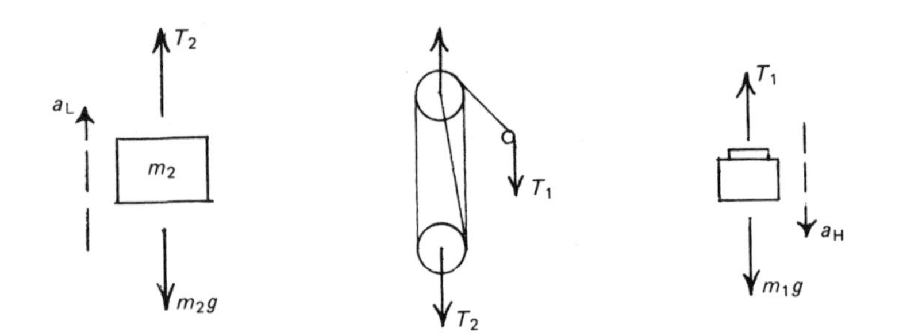

Working in algebra, the equations of motion are

for 11 kg, $\quad m_1g - T_1 = m_1a_H;$ (1)

for 60 kg, $\quad T_2 - m_2g = m_2a_L.$ (2)

[a] $\quad a_H = Ga_L;$ (3)

gearbox $\quad T_2 = GT_1.$ (4)

By eliminating T_1, T_2 and a_L from these equations we are, in effect, 'referring' the system to the high-speed rope.

Substituting for T_1 and T_2 in equation 4:

$$m_2a_L + m_2g = G(m_1g - m_1a_H).$$

$$m_2\left(\frac{a_H}{G}\right) + m_2g = Gm_1g - Gm_1a_H.$$

$$\therefore\ m_1g - \frac{m_2g}{G} = a_H\left(m_1 + \frac{m_2}{G^2}\right). \qquad (5)$$

Examine this result carefully. It is analogous to the result of Example 4.1 part (a). The left-hand side is the resultant force on the system, i.e. the difference between the weight of m_1 and the effective weight at this point of m_2; the bracketed term on the right-hand side is the effective mass of the system referred to the high-speed side of the pulley.

Substituting numbers, recalling that $m_1 = (10 + 1) = 11$ kg.

$$g\left(11 - \frac{60}{6}\right) = a_H\left(11 + \frac{60}{6^2}\right).$$

$$\therefore\ a_H = \frac{9.81 \times 1}{12.67} = 0.7748 \text{ m s}^{-2}.$$

$$\therefore\ a_L = \frac{0.7748}{6} = 0.1291 \text{ m s}^{-2}.$$

The required time is obtained from

$$x = ut + \tfrac{1}{2}at^2.$$

$$\therefore\ 1 = 0 + \tfrac{1}{2} \times 0.1291t^2.$$

$$\therefore\ t = \underline{3.936 \text{ s}}.$$

Part (b)

We shall not repeat the detailed analysis. Instead, we may write the system equation straight away, using equation 5 as a guide.

The equivalent mass referred to the low-speed rope will be

$$m_{eL} = (m_2 + G^2 m_1)$$

and the 'nett' or resultant force at this point will be $(m_2 g - G m_1 g)$.

The equation is

$$m_2 g - G m_1 g = a_L (m_2 + G^2 m_1).$$

Substituting,

$$g(61 - 6 \times 10) = a_L (61 + 36 \times 10).$$

$$\therefore a_L = \frac{9.81 \times 1}{421} = 0.0233t^2.$$

$$x = ut + \tfrac{1}{2}at^2.$$

$$\therefore 1 = 0 + \tfrac{1}{2} \times 0.0233t^2.$$

$$\therefore t = \underline{9.265 \text{ s.}}$$

Example 4.4

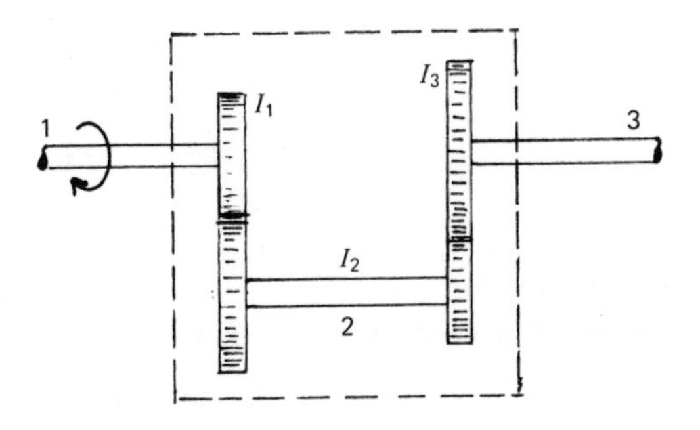

The diagram shows in schematic form a two-stage gearbox with an input shaft carrying a wheel of moment of inertia I_1, an intermediate shaft carrying wheels of total moment of inertia I_2 and an output (low-speed) shaft carrying a wheel of moment of inertia I_3. The high-speed shaft 1 turns at G_1 times the speed of the intermediate shaft 2; this shaft turns at G_2 times the speed of the low-speed shaft 3.

I_1 has a moment of inertia of 0.02 kg m^2; I_2, 0.18 kg m^2; I_3, 2.8 kg m^2. G_1 is 5.6 and G_2 is 8.4. A torque of 0.2 N m is applied to shaft 1. Calculate the speed of shaft 3 after 5 seconds.

If, after this time, a second torque of magnitude 10 N m acts on shaft 3 to bring the system to rest, assuming the first torque is still being applied, how long will it take the box to come to rest?

Solution 4.4

For the first part we require the equivalent inertia of the system, referred to shaft 1. This takes two stages. First we must reduce I_2 and I_3 to a single equivalent inertia referred to shaft 2.

Using the result of Example 4.1 part (a):

For I_2 and I_3 only,

$$I_{e2} = I_2 + \frac{I_3}{G_2^2}.$$

We now refer this to shaft 1:

$$I_{e1} = I_1 + \left(\frac{1}{G_1^2}\right)\left(I_2 + \frac{I_3}{G_2^2}\right)$$

$$= I_1 + \frac{I_2}{G_1^2} + \frac{I_3}{G_1^2 G_2^2}.$$

Substituting numbers,

$$I_{e1} = 0.02 + \frac{0.18}{(5.6)^2} + \frac{2.8}{(5.6)^2(8.4)^2}$$

$$= 0.027 \text{ kg m}^2.$$

The system equation of motion is

$$M_1 = I_{e1}\alpha_1.$$

$$\therefore\ 0.2 = 0.027\alpha_1.$$

$$\therefore\ \alpha_1 = 7.407 \text{ rad s}^{-2}.$$

The corresponding angular acceleration α_3 of the output shaft is

[a]

$$\alpha_3 = \frac{7.407}{5.6 \times 8.4} = 0.1575 \text{ rad s}^{-2}.$$

$$\omega_2 = \omega_1 + \alpha t$$

$$= 0 + 0.1575 \times 5$$

$$= \underline{0.787 \text{ rad s}^{-1}}.$$

For the second part we refer inertias to shaft 3. Use the result of Example 4.1 part (b). The equivalent inertia is

$$I_{e3} = I_3 + G_2^2 I_2 + (G_1 G_2)^2 I_1$$

$$= 2.8 + (8.4)^2(0.18) + (8.4 \times 5.6)^2(0.02)$$

$$= 59.76 \text{ kg m}^2.$$

The effective torque at this point is the torque of 10 N m less the multiplied torque of 0.02 N m still acting to accelerate the box.

$$\text{Total effective torque} = 10 - (0.2 \times 5.6 \times 8.4)$$

$$= 0.592 \text{ N m}.$$

$$\text{Equation of motion: } M_3 = I_{e3}\alpha_3.$$

$$0.592 = 59.76\alpha_3.$$

$$\therefore\ \alpha_3 = 0.009\,906 \text{ rad s}^{-2}.$$

$$\omega_2 = \omega_1 + \alpha t.$$

$$\therefore\ 0 = 0.787 - 0.009\,906t.$$

$$\therefore\ t = 79.45 \text{ s}.$$

Example 4.5

Bodies having mass m_1 and m_2 ($m_2 > m_1$) are connected by a light string which passes over a pulley of radius R and moment of inertia I which turns about a frictionless fixed axis, as shown in the diagram. The string does not slip on the pulley. Obtain an expression for the linear acceleration of the body of mass m_2.

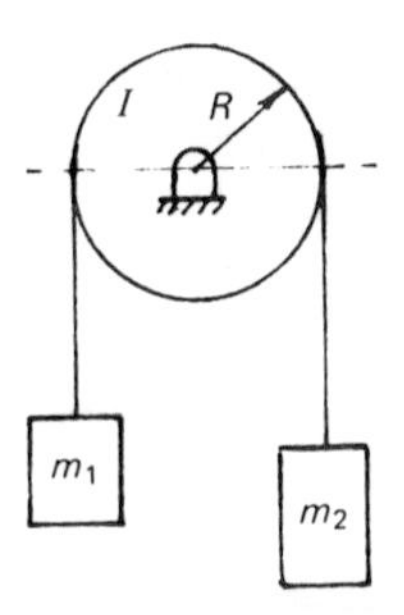

If m_1 is 2 kg, m_2 is 2.4 kg, I is 1.4 kg m^2 and R is 0.2 m, calculate the magnitude of the torque M required at the pulley axis to cause m_2 to be lifted a distance of 1 m from rest in 1 second.

Solution 4.5

We draw a free-body diagram (as here) for each of the three 'elements' of the system. The string tension forces are shown as T_1 and T_2. Note that these are different, although they exist in the same string. The reason is that if no slip occurs, as stated, there must be a friction force between string and pulley, causing a difference of tension between the two ends.

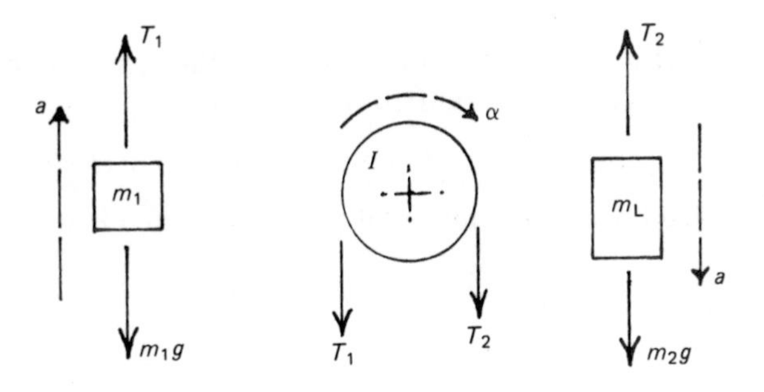

Because both bodies are connected to the same string, the linear acceleration a is the same for each. This is also the same as the peripheral acceleration of the pulley rim. Thus

$$a = \alpha R. \tag{1}$$

The three equations of motion, having regard to the signs of the accelerations, are as follows:

$$T_1 - m_1 g = m_1 a. \tag{2}$$

$$T_2 R - T_1 R = I\alpha. \tag{3}$$

$$m_2 g - T_2 = m_2 a. \tag{4}$$

Substitute in equation 3 for T_1 and T_2:

$$R\,(m_2 g - m_2 a) - R\,(m_1 a + m_1 g) = I\alpha = I\,\frac{a}{R}\,.$$

$$\therefore\ m_2 g - m_2 a - m_1 a - m_1 g = I\,\frac{a}{R^2}\,.$$

$$\therefore\ m_2 g - m_1 g = a\left(\frac{I}{R^2} + m_1 + m_2\right)\,. \tag{5}$$

This last equation, equation 5, has the general form $\Sigma F = ma$. The left-hand side is seen to be the resultant force on the system, i.e. the excess or unbalanced weight of the two masses. The quantity in brackets should now be recognised as the equivalent mass of the system, referred to the location of m_2.

The required expression for a is thus

$$a = \frac{m_2 g - m_1 g}{I_e},$$

where I_e is the group of terms in brackets in equation 5.

The remainder of the problem requires the equation in terms of an equivalent inertia, referred to the pulley axis. A complete reanalysis is not necessary. We may manipulate equation 5.

Multiplying throughout by R^2 and writing $a = \alpha R$,

$$R^2(m_2 g - m_1 g) = \alpha R\,(I + m_1 R^2 + m_2 R^2).$$

$$\therefore\ m_2 gR - m_1 gR = \alpha\,(I + m_1 R^2 + m_2 R^2). \qquad (6)$$

Now the left-hand side is seen to be the resulting *torque* due to the unbalance of the weights, and the bracketed quantity is the equivalent *inertia* referred to the pulley axis.

If a torque M is now added, in the sense to raise m_2 (i.e. anticlockwise), equation 6 must be modified thus:

$$M - m_2 gR + m_1 gR = \alpha\,(I + m_1 R^2 + m_2 R^2). \qquad (7)$$

The linear acceleration of m_2 is calculated from

$$x = ut + \tfrac{1}{2}at^2.$$

$$\therefore\ 1 = 0 + \tfrac{1}{2}a(1)^2.$$

$$\therefore\ a = 2\ \text{m s}^{-2}.$$

$$\therefore\ \alpha = \frac{a}{R} = \frac{2}{0.2} = 10\ \text{rad s}^{-2}.$$

Substituting in equation 7,

$$M - 2.4g \times 0.2 + 2g \times 0.2 = 10[1.4 + 2(0.2)^2 + 2.4(0.2)^2].$$

$$\therefore\ M = 10 \times 1.576 + 0.4g \times 0.2.$$

$$= \underline{16.54\ \text{N m}}.$$

Example 4.6

A simple hoist consists of a drum of radius R = 1.4 m and moment of inertia I = 240 kg m^2, with a rope attached to the rim, which supports a load of mass m = 2000 kg as shown in diagram (a). The load is raised by a torque M applied to the drum axle. Calculate the value of the torque M sufficient to produce an upward acceleration of the load of 1 m s^{-2}.

A counter-weight, also of mass m = 2000 kg, is then added to the system, such that as the load rises, the counterweight falls. (See diagram (b).) Calculate the value of torque M now required to produce the same upward acceleration of the load as before.

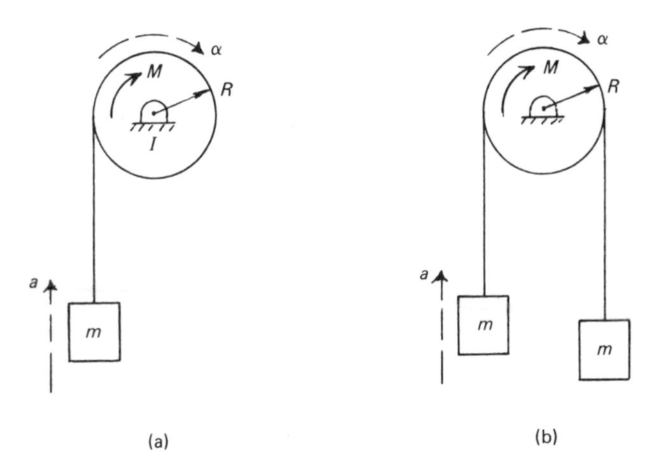

Solution 4.6

If you have worked through Example 4.5 you should realise that it will not be necessary to draw free-body diagrams and write separate equations of motion. We can make use of equation 7 of that example to write the system equation straight down. For the first part of the question, $m_2 = m$ and $m_1 = 0$. Therefore

$$M - mgR = \alpha(I + mR^2).$$

$$\alpha = \frac{a}{R} = \frac{1}{1.4}.$$

$$\therefore\ M = 2000g \times 1.4 + \frac{1}{1.4}[240 + 2000 \times (1.4)^2]$$

$$= 27\,468 + \frac{1}{1.4} \times 4160$$

$$= \underline{30.44 \text{ kN m}}.$$

For the second part of the question, $m_1 = m$, $m_2 = m$. The equation is therefore

$$M - mgR + mgR = \alpha(I + mR^2 + mR^2).$$

$$\therefore\ M = \frac{1}{1.4}[240 + 2 \times 2000 \times (1.4)^2]$$

$$= \frac{1}{1.4} \times 8080$$

$$= \underline{5.771 \text{ kN m}}.$$

Now notice that although including a second load of mass m has added considerably to the effective inertia of the system (an increase from 4160 to 8080 kg m^2), we have more than compensated for this by getting rid of the large figure of 27 468 N m. This figure represents the torque required at the drum axis merely to support the static weight of the load. We may call this the 'static torque'. The remainder is what is required to accelerate the whole system. By adding a counterweight, we have eliminated the static torque altogether, leaving only the 'dynamic torque', or the accelerating torque. Industrial and commercial hoists such as the familiar 'lift', or elevator, incorporate a counterweight of approximately the same weight as the load, in order that the driving motor can be reasonably small and energy not be used unnecessarily.

Example 4.7

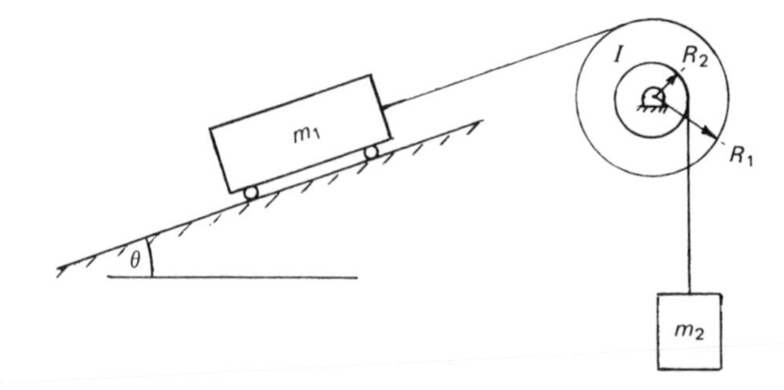

The diagram shows an inclined hoist consisting of a compound drum of radii R_1 and R_2 on which ropes are attached to connect to loads of masses m_1 and m_2. m_2 hangs vertically and m_1 lies on a frictionless plane inclined at angle θ to the horizontal.

Use the concept of equivalent inertia to determine the value of the acceleration of load m_1. $m_1 = 200$ kg; $m_2 = 300$ kg; $I = 280$ kg m^2; $R_1 = 1.2$ m; $R_2 = 0.45$ m; $\theta = 40°$.

Solution 4.7

From earlier examples we know that the equivalent inertia of a mass at radius R on a drum or pulley is (mR^2). The equivalent inertia of the system referred to the drum axis, is thus

$$I_e = I + m_1 R_1^2 + m_2 R_2^2$$

(substituting)

$$= 280 + 200 \times (1.2)^2 + 300 \times (0.45)^2$$

$$= 628.75 \text{ kg m}^2.$$

The direction of acceleration is not stated. We assume that the drum acceleration is clockwise; thus, m_1 is assumed to accelerate up the plane and m_2 to accelerate downwards.

In determining the resultant torque on the drum, we note that the weight of m_2 exerts a positive torque of $(m_2 g R_2)$. The torque due to the weight of m_1 is negative. But only the weight component along the direction of motion contributes to the effective torque. Thus the retarding torque due to m_1 is $(m_1 g R_1 \sin\theta)$.

We may now write the system equation:

$$m_2 g R_2 - m_1 g R_1 \sin\theta = \alpha I_e.$$

$$\therefore \alpha = \frac{300g \times 0.45 - 200g \times 1.2 \sin 40°}{628.75}$$

$$= -0.3006 \text{ rad s}^{-2}.$$

Relating the linear acceleration of m_1 to the drum acceleration,

$$a_1 = \alpha R_1 = (-0.3006) \times 1.2.$$

$$\therefore a_1 = \underline{-0.3607 \text{ m s}^{-2}}.$$

The negative sign indicates that the acceleration will be down the plane instead of up as assumed.

Example 4.8

An electric motor drives the drum of a hoist through a speed-reducing gearbox of ratio 144. The motor output torque is 2.6 N m. The rotating parts on the high-speed input shaft to the gearbox have a total moment of inertia of 0.02 kg m^2. The rotating parts on the low-speed output shaft, including the drum, have a moment of inertia of 26.4 kg m^2. A rope around the drum at a radius of 0.6 m supports a hanging load of mass 48 kg. The gearbox has a transmission efficiency of 82 per cent and the drum is subjected to a friction torque at the bearings of 6.4 N m. Calculate the linear upward acceleration of the load.

Solution 4.8

When friction losses are introduced, the method of equivalent mass or inertia does not offer the simpler solution illustrated by earlier examples. It is better to return to drawing free-body diagrams and writing equations for each element. Care spent on a good clear diagram will help to avoid mistakes.

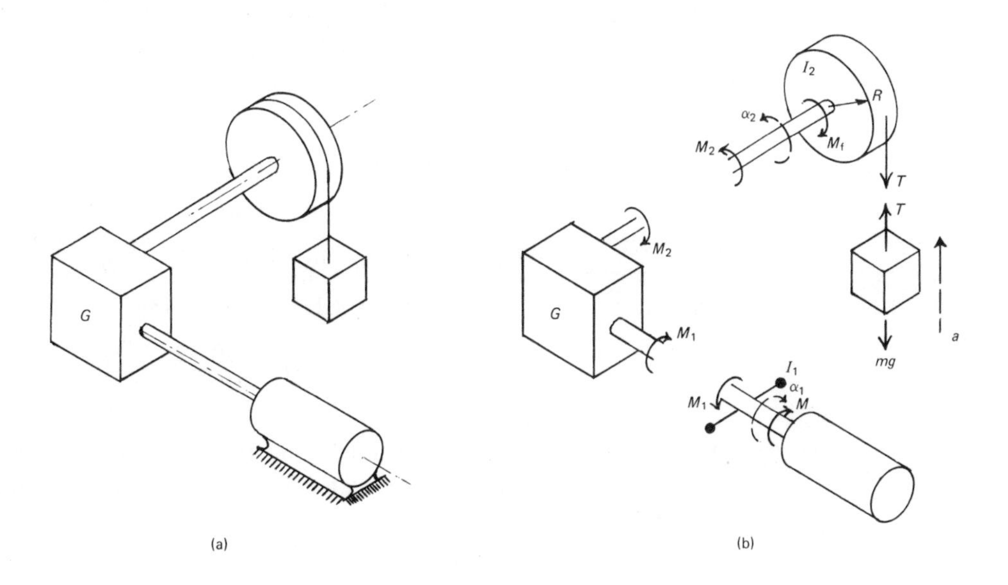

Diagram (a) shows a diagrammatic representation of the system; (b) 'breaks down' the system into the four elements. M_1 is the intermediate torque between inertia I_1 and the gearbox; M_2 is the intermediate torque between gearbox and inertia I_2 (the drum); T is the intermediate rope tension force between drum and load; M_f is the friction torque on the drum axle. All data are given, the only unknown terms being the three accelerations. We begin by relating these to a, the required acceleration:

$$\alpha_2 = \frac{a}{R} = \frac{a}{0.6} = 1.667a;$$

$$\alpha_1 = G\alpha_2 = 144\alpha_2 = 240a.$$

Now we write equations for each element, starting at the motor, and arranging each equation so that the 'subject' is the torque or force required for the next equation. This example is an exception to the usual rule of working in algebra where possible; this would result in six equations, and some trouble in manipulating them. In this solution, we will complete each calculation as it arises.

The equation of motion for I_1 is

$$M - M_1 = I_1 \alpha_1.$$

$$2.6 - M_1 = 0.02\,(240a).$$

$$\therefore\ M_1 = 2.6 - 4.8a.$$

The equation for the gearbox is

[c]

$$M_2 = M_1 GE$$

$$= (2.6 - 4.8a)\,144 \times 0.82$$

$$= 307.0 - 566.8a.$$

The equation of motion for drum I_2 is

$$M_2 - M_f - TR = I_2 \alpha_2.$$

$$\therefore\ T = \frac{1}{R}(M_2 - M_f - I_2 \times 1.667a)$$

$$= \frac{1}{0.6}(307.0 - 556.8a - 6.4 - 26.4 \times 1.667a)$$

$$= 501 - 1001.3a.$$

The equation of motion for the load m is

$$T - mg = ma.$$

$$\therefore\ 501 - 1001.3a - 48g = 48a.$$

$$\therefore\ a(48 + 1001.3) = 501 - 48g.$$

$$\therefore\ a = \frac{30.12}{1049.3} = \underline{0.0287 \text{ m s}^{-2}}.$$

The solution begins with the motor and the equation for I_1 because the motor torque is given. If the question stated the acceleration of the load and asked for the corresponding motor torque, we would have begun with the equation of the load and worked backwards. (See Problem 4.9 at the end of this chapter.)

Example 4.9

A simple hoist consists of a drum of radius 0.6 m driven by a motor through a reduction gearbox of ratio 14.2. The hoist is used to raise a load of 200 kg which hangs from a rope wrapped round the drum. When the torque from the motor is 122 N m the load is lifted with an upward acceleration of 0.93 m s^{-2}. When the motor torque is increased to 150 N m the acceleration of the load is increased to 2.33 m s^{-2}. Estimate the efficiency of the gearbox. Neglect all other energy losses.

Solution 4.9

We first calculate the output torque of the gearbox. Call this M_2.

[c]

$$M_2 = (122 \times 14.2\eta) \text{ N m}$$

where η is the required efficiency. 'Referring' all torques and forces to the load-point, the effective force due to this torque, F_e, is

$$F_e = \frac{\text{Torque}}{\text{Drum radius}}$$

$$= \frac{122 \times 14.2\eta}{0.6} = 2887.3\eta.$$

In writing the system equation of motion we must recognise that the 'equivalent mass' is the mass of 200 kg plus an equivalent mass of the drum. Since no data are given we call this equivalent mass m_e. The system equation of motion, referred to the load, is

$$2887.3\eta - 200g = m_e a = 0.93 m_e. \qquad (1)$$

For the increased motor torque the corresponding effective force at the load is

$$F'_e = \frac{150 \times 14.2\eta}{0.6} = 3550.0\eta$$

and the system equation of motion is therefore

$$3550.0\eta - 200g = 2.33 m_e. \qquad (2)$$

Equations 1 and 2 may be used to solve for m_e and η. Since η is required we may eliminate m_e by dividing equation 2 by equation 1.

$$\frac{3550.0\eta - 200g}{2887.3\eta - 200g} = \frac{2.33\, \cancel{m_e}}{0.93\, \cancel{m_e}}.$$

$$3550.0\eta - 200g = 2.505\,(2887.3\eta - 200g).$$

$$\therefore \; \eta\,(2.505 \times 2887.3 - 3550) = 200g\,(2.505 - 1).$$

$$\therefore \; \eta = \frac{200g \times 1.505}{3682.6} = \underline{0.802.}$$

Example 4.10

A solid uniform cylinder of mass m and radius R rolls without slipping down an inclined plane of slope θ. Obtain an expression for the linear acceleration of the cylinder centre.

A uniform hollow metal ring of outer radius 20 mm is released from rest at the top of an inclined plane of slope 5°. It is observed that it takes 1.95 seconds to travel 1 metre down the plane. Calculate the internal diameter of the ring.

The moment of inertia of a uniform hollow ring of outer and inner radii R_o and R_i is

$$I = \tfrac{1}{2} m\,(R_o^2 + R_i^2).$$

Solution 4.10

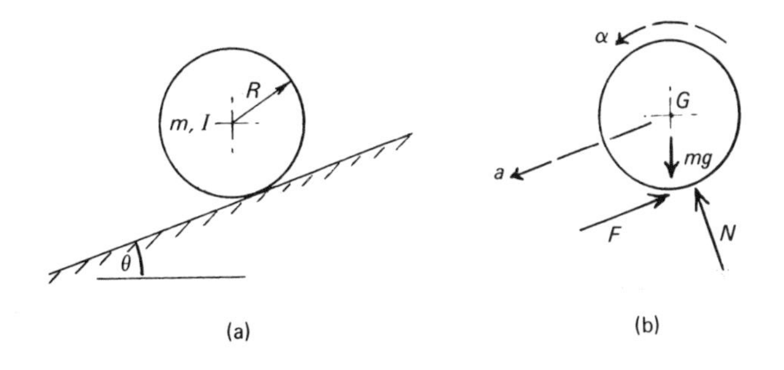

Diagram (a) shows the cylinder on the plane. (b) shows the free-body diagram. This is an instance of combined translation and rotation. The centre of the cylinder has a linear translation, with associated velocity and acceleration. At the same time, the cylinder rotates about its own centre with associated angular velocity and acceleration. Two equations of motion will therefore be needed.

Because the cylinder rolls without slip there must be a friction force between plane and cylinder. This must have anticlockwise moment about the cylinder centre to cause it to roll downwards. This is force *F* on the free-body diagram.

For the equation of motion of translation, mg is resolved into components along the plane and perpendicular to it. The equation is

$$\Sigma F = ma.$$

$$mg \sin \theta - F = ma. \quad (1)$$

For the equation of motion of rotation, F is the only force having a moment about the mass centre G. The equation is

$$\Sigma M = I\alpha.$$

$$\therefore\ F \times R = I_g \alpha. \quad (2)$$

There is a kinematic relation between the linear acceleration *a* and the angular acceleration α. If the cylinder turned with angular acceleration α about a fixed centre, then, for no slip to occur, the track would have to move upwards in the direction of *F* with a linear acceleration equal to the peripheral linear acceleration of the cylinder. Since the plane does not move, the cylinder centre moves down the plane with this acceleration. Thus

$$a = \alpha R. \quad (3)$$

Eliminating F and α from the three equations,

$$mg \sin \theta - \frac{I_g}{R}\left(\frac{a}{R}\right) = ma.$$

$$\therefore\ mg \sin \theta = a\left(m + \frac{I_g}{R^2}\right). \quad (4)$$

The quantity in the brackets should by now be recognised as the equivalent mass of the rolling cylinder.

Replacing I_g by $\frac{1}{2}mR^2$ (using the formula given, with $R_i = 0$),

$$mg \sin \theta = a(m + \tfrac{1}{2}m).$$

$$\therefore\ g \sin \theta = 1\tfrac{1}{2}a.$$

$$\therefore\ \underline{a = \tfrac{2}{3}g \sin \theta}.$$

For the second part of the question it is not necessary to repeat the analysis. Equation 4 is valid for any cylinder. First we may calculate the linear acceleration *a*.

$$x = ut + \tfrac{1}{2}at^2.$$

$$1 = 0 + \tfrac{1}{2}a(1.95)^2.$$

$$\therefore\ a = 0.526 \text{ m s}^{-2}.$$

Substituting data in equation 4,

$$mg \sin 5° = 0.526 \left(m + \frac{\frac{1}{2}m(R_o^2 + R_i^2)}{R_o^2}\right).$$

Cancel m right through:

$$\frac{g \sin 5°}{0.526} = 1 + \tfrac{1}{2} + \tfrac{1}{2}\left(\frac{R_i^2}{R_o^2}\right).$$

$$\therefore \frac{R_i^2}{R_o^2} = 2(1.625 - 1.5) = 0.25.$$

$$\therefore R_i = R_o\sqrt{0.25}$$

$$= 20 \times 0.5$$

$$= \underline{10 \text{ mm.}}$$

Example 4.11

A wagon has a total mass m and four wheels, all of radius R, radius of gyration k and mass $m\omega$. It is released from rest at the top of a slope of angle θ and travels down a distance L. Derive an expression for the velocity of the truck at this point. Neglect any energy loss due to friction or air resistance.

A wagon of total mass 1650 kg has four identical wheels, each of mass 70 kg and rolling radius 0.4 m. It is released from rest and allowed to roll down a track having a slope of $\sin^{-1}$ 0.02. After 10 seconds it has travelled 9.2 m. Calculate the radius of gyration of the wheels.

Solution 4.11

The solution may be obtained by applying the principles of conservation of energy. (See Chapter 5.) No work is done on the wagon or by it. The energy equation is therefore

Loss of potential energy = Gain of kinetic energy.

For a track distance L, vertical drop = $L \sin \theta$.

$$\therefore \text{Loss of p.e.} = mgL \sin \theta.$$

(Note that m is the total mass, including that of the wheels.)

The kinetic energy consists of energy of translation ($\frac{1}{2}mv^2$) for the whole wagon, including the wheels, and energy of rotation ($\frac{1}{2}I\omega^2$) for the wheels themselves.

The linear speed v of the wagon is equal to the peripheral speed (due to rotation) of the wheels; i.e.

$$v = \omega R.$$

$$\therefore \text{k.e.} = \tfrac{1}{2}mv^2 + \tfrac{1}{2}I\omega^2$$

$$= \tfrac{1}{2}mv^2 + \tfrac{1}{2}I\left(\frac{v}{R}\right)^2$$

$$= \tfrac{1}{2}v^2\left(m + \frac{I}{R^2}\right).$$

The term in brackets is recognised as the equivalent mass of the wagon and wheel system.

The energy equation is thus

$$mgL \sin\theta = \tfrac{1}{2}v^2 \left(m + \frac{I}{R^2}\right).$$

Replacing I by $4m_w k^2$ (recalling that $I = mk^2$, where k is the radius of gyration),

$$mgL \sin\theta = \tfrac{1}{2}v^2 \left(m + 4m_w \frac{k^2}{R^2}\right).$$

$$\therefore\ v = \left(\frac{2mgL \sin\theta}{m + 4m_w \dfrac{k^2}{R^2}}\right)^{1/2}. \qquad (1)$$

The velocity may be calculated from

$$x = \tfrac{1}{2}(u + v)\,t.$$

$$9.2 = \tfrac{1}{2}(0 + v)\,10.$$

$$\therefore\ v = 1.84 \text{ m s}^{-1}.$$

Substitute in equation 1, writing the equivalent mass as m_e for the present:

$$(1.84)^2 = \frac{2 \times 1650 \times 9.81 \times 9.2 \times 0.02}{m_e}.$$

$$\therefore\ m_e = 1759.4 \text{ kg}.$$

$$\therefore\ 1650 + 4 \times 70 \left(\frac{k^2}{(0.4)^2}\right) = 1759.4.$$

$$\therefore\ k^2 = (1759.4 - 1650)\frac{(0.4)^2}{4 \times 70} = 0.0625.$$

$$\therefore\ k = \underline{0.25 \text{ m}}.$$

Example 4.12

A motor vehicle has a total mass of 1200 kg. It has wheels of 0.74 m diameter, the total moment of inertia of all four wheels being 11.5 kg m^2. The output torque of the engine has a constant value of 180 N n. The gearbox ratio is 6.2. Rotating parts on the engine or input shaft to the gearbox have a total moment of inertia of 0.62 kg m^2. There is a friction torque on the engine shaft of 35 N m and a friction torque on the gearbox output shaft and wheel bearings of 25 N m. Resistance to the vehicle due to air is estimated at 440 N. Determine the acceleration of the vehicle (a) on a straight level road; (b) up an incline of slope $\sin^{-1}$ 0.05.

Solution 4.12

This is similar to Example 4.8 in that a simpler solution is obtained by writing an equation for each element in turn, and calculating as we go. Again, a good clear diagram is necessary. The illustration shows a schematic representation of the vehicle, each element shown as a free-body diagram. The wheels have been represented as a single rotor. M is the engine output torque; M_1 and M_2 are respectively the gearbox input and output torques; M_{f1} and M_{f2} are the friction torques on engine shaft and gearbox output shaft, and wheel bearings respectively; F is the tangential force between driving wheels and road (the tractive force). Observe that on the wheel diagram F is shown also as an equal and opposite force at the axle. Strictly speaking, this cannot be correct, as the wheels have mass and need a resultant force to accelerate them. But this is taken account of by adding the mass of the wheels to that of the vehicle.

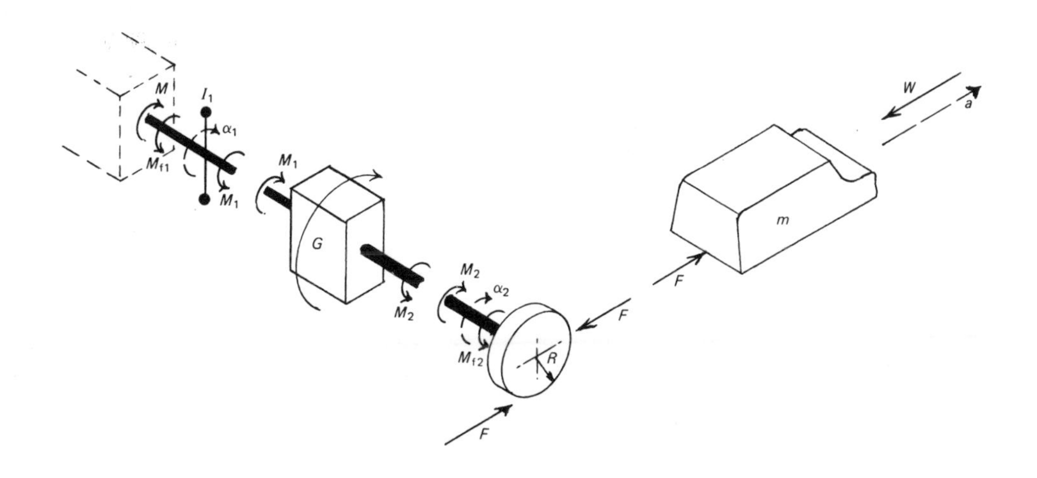

First we relate all accelerations to a:

$$\alpha_2 = \frac{a}{R} = \frac{a}{0.37} = 2.702a.$$

$$\alpha_1 = G\alpha_2 = \frac{6.2a}{0.37} = 16.7a.$$

The equation of motion for I_1 is

$$M - M_1 - M_{f1} = I_1\alpha_1.$$

$$\therefore\ M_1 = M - M_{f1} - I_1\alpha_1.$$

$$= 180 - 35 - 0.62 \times 16.76a$$

$$= 145 - 10.39a.$$

For the gearbox,

$$M_2 = GM_1$$

$$= 6.2\,(145 - 10.39a)$$

$$= 899 - 64.42a.$$

The equation of motion for I_2 (the wheels) is

$$M_2 - M_{f2} - FR = I_2\alpha_2.$$

$$\therefore\ F = \frac{1}{R}(M_2 - M_{f2} - I_2\alpha_2).$$

$$= \frac{1}{0.37}\,(899 - 64.42a - 25 - 11.5 \times 2.702a)$$

$$= 2362 - 258.1a.$$

Part (a)

The vehicle equation of motion for a straight level road is

$$F - W = ma.$$

$$2362.2 - 258.1a - 440 = 1200a.$$

$$\therefore\ a = \frac{2362 - 440}{1200 + 258.1} = \underline{1.318\ \text{m s}^{-2}}.$$

Part (b)

For the upward incline the additional retarding component of weight down the slope must be included:

$$F - W - mg \sin \theta = ma.$$

$$2362.2 - 258.1a - 440 - 1200g \times 0.05 = 1200a.$$

$$\therefore a = \frac{2362.2 - 440 - 1200 \times 9.81 \times 0.05}{1200 + 258.1}$$

$$= \underline{0.915 \text{ m s}^{-2}}.$$

Example 4.13

A road vehicle has a mass of 1140 kg. It has four wheels, each of mass 21 kg, radius 0.34 m and radius of gyration 0.28 m. The engine output torque has a constant value of 165 N m. The gearbox ratio is G and energy losses are given in terms of a transmission efficiency of 0.85. The engine and gearbox input shaft rotating parts have a total moment of inertia of 0.53 kg m^2; the inertia of the gearbox output shaft rotating parts is negligible in comparison with the wheels. At a certain speed along a straight level road, resistance due to air and friction is constant at 580 N.

Show that the resulting acceleration a of the vehicle is given by

$$a = \frac{G - A}{B + CG^2}$$

and calculate the value of A, B and C. Find the value of G corresponding to maximum acceleration of the vehicle under the conditions stated, and calculate this acceleration.

Solution 4.13

The method of solution may be generally the same as in Example 4.12. We may make use of the free-body diagrams for that example, but notice that in *this* example losses are given in terms of transmission efficiency, and not as friction torques. M_{f1} and M_{f2} will therefore both be zero.

Relating accelerations as previously,

$$\alpha_2 = \frac{a}{R} = \frac{a}{0.34} = 2.941a; \qquad \alpha_1 = G\alpha_2 = 2.941Ga.$$

The equation of motion for I_1 is

$$M - M_1 = I_1 \alpha_1.$$

$$\therefore M_1 = M - I_1 \alpha_1$$

$$= 165 - 0.53 \times 2.941Ga$$

$$= 165 - 1.559Ga.$$

For the gearbox,

$$M_2 = M_1 GE = 0.85G(165 - 1.559Ga)$$

$$= 140.3G - 1.325G^2 a.$$

The equation of motion for the wheels, I_2, is

$$M_2 - FR = I_2 \alpha_2.$$

$$\therefore F = \frac{1}{R}(M_2 - I_2 \alpha_2).$$

Recall that $I = mk^2$.

$$\therefore\ I_2 = 4 \times 21 \times (0.28)^2 = 6.586 \text{ kg m}^2.$$

$$\therefore\ F = \frac{1}{0.34}(140.3G - 1.325G^2 a - 6.586 \times 2.941a)$$

$$= 412.6G - 3.897G^2 a - 56.97a.$$

The vehicle equation of motion is

$$F - W = ma.$$

$$412.6G - 3.897G^2 a - 56.97a - 580 = 1140a.$$

$$\therefore\ a(1140 + 56.97 + 3.897G^2) = 412.6G - 580.$$

$$\therefore\ a = \frac{412.6G - 580}{1196.97 + 3.897G^2}.$$

To arrange in the required form we divide throughout by 412.6:

$$a = \underline{\frac{G - 1.406}{2.901 + 0.009\,44G^2}}.$$

The required values are therefore

$$\underline{A = 1.406\ :\ B = 2.901\ :\ C = 0.009\,44.}$$

To determine maximum acceleration we differentiate a with respect to G. Using the quotient formula,

$$\frac{\mathrm{d}a}{\mathrm{d}G} = 0 = \frac{v\ \mathrm{d}u - u\ \mathrm{d}v}{v^2}.$$

Reverting to using the terms A, B and C,

$$0 = \frac{(B + CG^2) \times 1 - (G - A)(2CG)}{v^2}.$$

(We do not bother to substitute for v^2 on the bottom line; if the expression is equal to 0 then the top line must equal 0.)

$$B + CG^2 - 2CG^2 + 2CGA = 0.$$

$$\therefore\ CG^2 - 2CGA - B = 0.$$

$$\therefore\ G^2 - G(2A) - (B/C) = 0.$$

$$\therefore\ G = \frac{2A \pm \sqrt{4A^2 + 4(B/C)}}{2}$$

$$= A \pm \sqrt{A^2 + (B/C)}$$

$$= 1.406 \pm \sqrt{1.977 + 307.3}$$

$$= 1.406 \pm 17.59$$

$$= \underline{18.996 \approx 19.0.}$$

Substituting in expression for a,

$$a_{max} = \frac{19 - 1.406}{2.901 + 0.00944 \times 19^2}$$

$$= \underline{2.789 \text{ m s}^{-2}}.$$

Example 4.14

A car has a total mass of 980 kg. The gear ratio (engine speed : wheel speed) is 1.6. The engine produces a constant output torque of 790 N m. The efficiency of the transmission is 0.8 and the driving wheels have a radius of 0.36 m. When the car is driven along a horizontal level road against a constant resistance due to air of 600 N the car has a forward acceleration of 1.7 m s^{-2}. Determine the acceleration when the car is driven (a) up a slope of $\sin^{-1}$ 0.05; (b) down a slope of $\sin^{-1}$ 0.05. Assume the same value of air resistance for all cases.

Solution 4.14

In this question no details are given of inertia of the elements. This suggests the use of the concept of equivalent mass. We will treat the problem as before, writing equations for each element, but this time retaining algebraic symbols. The illustration shows the system in the form of free-body diagrams. The notation in the diagrams corresponds to the equations. The linear force *mg* sin θ is shown acting both ways; when ascending the slope, the force will be against the motion, and when descending it will be in the same direction. Along the horizontal track this force will of course be absent.

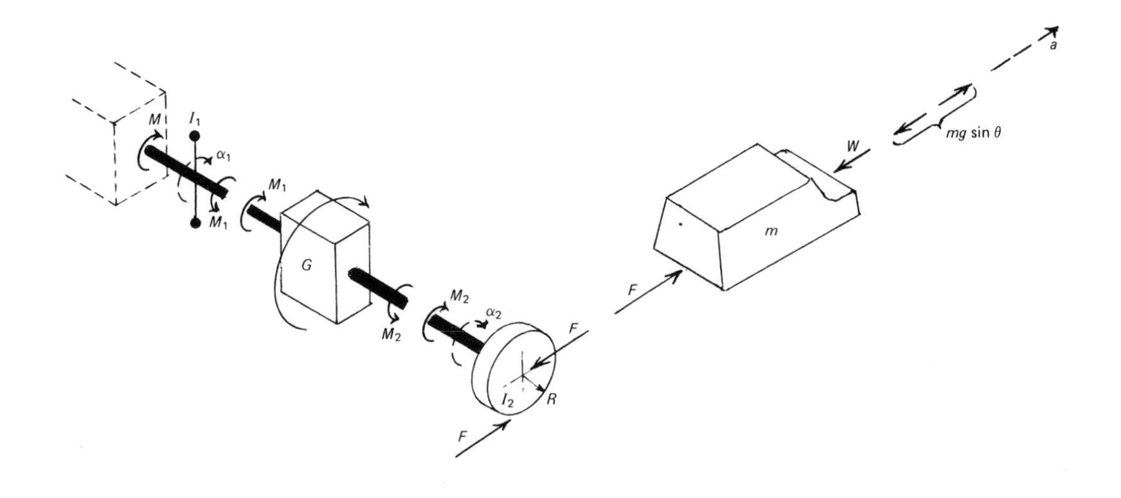

We first relate the accelerations, as before:

$$\alpha_2 = \frac{a}{R}.$$

$$\alpha_1 = \frac{Ga}{R}.$$

The equation of motion for I_1 is

$$M - M_1 = I_1\alpha_1 = I_1\frac{Ga}{R}.$$

$$\therefore\ M_1 = M - I_1\frac{Ga}{R}.$$

For the gearbox

[c] $$M_2 = M_1 G\eta = MG\eta - I_1 G^2\eta\frac{a}{R}.$$

The equation of motion for I_2 is

$$M_2 - FR = I_2 \alpha_2 = I_2 \left(\frac{a}{R}\right).$$

$$\therefore\ F = \frac{1}{R}\left(M_2 - I_2 \frac{a}{R}\right)$$

$$= \frac{MG\eta}{R} - \frac{I_1 G^2 \eta a}{R^2} - \frac{I_2 a}{R^2}.$$

For the car along the level road,

$$F - W = ma.$$

$$\therefore\ \frac{MG\eta}{R} - \frac{I_1 G^2 \eta a}{R^2} - \frac{I_2 a}{R^2} - W = ma.$$

Collecting the 'a' terms on the right-hand side,

$$\frac{MG\eta}{R} - W = a\left(m + \frac{I_1 G^2 \eta}{R^2} + \frac{I_2}{R^2}\right). \qquad (1)$$

I_1 and I_2 are not known, but we may replace the whole of the contents of the bracket by m_e, the equivalent mass of the car.

$$\frac{MG\eta}{R} - W = am_e. \qquad (2)$$

Substituting the numbers,

$$\frac{790 \times 1.6 \times 0.8}{0.36} - 600 = 1.7m_e.$$

$$\therefore\ m_e = \frac{2808.9 - 600}{1.7} = 1299.4 \text{ kg}.$$

For ascending the slope, we modify equation 2:

$$\frac{MG\eta}{R} - W - mg \sin\theta = am_e.$$

$$\therefore\ a = \frac{2808.9 - 600 - 980 \times 9.81 \times 0.05}{1299.4}$$

$$= \underline{1.33 \text{ m s}^{-2}}$$

and for descending the slope,

$$\frac{MG\eta}{R} - W + mg \sin\theta = am_e.$$

$$a = \frac{2809.9 - 600 + 980 \times 9.81 \times 0.05}{1299.4}$$

$$= \underline{2.07 \text{ m s}^{-2}}.$$

Example 4.15

The diagram shows a gearbox having a velocity ratio, $\frac{\text{input speed}}{\text{output speed}}\left(\frac{\omega_i}{\omega_o}\right)$, of G. Rotors of moments of inertia I_a and I_b are attached to the input and output shafts as shown. $I_a = 0.6$ kg m^2 and $I_b = 24$ kg m^2.

(a) Given that $G = 7\frac{1}{2}$, calculate the torque required at point A such that I_a reaches a speed of 600 rev min^{-1} from rest in 1.5 seconds.

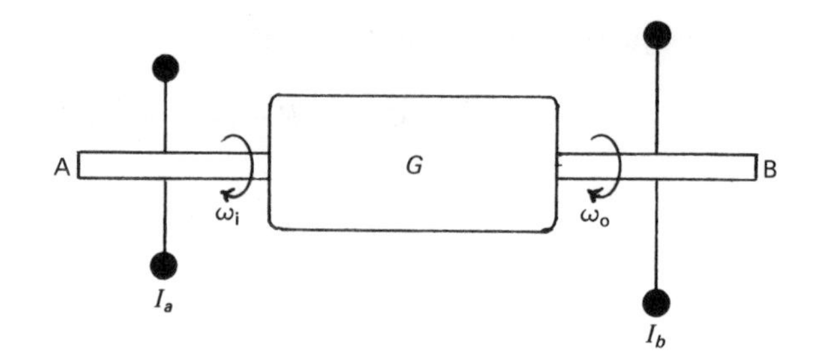

(b) With the system running at this speed, calculate the torque then required at point B to bring I_b to rest in 10 revolutions.
(c) Determine the value of G such that a torque applied at point A produces the maximum acceleration of rotor I_b.

The inertial effects of the components of the gearbox itself may be considered negligible.

PCL

Solution 4.15

Part (a)

We clearly require to 'refer' all inertias to point A. We will make use of the expression for equivalent inertia derived in Example 4.1. Using equation 5 from that example, which gives the equivalent inertia referred to the high-speed shaft,

$$I_{ea} = I_H + \frac{I_L}{G^2 \eta}$$

η is not stated; we therefore assume no losses, and $\eta = 1$.

$$\begin{aligned} I_{ea} &= I_a + \frac{I_b}{G^2} \\ &= 0.6 + \frac{24}{(7\frac{1}{2})^2} \\ &= 1.0267 \text{ kg m}^2. \end{aligned}$$

The acceleration α_a is given by

$$\begin{aligned} \omega_2 &= \omega_1 + \alpha t. \\ 600 \times \frac{2\pi}{60} &= 0 + 1.5\alpha_a. \\ \therefore \alpha_a &= 41.89 \text{ rad s}^{-2}. \\ M_a &= I_{ea}\alpha_a \\ &= 1.0267 \times 41.89 \\ &= \underline{43.01 \text{ N m.}} \end{aligned}$$

Part (b)

The problem now requires referring to the low-speed side of the gearbox. Equation 10 of Example 4.1 applies now.

$$\begin{aligned} I_{eb} &= I_b + G^2 I_a \\ &= 24 + (7\tfrac{1}{2})^2 \times 0.6 \\ &= 57.75 \text{ kg m}^2. \end{aligned}$$

The initial speed of shaft B is $\left(\frac{600}{7\frac{1}{2}}\right)$ rev min^{-1}. Since the retardation is given in terms of shaft displacement instead of time, we use:

$$\omega_2^2 = \omega_1^2 + 2\alpha\theta.$$

$$0 = \left(\frac{600}{7\frac{1}{2}} \times \frac{2\pi}{60}\right)^2 + 2\alpha_b(10 \times 2\pi).$$

(Always remember to express all terms in radians and the corresponding derivatives when using these formulae.)

$$\therefore\ \alpha_b = 0.5585 \text{ rad s}^{-2}.$$

$$M_b = I_{eb}\alpha_b$$

$$= 57.75 \times 0.5585$$

$$= \underline{32.25 \text{ N m.}}$$

Part (c)

Example 4.13 was concerned with an 'optimum' gear ratio designed to provide the maximum acceleration at the output end of a gearbox. The equation already written in part (a) can be used again, substituting for I_{ea}.

$$M_a = \alpha_a\left(I_a + \frac{I_b}{G^2}\right).$$

[a] But

$$\alpha_a = G\alpha_b.$$

$$\therefore\ M_a = G\alpha_b\left(I_a + \frac{I_b}{G^2}\right).$$

$$\therefore\ \alpha_b = \frac{M_a}{G\left(I_a + \frac{I_b}{G^2}\right)}.$$

$$\therefore\ \alpha_b = \frac{M_a G}{I_a G^2 + I_b}.$$

For a maximum value of α_b, $\frac{d(\alpha_b)}{d(G)} = 0$. Using the 'quotient' formula,

$$\frac{(I_a G^2 + I_b)(M_a) - (M_a G)(2I_a G)}{(I_a G^2 + I_b)^2} = 0.$$

It is sufficient to equate the top line to zero. Expanding,

$$M_a I_a G^2 + M_a I_b - 2M_a I_a G^2 = 0,$$

giving

$$G^2 = \frac{I_b}{I_a}.$$

$$\therefore\ G = \sqrt{\frac{24}{0.6}} = \underline{6.325.}$$

Example 4.16

A simple hoist device comprises a truck of mass m = 2200 kg hauled along a horizontal frictionless rail by a light rope wound round a drum of radius R = 0.8 m and moment of inertia 176 kg·m^2. The drum is mounted directly on the low-speed shaft of a 16 to 1 ratio

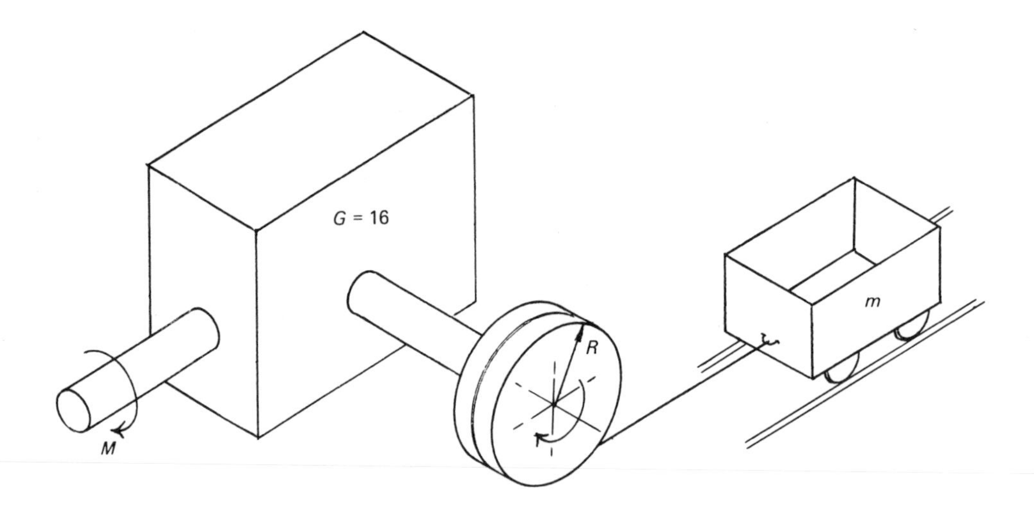

gearbox, and power to the hoist is supplied by a motor which applies a torque M to the high-speed shaft of the gearbox. It may be assumed that there is no energy loss within the gearbox.

(a) Calculate the angular acceleration of the high-speed shaft of the gearbox corresponding to a linear acceleration of the truck of 2.5 m s^{-2}.
(b) Calculate the effective moment of inertia of the system, referred to the high-speed shaft of the gearbox.
(c) Hence calculate the required torque M at the high-speed shaft to produce an acceleration of the truck of 2.5 m s^{-2} towards the drum.
(d) Calculate the power output of the driving motor at the instant the truck is moving at 5 m s^{-1} with the stated acceleration.

PCL

Solution 4.16

Part (a)

The drum acceleration α_o is given by

$$\alpha_o = \frac{a}{R}$$

$$= \frac{2.5}{0.8} \text{ rad s}^{-2}.$$

Hence the motor acceleration is given by

$$\alpha_i = \frac{2.5}{0.8} \times 16 = \underline{50 \text{ rad s}^{-2}}.$$

Part (b)

In Examples 4.5 and 4.6 we dealt with masses attached to rotating drums, and we will use the results obtained to apply here. The effective inertia of truck and drum is

$$I_{eo} = I + mR^2$$

$$= 176 + 2200 \times (0.8)^2$$

$$= 1584 \text{ kg m}^2$$

and the corresponding effective inertia of this referred to the gearbox input is

$$I_{ei} = \frac{I_{eo}}{G^2}$$

$$= \frac{1584}{(16)^2}$$

$$= \underline{6.1875 \text{ kg m}^2}.$$

Part (c)

$$M_i = I_{ei}\alpha_i$$

$$= 6.1875 \times 50$$

$$= \underline{309.375 \text{ N m.}}$$

Part (d)

The angular velocity of the drum, ω_o, is

$$\omega_o = \frac{v}{R} = \frac{5}{0.8}.$$

The angular velocity of the motor, ω_i, is thus

$$\omega_i = G\omega_o = 16\left(\frac{5}{0.8}\right) = 100 \text{ rad s}^{-1},$$

and the motor power is

$$W = M_i\omega_i$$

$$= 309.375 \times 100$$

$$= \underline{30.9375 \text{ kW.}}$$

This question is a reminder of the importance of reading a question carefully. Many candidates are frequently daunted by a question which appears very long, and contains a number of separate parts, as does this one. But if you study the wording, and also the solution, you will see that the examiner has actually carefully laid out the order of the work you need to do. Thus, if (a) and (b) had been omitted, and the instruction had begun with (c), you would still have had to go through the same working. So what at first appears to be a drawback to the question is actually an asset. The examiner has helped the candidate by telling him in effect how to solve the question.

Example 4.17

Masses m_1 of 8 kg and m_2 of 3 kg hang by light strings from two pulleys of radius 15 cm and 35 cm respectively, the pulleys being attached to a drum of moment of inertia 0.36 kg m^2.

(a) Calculate the torque M required at the axis of the drum to raise m_1 a distance of 4 m from rest in 10 seconds.

(b) If the torque is then removed, in what additional distance will m_1 come to rest?

Neglect all frictional effects.

PCL

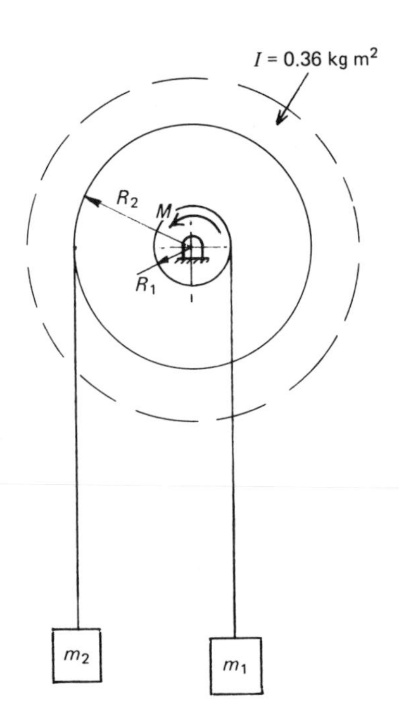

Solution 4.17

Part (a)

Whereas this question could be answered by the use of equivalent inertias, the method would offer very little advantage over that of analysing the motion of each element separately. A solution to part (a) using the first method will be found at the end.

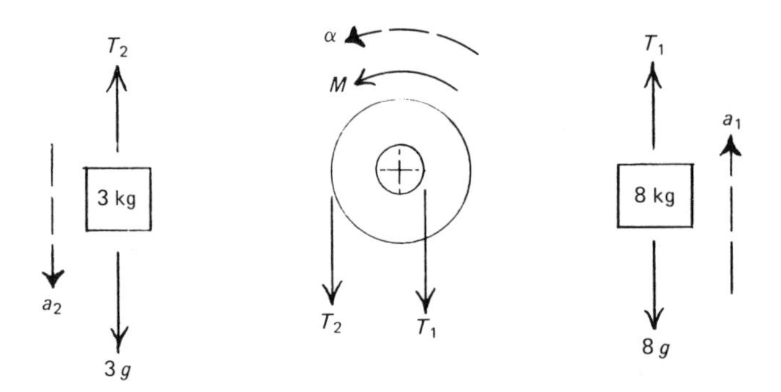

Shown here are free-body diagrams of the three elements of the system. The three equations of motion, paying due regard to sign, are as follows:

$$3g - T_2 = 3a_2. \qquad (1)$$

$$T_1 - 8g = 8a_1. \qquad (2)$$

$$M + T_2 \times 0.35 - T_1 \times 0.15 = I\alpha = 0.36\alpha. \qquad (3)$$

We do not need to write the kinematic equations relating a_1, a_2 and α because a_1 can be calculated directly from the data, and thus a_2 and α can also be found.

$$x = ut + \tfrac{1}{2}at^2.$$

$$\therefore\ 4 = 0 + \tfrac{1}{2}(a_1)(10)^2.$$

$$\therefore\ a_1 = 0.08 \text{ m s}^{-2}.$$

$$\alpha = \frac{a_1}{R_1} = \frac{0.08}{(15 \times 10^{-2})} = 0.5333 \text{ rad s}^{-2}.$$

$$a_2 = \alpha R_2 = 0.5333 \times (35 \times 10^{-2})$$

$$= 0.1867 \text{ m s}^{-2}.$$

Use equations 1 and 2 to substitute in equation 3 for T_1 and T_2. Rearranging to make M the 'subject',

$$M = 0.36 \times 0.5333 - 0.35\,(3g - 3 \times 0.1867) + 0.15\,(8 \times 0.08 + 8g)$$

$$= 0.1920 - 10.104 + 11.868$$

$$= \underline{1.956 \text{ N m.}}$$

Part (b)

Since the accelerations are now unknown, we shall need to use the kinematic equations:

$$a_1 = \alpha R_1 = 0.15\alpha;$$

$$a_2 = \alpha R_2 = 0.35\alpha.$$

The equations of motion will be the same as for part (a) except that M will now be zero. And, of course, we expect to obtain a negative answer for α.

Substituting for T_1 and T_2 in equation 3 as before,

$$0.35\,(3g - 3\alpha \times 0.35) - 0.15\,(8\alpha \times 0.15 + 8g) = 0.36\alpha.$$

Collecting all terms containing α to the right-hand side,

$$0.35 \times 3g - 0.15 \times 8g = \alpha\,[0.36 + 8\,(0.15)^2 + 3\,(0.35)^2\,].$$

$$\therefore\ \alpha = \frac{-0.15g}{0.9075}$$

$$= -1.6215 \text{ rad s}^{-2}.$$

$$a_1 = \alpha R_1$$

$$= -1.6215 \times 0.15$$

$$= -0.2432 \text{ m s}^{-2}.$$

Before solving, we need to know the velocity attained by m_1 at the instant the torque was removed.

$$x = \tfrac{1}{2}(u + v)\,t.$$

$$4 = \tfrac{1}{2}(0 + v)\,10.$$

$$\therefore\ v = 0.8 \text{ m s}^{-1}.$$

For the distance to come to rest from this velocity,

$$v^2 = u^2 + 2ax.$$

$$0 = (0.8)^2 + 2(-0.2432)x.$$

$$\therefore\ x = \frac{0.64}{0.4846} = \underline{1.321\ \text{m.}}$$

Alternative solution to part (a)

The equivalent inertia of the system referred to the drum axis is

$$I_e = 0.36 + 8 \times (0.15)^2 + 3 \times (0.35)^2$$
$$= 0.9075\ \text{kg m}^2.$$

The net effective driving torque, M_e, is given by

$$M_e = M + 3g(0.35) - 8g(0.15)$$
$$= (M - 1.4715)\ \text{N m}.$$

As previously, angular acceleration α is 0.5333 rad s^{-2}.
The system equation of motion ($\Sigma M = I\alpha$) is

$$M - 1.4715 = 0.9075 \times 0.5333$$
$$\therefore\ M = 0.4840 + 1.4715$$
$$= \underline{1.956\ \text{N m.}}$$

Example 4.18

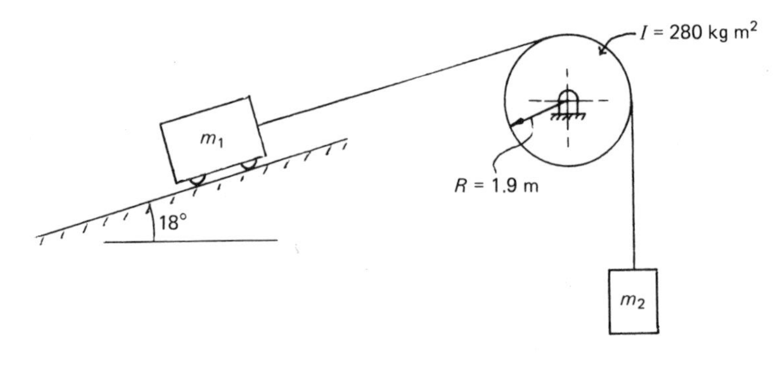

The elements of a simple hoist are shown in the diagram. A truck of total mass $m_1 = 1100$ kg has four wheels, each of radius 0.3 m and moment of inertia 20 kg m^2. A rope attached to the truck passes round a winding drum such that there is no slip of the rope on the drum. The drum radius is 1.9 m and its moment of inertia is 280 kg m^2. The other end of the rope hangs vertically, and a counterweight, mass $m_2 = 260$ kg, hangs from it. The truck is released from rest at the top of a slope of 18°. When in motion, the truck is subjected to a frictional resistance F which may be assumed constant. Calculate the magnitude of this force given that the truck attains a velocity of 0.5 m s^{-1} when it has travelled 120 m down the slope.

An energy equation is suggested.

PCL

Solution 4.18

Although an energy equation is suggested, it is not compulsory. The solution given below is by free-body diagrams and equations of motion. An energy solution to the same problem will be found in Chapter 5, Example 5.21.

The diagram shows free-body diagrams for the three elements of the connected system. Before writing the equations of motion, note that the distance and final velocity of the truck are given. All accelerations can therefore be calculated.

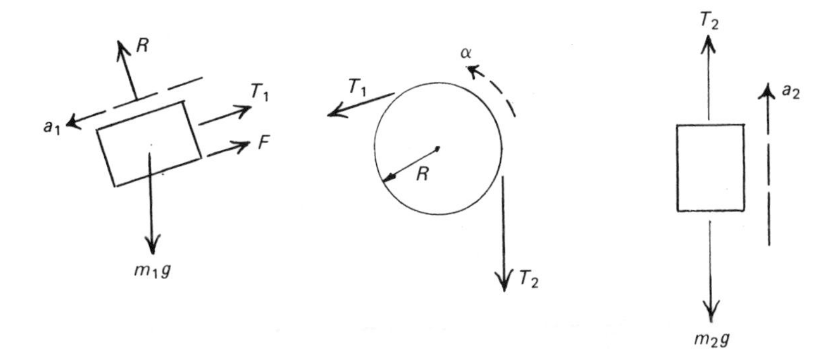

(When this can be done, the algebra for dealing with connected systems is much simplified.)

We begin by calculating the accelerations.

For the truck,
$$v^2 = u^2 + 2ax.$$
$$(0.5)^2 = 0 + 2a_1 \times 120.$$
$$\therefore\ a_1 = 0.001\,042\ \text{m s}^{-2}.$$

For the drum, $\alpha = \dfrac{a_1}{R} = \dfrac{0.001\,042}{1.9} = 0.000\,548\,2\ \text{rad s}^{-2}.$

For the hanging weight, $a_2 = 0.001\,042\ \text{m s}^{-2}$ (as for truck).

The moments of inertia of the truck wheels are stated; the equivalent mass of truck plus wheels must be calculated.

$$m_e = m + \frac{I}{R^2}$$
$$= 1100 + \frac{4 \times 20}{(0.3)^2}$$
$$= 1988.9\ \text{kg}.$$

Do not forget that while rotating wheels add to the effective *mass* of a body, they do not add to its *weight*.

We now write the three equations:

Truck: $$m_1 g \sin\theta - F - T_1 = m_e a_1. \quad (1)$$

Drum: $$T_1 R - T_2 R = I\alpha. \quad (2)$$

Weight: $$T_2 - m_2 g = m_2 a_2. \quad (3)$$

T_2 can be calculated directly from equation 3:

$$T_2 = m_2 a_2 + m_2 g$$
$$= 260\,(0.001\,042 + 9.81)$$
$$= 2550.9\ \text{N}.$$

Substituting this in equation 2 enables T_1 to be calculated.

$$T_1 = \frac{I\alpha}{R} + T_2$$

$$= \frac{280 \times 0.000\,548\,2}{1.9} + 2550.9$$

$$= 2551.0 \text{ N.}$$

Making F the 'subject' of equation 1,

$$F = m_1 g \sin\theta - T_1 - m_e a_1$$

$$= 1100g \sin 18° - 2551.0 - 1988.9 \times 0.001\,042$$

$$= 3334.60 - 2551.0 - 2.1$$

$$= \underline{781.5 \text{ N.}}$$

It can be seen that the 'inertia' forces in this example are extremely small when compared with the weights and rope tensions. In practice, they might well be neglected. But in an examination you must of course include them. It would be quite wrong to assume that because you left something out, the marks deducted would be in proportion to the magnitude of the quantity omitted!

Example 4.19

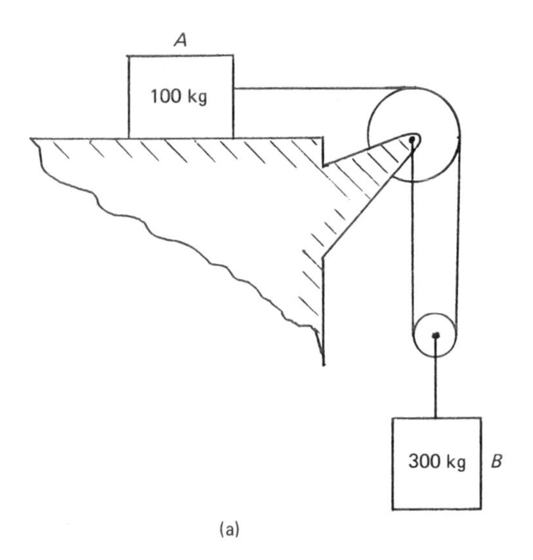

(a)

Two blocks shown in diagram (a) are initially at rest. The horizontal plane and the pulley are frictionless, and the pulley is assumed to be of negligible mass. Determine the acceleration of each block and the tension in each cord.

KP

Solution 4.19

The two free-body diagrams are shown here as (b) and (c). In addition to the two blocks, we also show the hanging pulley (d), although we are told that its mass is negligible. Before writing equations, we will first determine the relationship

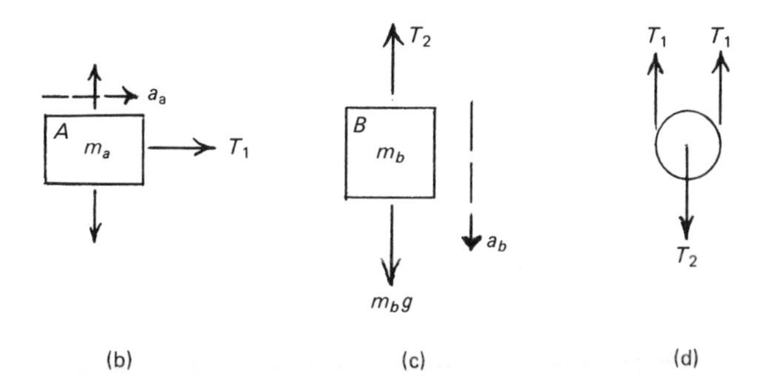

between the two accelerations. You may recognise the pulley system as a simple 'two-to-one ratio' system, which means that block B moves half the distance that block A moves. A simple way to understand this is to imagine first that we raise B by a vertical distance, say y. The pulley will be lifted off the cord, leaving a loop, or 'bight', of cord slack. Diagram (e) shows this clearly. In order to take up this slack, it is clear that we should have to move A to the left by a distance of $2y$.

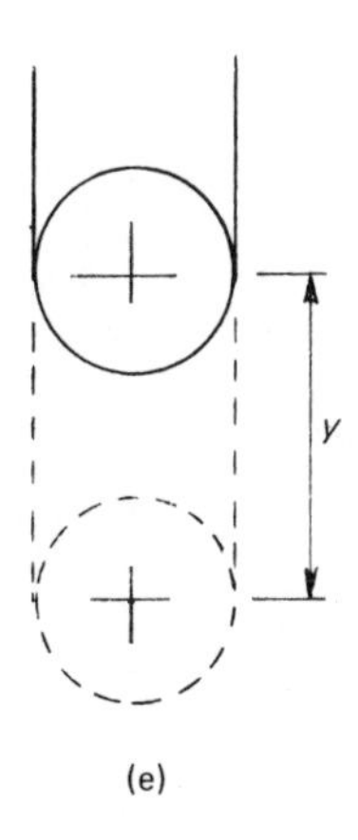

(e)

As the relative displacements are related, so also are the relative velocities and accelerations. (See Fact Sheet (a).)

We write the equations of motion.

For A,

$$T_1 = m_a a_a$$

$$= 100a_a. \tag{1}$$

For B,

$$m_b g - T_2 = m_b a_b$$

$$= 300a_b. \tag{2}$$

Since the pulley has negligible mass, the resultant force on it must be zero.

$$2T_1 = T_2. \tag{3}$$

And we have already shown that

$$a_a = 2a_b. \tag{4}$$

Substituting for T_2 and a_b in equation 2,

$$300g - 2T_1 = 300\,(\tfrac{1}{2}a_a).$$

$$300g - 2(100a_a) = 150a_a.$$

$$300g = 350a_a.$$

$$\therefore\ a_a = 8.409 \text{ m s}^{-2}.$$

$$\therefore\ a_b = \underline{4.205 \text{ m s}^{-2}}.$$

Substituting in equation 1,

$$T_1 = \underline{840.9 \text{ N}},$$

and in equation 3, $T_2 = \underline{1681.8 \text{ N.}}$

Some explanation is due in respect of the diagram of the small pulley. Why are we justified in calling both the upward tensions T_1? The answer is that the pulley is stated to have negligible moment of inertia. If this were not so, the pulley, which clearly is accelerating angularly as well as linearly, would be subjected to a resultant torque, which could only come from a difference of tensions of the two strings. For an example involving a pulley of non-negligible inertia and mass, see Example 4.20.

Example 4.20

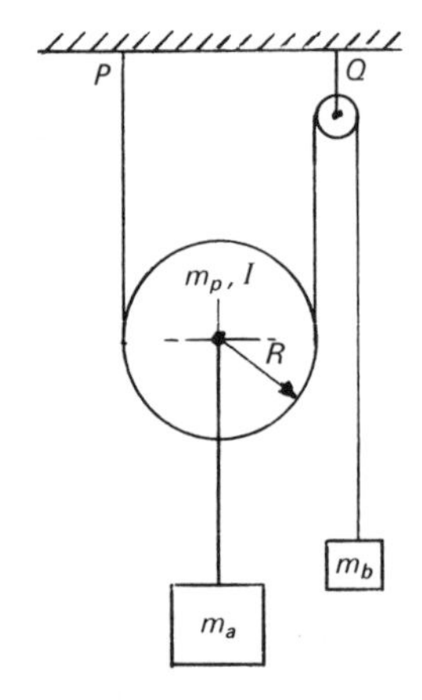

The diagram shows a pulley system. The large pulley has a mass m_p and moment of inertia about its central axis of I. A weight of mass m_a hangs from it. A single cord, attached to a fixed point at P, passes round the pulley, and round a second pulley at Q of negligible inertia, and a second weight of mass m_b hangs vertically on the end of the cord. The cord does not slip on the large pulley. The parts of the cord supporting the large pulley are vertical. Friction everywhere is negligible. Obtain an expression for the acceleration of m_b, assuming this to be downwards.

Solution 4.20

This pulley system should now be recognised as a simple 2-to-1 ratio system. (See Example 4.19.) Calling 'a' the required acceleration of m_b, this permits us to indicate the linear acceleration of the pulley and of m_a on the free-body diagrams as $\frac{1}{2}a$ straight away, saving the necessity of writing kinematic compatibility equations. We can now draw the free-body diagrams, as here.

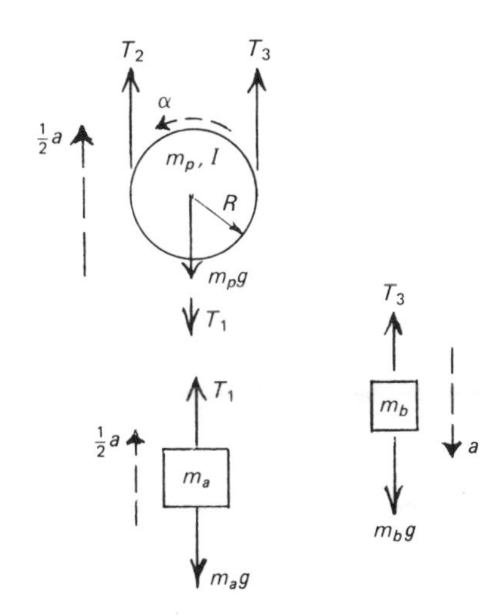

When the mass and moment of inertia of the pulley are not negligible, you have to recognise that the motion is both translational – the mass centre of the pulley moves bodily – and rotational – the pulley turns about its own centre. In such cases, *two* equations of motion are needed, one for each mode of motion. (See Chapter 2, Fact Sheet (a).) Before writing the equations, we first require to complete the kinematic analysis, and determine the angular acceleration α of the pulley in terms of the acceleration a. To do this, we first find the relation between linear and angular displacement, recalling that the acceleration ratio is the same as the displacement ratio. (This chapter, Fact Sheet (a).)

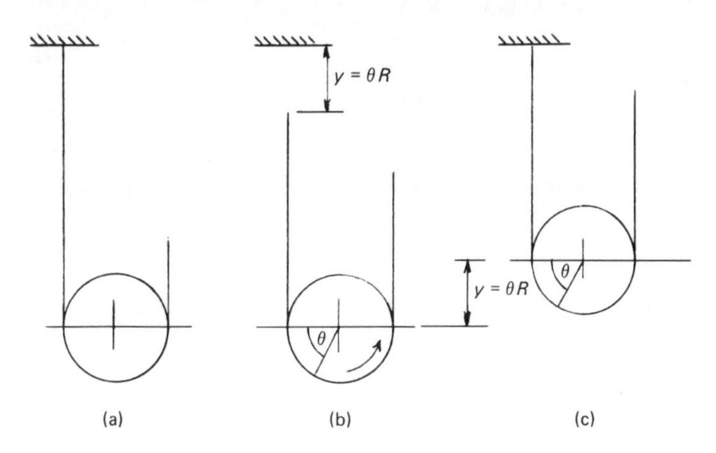

Diagrams (a) to (c) show a method for relating these displacements. In (a) the pulley is shown with the cord connected to the fixed point. We now imagine the pulley to be turned through a small angle θ anticlockwise. Since the cord does not slip on the pulley, the cord must be pulled down by a length $y = \theta R$ as illustrated in (b). But since the upper end of the cord is actually fixed, this can only be compatible with raising the whole pulley upwards by the same distance y, as we have shown in (c). Thus

$$y = \theta R$$

and differentiating twice,

$$\frac{d^2 y}{dt^2} = \frac{d^2(\theta R)}{dt^2}.$$

Hence

$$\tfrac{1}{2}a = \alpha R.$$

Diagrams (a) to (c) incidentally make it quite clear that an upward acceleration of the pulley is compatible with an *anticlockwise* angular acceleration.

We may now write the four equations of motion. Take care to note that the two cord tension forces T_2 and T_3 are now different. (They were the same in Example 4.19 when the moment of inertia was negligible.)

For m_a,
$$T_1 - m_a g = m_a(\tfrac{1}{2}a). \tag{1}$$

For m_b,
$$m_b g - T_3 = m_b a. \tag{2}$$

For pulley ($\Sigma F = ma$):
$$T_2 + T_3 - T_1 - m_p g = m_p(\tfrac{1}{2}a). \tag{3}$$

For pulley ($\Sigma M = I\alpha$):
$$T_3 R - T_2 R = I\alpha = I\left(\frac{\frac{1}{2}a}{R}\right). \tag{4}$$

Handling algebra is often a source of errors, and there is unfortunately no substitute for lots of practice. But some points may be made which might help. First, our job is to reduce these four equations of motion to one single equation for the whole system. To do this, we have to 'eliminate' the system internal forces, that is, the cord tensions. Now you can see that T_1 and T_3 are respectively the only tension forces in equations 1 and 2. We can therefore begin by substituting for T_1 and T_3 in the second two equations, thus reducing our four equations to two.

From equation 1,
$$T_1 = m_a(\tfrac{1}{2}a) + m_a g.$$

And from equation 2,
$$T_3 = m_b g - m_b a.$$

Substituting these values in equations 3 and 4,
$$T_2 + m_b g - m_b a - m_a(\tfrac{1}{2}a) - m_a g - m_p g = m_p(\tfrac{1}{2}a). \tag{5}$$

It is convenient to divide equation 4 throughout by R before substituting.

$$T_3 - T_2 = I\left(\frac{\frac{1}{2}a}{R^2}\right).$$

$$\therefore\ m_b g - m_b a - T_2 = I\left(\frac{\frac{1}{2}a}{R^2}\right). \tag{6}$$

Making T_2 the 'subject' in equations 5 and 6 and equating reduces to a single equation.

$$T_2 = m_p(\tfrac{1}{2}a) - m_b g + m_b a + m_a(\tfrac{1}{2}a) + m_a g + m_p g$$
$$= m_b g - m_b a - I\left(\frac{\frac{1}{2}a}{R^2}\right).$$

Collect all terms in a to the left-hand side and the others to the right-hand side:

$$a\left[\tfrac{1}{2}m_p + m_b + \tfrac{1}{2}m_a + m_b + I\left(\frac{\frac{1}{2}}{R^2}\right)\right] = m_b g + m_b g - m_a g - m_p g,$$

giving the required expression:

$$a = \frac{g(2m_b - m_a - m_p)}{\left[2m_b + \frac{1}{2}m_a + \frac{1}{2}m_p + I \frac{\frac{1}{2}}{R^2}\right]}$$

$$\therefore\ a = \frac{g[m_b - \frac{1}{2}(m_a + m_p)]}{[m_b + \frac{1}{4}(I/R^2 + m_a + m_p)]}.$$

In the last line, the expression has deliberately been manipulated to make the first term on the bottom line m_b, since a is the acceleration at this point. Thus, the bottom line is the equivalent mass of the system, referred to this point in the system, and of course, the top line is the net effective force at this point. The factor of $\frac{1}{4}$ (= $(\frac{1}{2})^2$) recalls the 'G^2' term for equivalent inertia in Example 4.1 and several subsequent examples.

Example 4.21

An electric motor develops torque T_m and has a moment of inertia I_m. It drives a load of moment of inertia I_o, through a step-down gearbox. The 'no load' condition is that in which the motion of the load is not resisted. The 'loaded' condition is that in which there is a constant resistance torque T. Show that if $I_o = 64I_m$ and the load torque is six times the motor torque, a change of gearing of 2 : 1 will be necessary to maintain maximum acceleration for both 'no load' and 'loaded' conditions. What is the ratio of these maximum accelerations?

U. Lond. U.C.

Solution 4.21

First consider 'no load' conditions. The diagram shows the arrangement.

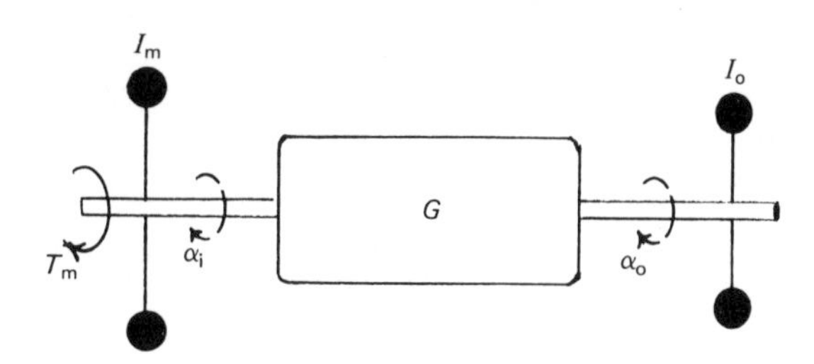

We are concerned with maximum acceleration of the output shaft. The equivalent inertia of the system referred to the input shaft is:

$$I_{ei} = I_m + \frac{I_o}{G^2}.$$

So the equation of motion will be

$$T_m = \alpha_i \left(I_m + \frac{I_o}{G^2}\right)$$

$$= G\alpha_o \left(I_m + \frac{I_o}{G^2}\right).$$

$$\therefore \ \alpha_o = \frac{T_m G}{I_m G^2 + I_o} \ . \qquad (1)$$

(This calculation is exactly the same as for part (c) of Example 4.15.)

For the condition that α_o has a maximum value, we differentiate, using the 'quotient' formula. Since we then equate the expression to zero, we do not need to worry about the bottom line; we simply equate the top line to zero.

$$\frac{d(\alpha_o)}{d(G)} = \frac{(I_m G^2 + I_o)T_m - (T_m G)(2I_m G)}{(\qquad)^2} \ .$$

Equating the top line to 0 and cancelling T_m,

$$I_m G^2 + I_o = 2I_m G^2 .$$

$$\therefore \ G = \sqrt{\frac{I_o}{I_m}} = \sqrt{64} = 8.$$

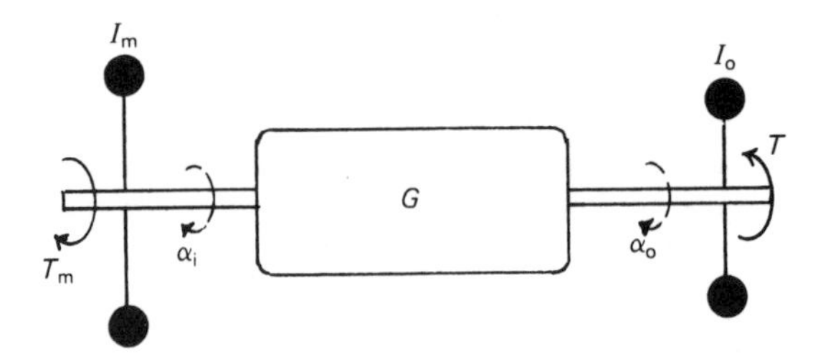

The diagram shows the 'loaded' condition. In equation 1 the top line is the effective torque at the output. This must now be reduced by the amount of load torque $T = 6T_m$. Thus,

$$\alpha_o = \frac{T_m G - 6T_m}{I_m G^2 + I_o} \ . \qquad (2)$$

For maximum α_o, differentiate as before:

$$\frac{d(\alpha_o)}{d(G)} = \frac{(I_m G^2 + I_o)T_m - (T_m G - 6T_m)(2I_m G)}{(\qquad)^2} \ .$$

Equating the top line to 0 and cancelling T_m,

$$I_m G^2 + I_o = 2I_m G^2 - 12I_m G.$$

$$\therefore \ I_m G^2 + 64I_m = 2I_m G^2 - 12I_m G,$$

which reduces to

$$G^2 - 12G - 64 = 0.$$

$$\therefore \ (G - 16)(G + 4) = 0,$$

giving $G = +16$ as the obvious positive solution. This is twice the ratio obtained for the 'no load' condition.

The ratio of accelerations is determined by dividing equation 1 by equation 2, substituting the appropriate values of G in each.

$$\frac{\alpha\,(\text{no load})}{\alpha\,(\text{load})} = \left(\frac{\cancel{T}_{\mathrm{m}}\ 8}{\cancel{I}_{\mathrm{m}}\ 64 + 64\cancel{I}_{\mathrm{m}}}\right)\left(\frac{\cancel{I}_{\mathrm{m}}\ 256 + 64\cancel{I}_{\mathrm{m}}}{\cancel{T}_{\mathrm{m}}\ 16 - 6\cancel{T}_{\mathrm{m}}}\right)$$

$$= \left(\frac{8}{128}\right)\left(\frac{320}{10}\right)$$

$$= \underline{2 \text{ to } 1}.$$

Example 4.22

A motor drives a machine shaft through a double-reduction gear, the reduction ratios of the two stages being the same and the efficiency of transmission of each stage being 0.97.

The moments of inertia of the masses rotating with the motor shaft, intermediate shaft and machine shaft are, in this same order, 700 kg cm^2, 8000 kg cm^2 and 30 000 kg cm^2.

The motor develops a constant torque of 25 N m and there is a constant torque of 220 N m opposing the rotation of the machine shaft.

The machine shaft is to be accelerated from rest to a speed of 80 rev min^{-1} as rapidly as possible.

Determine
(a) the reduction ratio of each stage of the reduction gear;
(b) the greatest speed attained by the motor; and
(c) the greatest power the motor must be capable of developing.

U. Lond. K.C.

Solution 4.22

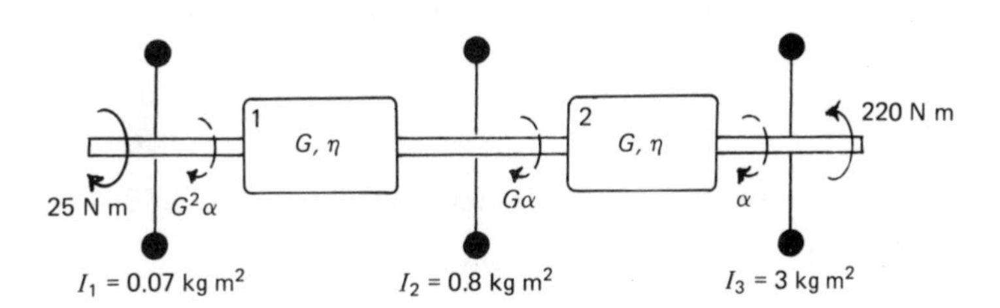

The illustration is a schematic diagram of the arrangement. As with earlier examples (e.g. Examples 4.8 and 4.12), when there are energy losses within the system it is advisable not to make use of the concepts of equivalent inertia but to work through the system, writing equations as we go through. In the diagram we have shown the angular acceleration of each section.

To draw free-body diagrams of the five elements of the system would, in this case, be tedious and unnecessary. It is seen that a torque of 25 N m acts at the input shaft. But 25 N m will not be the input to the first gearbox, because some of this torque is used in accelerating I_1. We may call this amount the 'inertia torque' of the inertia I_1.

$$\text{Input to gearbox 1} = 25 - I_1(G^2\alpha)$$
$$= 25 - 0.07G^2\alpha$$

(converting I_1 to kg m^2).

(Drawing separate free-body diagrams of I_1 and the gearbox would of course lead to the same result, but would be more tedious and time-consuming. An alternative approach is to make use of d'Alembert's principle, whereby an accelerating mass or inertia is ascribed a negative inertia force or torque, called a 'reversed effective'

force or torque, then treating the system as though it were in static equilibrium. This approach is referred to in Chapter 2. So you may think of the input to the first box as 25 N m less the reversed effective torque of I_1.)

$$\begin{aligned}
\text{Output from gearbox 1 } &= G\eta(25 - 0.07G^2\alpha) \\
&= 24.25G - 0.0679G^3\alpha \qquad (\text{taking } \eta = 0.97). \\
\text{Input to gearbox 2 } &= 24.25G - 0.0679G^3\alpha - 0.8G\alpha. \\
\text{Output from gearbox 2 } &= G\eta(24.25G - 0.0679G^3\alpha - 0.8G\alpha) \\
&= 23.52G^2 - 0.0659G^4\alpha - 0.776G^2\alpha. \\
\text{Input to shaft } &= 23.52G^2 - 0.0659G^4\alpha - 0.776G^2\alpha - 3\alpha = 220 \text{ N m}.
\end{aligned}$$

Collecting terms and rearranging gives

$$\alpha = \frac{23.52G^2 - 220}{3 + 0.776G^2 + 0.0659G^4}.$$

To determine the optimum value of G for maximum acceleration, we differentiate this expression with respect to G. We use the 'quotient' formula:

$$\frac{\mathrm{d}(u/v)}{\mathrm{d}G} = \frac{v\,\mathrm{d}u/\mathrm{d}G - u\,\mathrm{d}v/\mathrm{d}G}{v^2}.$$

Because the expression is equated to 0, we equate just the top line to 0. Hence

$$v\frac{\mathrm{d}u}{\mathrm{d}G} = u\frac{\mathrm{d}v}{\mathrm{d}G}.$$

$$\therefore\ (3 + 0.776G^2 + 0.0659G^4)\,47.04G = (23.52G^2 - 220)(1.552G + 0.2636G^3).$$

Cancelling G throughout and expanding gives:

$$141.2 + \cancel{36.50G^2} + 3.10G^4 = \cancel{36.50G^2} - 341.4 + 6.20G^4 - 58.0G^2.$$

Simplifying,

$$3.10G^4 - 58.0G^2 - 482.6 = 0$$

$$G^4 - 18.71G^2 - 155.7 = 0.$$

Solution of this quadratic in G^2 gives

$$G^2 = \frac{18.71 \pm \sqrt{350.1 + 4 \times 155.7}}{2}$$

$$= 24.95 \text{ (ignoring a negative answer).}$$

$$\therefore\ G = \underline{4.995.}$$

$$\begin{aligned}
\text{Maximum motor speed } &= \text{maximum shaft speed} \times (G)^2 \\
&= 80\,(4.995)^2 \\
&= \underline{1996 \text{ rev min}^{-1}}.
\end{aligned}$$

$$\begin{aligned}
\text{Maximum power: } W_{\max} &= \text{Torque} \times \text{Maximum speed} \\
&= 25 \times 1996 \times \frac{2\pi}{60} \\
&= \underline{5226 \text{ W.}}
\end{aligned}$$

The possibilities of error in answering a question of this type cannot be denied. This is the type of 'nightmare' question wherein a candidate may fill two sheets of calculations during three-quarters of an hour and get nowhere. It is the sort of question to leave alone unless you are really confident, having already solved similar questions before. But there may be consolation for the unfortunate ones who begin the question and cannot finish it. It is probable that a correct formulation of the equations at the beginning would earn at least a third of the total marks. And if you had reached the stage of writing the long equation in G, but failed to notice that it was a quadratic, then you might well earn two-thirds of the total marks.

This is once more a good place to stress the importance of neat and careful setting down of your work. A careless and untidy worker would almost certainly include several algebraic and arithmetical slips in an answer of this type.

Example 4.23

An electric motor drives a machine through a reduction gear of ratio 9 to 1. The masses rotating with the motor shaft have a moment of inertia of 0.5 kg m^2 and the masses rotating with the machine shaft have a moment of inertia of 38 kg m^2. A constant torque of 100 N m resists the rotation of the machine shaft and the efficiency of the reduction gear is 95 per cent. The motor can develop a maximum torque of 27 N m.

(a) Determine the power the motor develops when the machine shaft is driven at the uniform speed of 150 rev min^{-1}.

(b) Determine the least time required for the speed of the machine shaft to increase from zero to 150 rev min^{-1}.

(c) If the gear ratio were altered so as to give the machine shaft the greatest possible angular acceleration, what would then be the gear ratio?

U. Lond. K.C.

Solution 4.23

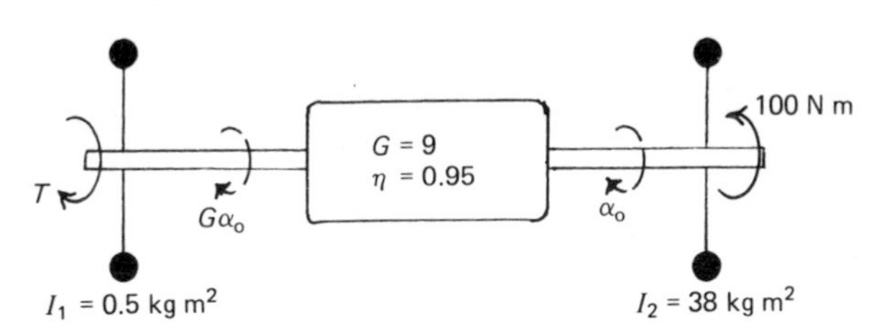

Figure 4.31 shows a schematic diagram of the system.

Part (a)

If the shaft runs at uniform speed, there is no acceleration. Hence

$$\text{Motor torque} = \frac{100}{9 \times 0.95} = 11.70 \text{ N m}.$$

$$\text{Power} = T\omega = 11.70 \times (150 \times 9)\ \frac{2\pi}{60}$$

$$= \underline{1654 \text{ W}}.$$

Part (b)

We will use the method of the previous Example 4.22 to determine the machine shaft acceleration. The motor will of course deliver the maximum torque of 27 N m.

Bearing in mind that in part (c) we shall need to calculate a different value of G, we shall obtain the general expression for α_o in terms of G.

Input to gearbox = $27 - 0.5\,(G\alpha_o)$.

Output from gearbox = $G\eta\,(27 - 0.5\alpha_o)$

$$= 25.65G - 0.475G^2\alpha_o$$

(taking $\eta = 0.95$).
Torque to machine shaft = $25.65G - 0.475G^2\alpha_o - 38\alpha_o = 100$ N m.

Rearranging gives

$$\alpha_o = \frac{25.65G - 100}{38 + 0.475G^2}. \qquad (1)$$

Substituting the value of $G = 9$,

$$\alpha_o = \frac{25.65 \times 9 - 100}{38 + 0.475 \times 81} = 1.711 \text{ rad s}^{-2}.$$

$$\omega_2 = \omega_1 + \alpha t.$$

$$\therefore\ 150 \times \frac{2\pi}{60} = 0 + 1.711t.$$

$$\therefore\ t = \frac{15.71}{1.711} = \underline{9.182 \text{ s.}}$$

Part (c)

As with Example 4.22, for the optimum value of G for maximum acceleration we differentiate equation 1 using the 'quotient' formula, and equating the top line to zero. We shall not bother to include the denominator of the differential.

$$v\frac{du}{dG} = u\frac{dv}{dG}.$$

$$(38 + 0.475G^2)\,25.65 = (25.65G - 100)\,(2 \times 0.475G).$$

$$\therefore\ 974.7 + 12.18G^2 = 24.37G^2 - 95G,$$

which reduces to the quadratic

$$G^2 - 7.793G - 79.96 = 0$$

giving

$$G = \frac{7.793 \pm \sqrt{60.73 + 319.84}}{2}$$

$$= \underline{13.65.}$$

(The arithmetic of solving a quadratic is simpler if the equation is first manipulated to make the coefficient of the first term 1.)

Example 4.24

Determine the mass of block A of the pulley system shown in the diagram so that when it is released from rest it moves block B a distance of 0.75 m up the inclined plane in two seconds. Neglect the mass of the pulleys and cords. Block B has a mass of 5 kg and the coefficient of friction between block B and the inclined plane = 0.25.

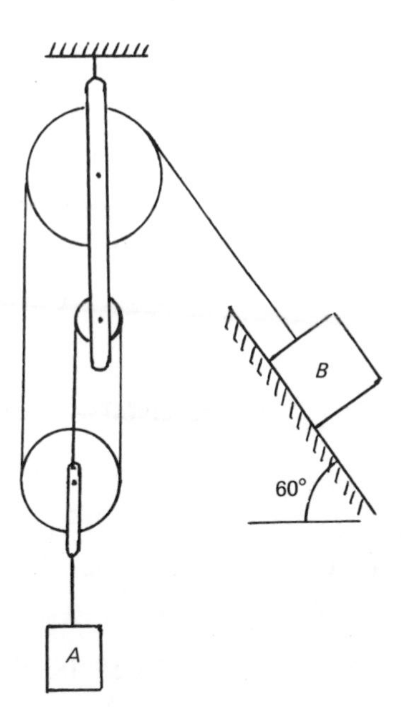

Thames Poly.

Solution 4.24

We may calculate the acceleration of B using

$$x = ut + \tfrac{1}{2}at^2 .$$
$$0.75 = 0 + \tfrac{1}{2}a\,(2)^2 .$$
$$\therefore\ a = 0.375 \text{ m s}^{-2} .$$

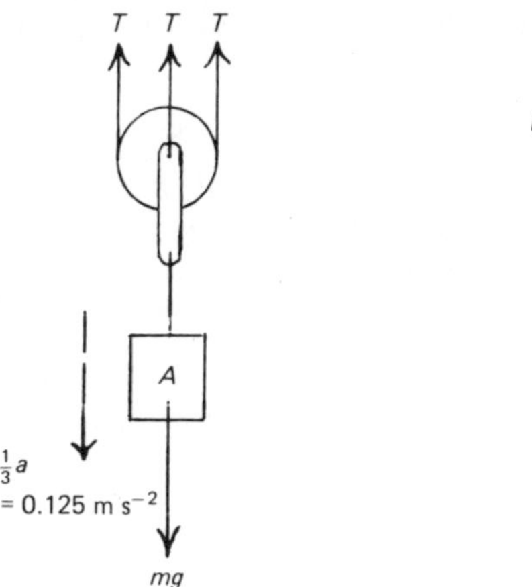

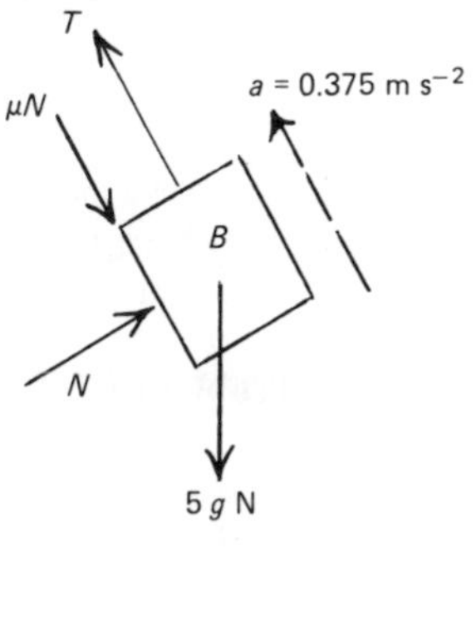

Here are shown free-body diagrams of the two blocks, with the hanging pulley included on block A. Because the same cord passes round the complete system, and because the inertia and friction of the pulleys is negligible, the same tension T acts in all parts of the cord. The velocity ratio of the pulley system is 3. (It is seen that raising the hanging pulley by, say, 1 m would produce 1 m slack in each of the three supporting cords.) Hence we may calculate the acceleration of block A:

$$a = \tfrac{1}{3} \times 0.375$$
$$= 0.125 \text{ m s}^{-2}.$$

The equation of motion for B ($\Sigma F = ma$) is

$$T - 5g \sin 60° - \mu N = 5 \times 0.375.$$

Across the plane (equilibrium),

$$N = 5g \cos 60°.$$

Substituting,

$$T - 5g \sin 60° - 0.25 \times 5g \cos 60° = 5 \times 0.375.$$
$$\therefore\ T = 1.875 + 1.25g \cos 60° + 5g \sin 60°$$
$$= 50.485 \text{ N}.$$

The equation of motion for block A is

$$mg - 3T = ma.$$
$$3 \times 50.485 = m(9.81 - 0.125).$$
$$\therefore\ m = \frac{3 \times 50.485}{9.685}$$
$$= \underline{15.64 \text{ kg}}.$$

4.3 Problems

Problem 4.1

Two wheels A and B are connected by a gearbox. The speed of A is G times that of B. The moments of inertia of the wheels are I_a and I_b.

If a torque M acts on wheel A to produce an angular acceleration on wheel B, show that this acceleration will have a maximum value for a given value of M, when G has the value

$$\sqrt{\frac{I_b}{I_a}},$$

A has a moment of inertia of 2.6 kg m^2, B has a moment of inertia of 72 kg m^2 and M is 24 N m. Calculate how long B takes to reach a speed of 250 rev min^{-1} starting from rest, assuming this optimum gear ratio.

Problem 4.2

A gearbox has an overall ratio of 12.6. When a torque of 1.8 N m acts on the high-speed shaft the low-speed shaft increases speed from 0 to 5 rev s^{-1} in 4 seconds. Calculate the torque required on the low-speed shaft to cause the high-speed shaft to accelerate at the same rate.

Hint: Call the respective inertias I_1 and I_2 and solve algebraically.

Problem 4.3

A pulley of diameter 1.6 m has a moment of inertia of 24 kg m^2. A rope round the pulley carries loads of mass 16 kg and 14 kg. A body of mass m is then added to the 14 kg load and the system released from rest. If the load moves down a distance of 6 m in 10 seconds, determine the value of m. If this third mass is then removed, how long will it take for the two loads to return to their original position?

Hint: When the third body is removed the system is *not* at rest.

Problem 4.4

A load of mass 50 kg hangs from the smaller of a pair of wheels of radii 50 mm and 120 mm as shown; the two wheels are fixed together. The moment of inertia of the compound wheel is 0.12 kg m^2. Gear teeth on the outer wheel drive a motor with an increasing speed ratio of 6 : 1. The moment of inertia of the motor is 0.09 kg m^2. If the system is released from rest, calculate how long it takes the motor to reach a speed of 200 rev min^{-1}. Neglect all friction effects.

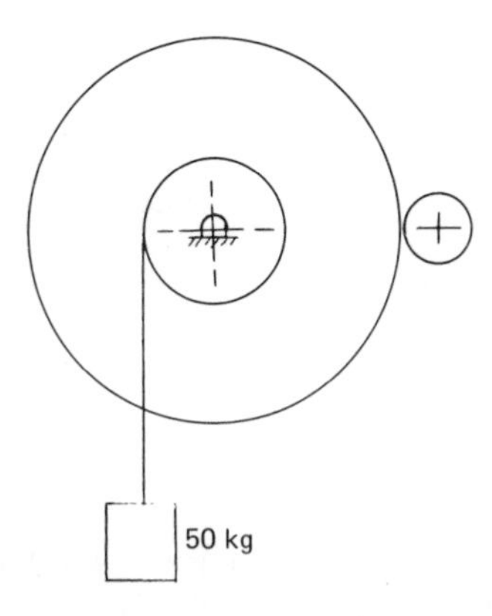

Problem 4.5

A hoist comprises a drum of radius 0.1 m and moment of inertia 1.8 kg m^2 carrying a rope which supports a load of mass 80 kg. The drum is driven by a motor via a reduction gearbox; a flywheel on the motor shaft has a moment of inertia of 0.6 kg m^2. The motor output torque is 20 N m. Neglecting all inertia effects other than those given and neglecting also all friction losses, calculate the ratio of the gearbox for the highest upward acceleration of the load, and calculate this acceleration.

Hint: Refer to Example 4.13.

Problem 4.6

A travelling crane comprises a drum of diameter 0.2 m driven by an electric motor through a reduction gearbox of overall ratio 120. The drum carries a rope which connects to a 6 : 1 reduction pulley system and a load of mass 2000 kg hangs from the pulley. The armature of the driving motor has a moment of inertia of 0.04 kg m^2. If the current to the motor is accidentally switched off, causing the load to drive the motor backwards, calculate the resulting downward acceleration of the load. If it is 4 m above the floor when it begins to descend, what will be its velocity when it reaches the floor? Inertias of the gearbox, drum and pulley, and also all friction forces and torques may be considered negligible.

Problem 4.7

A solid metal disc of radius 50 mm is fixed to a central spindle of diameter 3 mm. The spindle is placed on a pair of sloping rails so that the disc rolls down the slope without the spindle slipping. If the slope of the rails is 15°, calculate how long the disc will take, starting from rest, to travel 200 mm down the slope. What will be its angular velocity at this point? In calculating the moment of inertia of the disc, treat it as a uniform solid disc and neglect the inertia of the spindle.

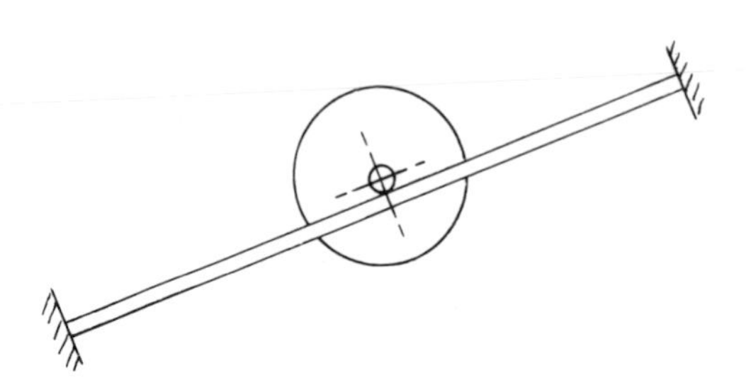

Problem 4.8

A drum of radius 1 m is connected by a light rope to a body of mass 40 kg which rests on a horizontal frictionless track. A force of 40 N applied to the mass causes it to attain a velocity of 5 m s^{-1} in 22 s. When this force is removed the system comes to rest after the mass has moved a further 129 m. Calculate the moment of inertia of the drum and the friction torque on the drum shaft.

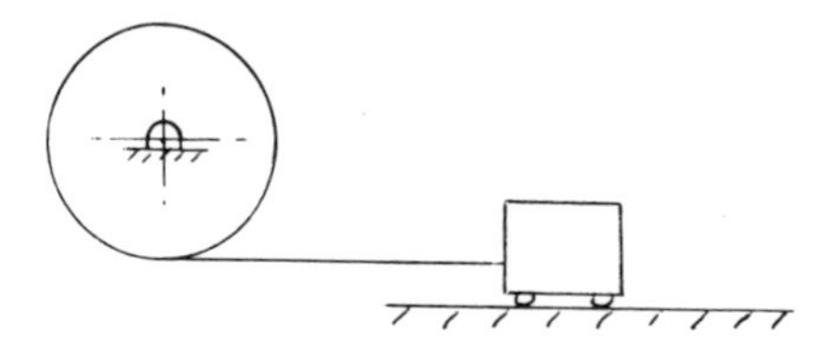

Problem 4.9

A car has a total mass of 1200 kg and four wheels, each having moment of inertia 14.6 kg m^2 and radius 0.19 m. The gearbox ratio is 5.2; the engine shaft has rotating parts having moment of inertia 2.2 kg m^2 and the transmission efficiency is 0.75. Calculate the engine torque required to drive the car up a slope of $\sin^{-1}$ 0.1 with a forward acceleration of 1.5 m s^{-2} against a wind resistance of 450 N.

Hint: Refer to Example 4.12 but write the equations and do the calculations from the road end of the system, not the engine.

Problem 4.10

A vehicle of mass m has wheels of radius R and total moment of inertia I_w. It is driven by a motor having a reduction gearbox of ratio G. The inertia of the high-speed input shaft rotating parts is I_e. The motor shaft output torque is M. The friction torque on the high-speed shaft is M_f. The vehicle is driven up an incline of slope θ against a wind resistance W. Show that the acceleration a of the vehicle is obtained from the system equation of motion

$$\frac{MG}{R} - \frac{M_f G}{R} - W - mg \sin\theta = a\left(m + \frac{I_w}{R^2} + \frac{G^2 I_e}{R^2}\right).$$

An electrically driven locomotive has a mass of 30 000 kg. It has eight wheels, each of radius 0.7 m and moment of inertia 180 kg m^2. The driving wheels are actuated by a motor driving through a reduction gearbox of ratio 96. The inertia of the high-speed rotating parts, including the motor, is 0.04 kg m^2. Friction in the transmission is equivalent to a torque at the high-speed shaft of 0.5 N m. Calculate the motor torque required to drive the locomotive up a slope of 1 in 120 with a forward acceleration of 1 m s^{-2} against a wind resistance of 4.5 kN.

4.4 Answers to Problems

4.1 29.85 s.
4.2 1.8 N m.
4.3 2.861 kg; 11.77 s.
4.4 0.496 s.
4.5 8.366; 0.199 m s^{-2}.
4.6 0.009 45 m s^{-2}; 0.275 m s^{-1}. (Equivalent mass (at load) = 2 075 600 kg.)
4.7 9.364 s; 28.47 rad s^{-1}.
4.8 83.4 kg m^2; 11.95 N m.
4.9 375.5 N m.
4.10 280.8 N m.

5 Work, Energy and Power

Work done by a force and a moment. Kinetic energy of translation and rotation. Potential energy of position and of strain. Power. Efficiency. Stiffness of a spring or elastic member. The general form of the energy equation for a system.

5.1 The Fact Sheet

(a) Work Done

Work in translation is defined as the product of a force and the distance moved by the force along its line of action.

$$E = Fx$$

where E is work done in joules (J),
F is force in newtons (N),
x is the distance moved along the line of action of F, in metres (m).

Work in rotation is defined as the product of moment or torque and the angle turned through.

$$E = M\theta$$

where E is work done in joules (J),
M is moment or torque in newton metres (N m),
θ is the angle turned through by M in radians (rad).

(b) Kinetic Energy

The kinetic energy of translation of a body is given by

$$E_k = \tfrac{1}{2}mv^2,$$

where E_k is kinetic energy in joules (J),
m is the mass of the body in kilograms (kg)
v is the velocity of the body in metres per second (m s^{-1}).

The kinetic energy of rotation of a body is given by

$$E_k = \tfrac{1}{2}I\omega^2$$

where E_k is kinetic energy in joules (J),
I is the moment of inertia of the body in kilogram metres squared (kg m^2),
ω is the angular velocity of the body in radians per second (rad s^{-1}).

(c) Potential Energy of Position

The potential energy of position of a body is given by

$$E_p = mgh,$$

where E_p is the potential energy in joules (J),
(mg) is the weight of the body in newtons (N),
h is the vertical height of the body above an arbitrary datum in metres (m).

(d) Potential Energy of Strain

Potential energy of strain, or strain energy, of a linear spring or elastic member is given by

$$E_s = \tfrac{1}{2}Fx = \tfrac{1}{2}kx^2 = \frac{F^2}{2k},$$

where E_s is strain energy in joules (J),
F is the tensile or compressive force in the member, in newtons (N),
x is the deformation of the member at the point of application of the force in newtons (N),
k is the stiffness of the spring or member in newtons per metre (N m^{-1}) as defined at (f) below.

Strain energy of a torsional spring or member is given by

$$E_s = \tfrac{1}{2}M\theta = \tfrac{1}{2}k_t\theta^2 = \frac{M^2}{2k_t}$$

where E_s is strain energy in joules (J),
M is the torque or moment carried by the member in newton metres (N m),
θ is the angular deformation in radians (rad),
k_t is the torsional stiffness of the spring or member in newton metres per radian (N m rad^{-1}) as defined at (f) below.

(e) Power

Power is the rate of work being done.
For a force in translation

$$W = Fv$$

where W is the power in watts (W),
F is the force in newtons (N),
v is the velocity of the force in metres per second (m s^{-1}).

For a torque or moment

$$W = M\omega$$

where W is the power in watts (W),
M is the moment or torque in newton metres (N m),
ω is the angular velocity in radians per second (rad s^{-1}).

For an electrical machine

$$W = IV$$

where W is the power in watts (W), I is the current in amperes (A),
V is the voltage in volts (V).

(f) Stiffness of a Spring or Elastic Member

The stiffness of a spring or elastic member is defined by

$$k = \frac{\text{Force}}{\text{Deformation}} = \frac{F}{x}$$

where k is the stiffness in newtons per metre (N m^{-1}),
F is the deforming force in newtons (N),
x is the linear deformation (stretch or compression) in the direction of the force at its point of application, in metres (m).

The stiffness of a torsional spring or elastic member is defined by

$$k_t = \frac{\text{Torque}}{\text{Angular deformation}} = \frac{M}{\theta}$$

where k_t is the torsional stiffness in newton metres per radian,
M is the torque or moment on the member in newton metres (N m),
θ is the angle of deformation in radians (rad).

(g) Efficiency

The efficiency of a machine or system is defined by

$$\eta = \frac{\text{Useful work done by the system}}{\text{Energy into the system}}$$

in a given time where η is the efficiency (dimensionless ratio).

(h) General Form of Energy Equation

The general energy equation for a system is
Initial energy + Work done on the system (gain of energy)
− Work done by the system (loss of energy)
= Final energy.
Work done on the system: e.g. a driving motor or engine, or pump.
Work done by the system: e.g. friction.

5.2 Worked Examples

Example 5.1

A body of mass 20 kg is projected up an inclined plane of slope 35° with an initial velocity of 15 m s^{-1}. There is a frictional resistance to motion of 24 N. Calculate (a) how far up the plane it will travel; (b) what its velocity will be when it returns to its starting point.

Solution 5.1

Part (a)

The energy equation appropriate to this problem is

[h] Initial energy (kinetic) – Loss of energy (work against friction) = Final energy (potential).

If the distance travelled up the plane is x:

[a] Work against friction = Fx.

[c] Gain of potential energy $= mgh = mgx \sin\theta$.

$$\therefore\ \tfrac{1}{2}mu^2 - Fx = mgx \sin\theta.$$

Substituting values,

$$\tfrac{1}{2} \times 20 \times 15^2 - 24x = 20gx \sin 35°.$$

$$\therefore\ x = \frac{\tfrac{1}{2} \times 20 \times 15^2}{20g \sin 35° + 24} = \underline{16.48 \text{ m.}}$$

Part (b)

The intermediate stage may be bypassed; we equate the difference between initial and final kinetic energy to the total work done against friction. Since the distance *x* is now known, this work is *F* × 2*x*.

$$\tfrac{1}{2}mu^2 - F \times 2x = \tfrac{1}{2}mv^2.$$

$$\therefore\ v^2 = u^2 - 2Fx \times \frac{2}{m}$$

$$= 15^2 - 24 \times 2 \times 16.48 \times \frac{2}{20} = 145.9.$$

$$\therefore\ v = \underline{12.08 \text{ m s}^{-1}}.$$

It is good practice for you to solve the two parts of this problem by drawing the free-body diagrams, writing the equations of motion, and using appropriate kinematic equations. You will find that more work is required. You should appreciate that both approaches are correct but that the energy approach is (in this case) the simpler one.

Example 5.2

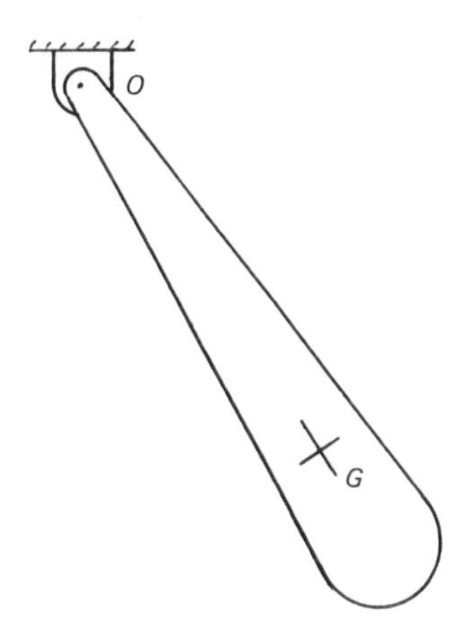

The diagram shows a compound pendulum comprising a non-uniform bar of mass 24 kg pivoted at a point O to swing in a vertical plane. The distance OG from pivot to mass centre is 0.64 m. The moment of inertia of the bar about a transverse axis through O is 12.6 kg m^2. The bar is swung to one side until G is level with O and released from rest. It comes to rest on the opposite swing 7° below the horizontal. Determine (a) the angular velocity of the pendulum at the instant OG is first vertical; (b) the angle OG makes with the horizontal when the pendulum again comes instantaneously to rest. It may be assumed that the resistance to the motion of the pendulum is a friction torque at O of constant magnitude and that resistance of the air is negligible.

Solution 5.2

Part (a)

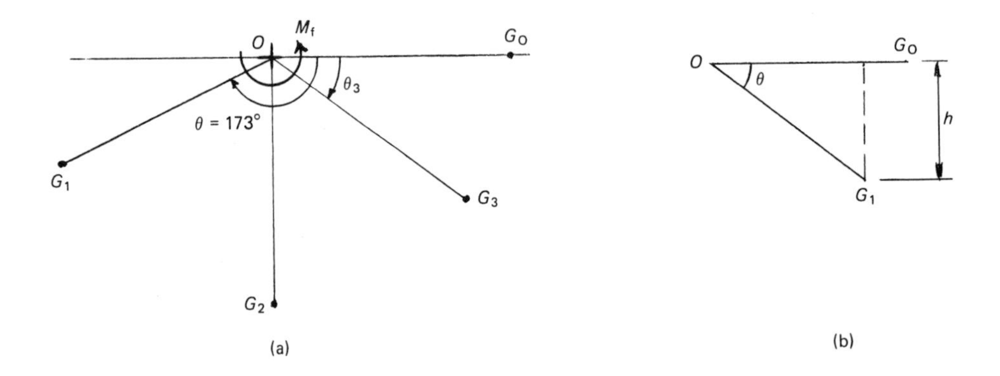

Due to the friction torque the pendulum loses 'height' in the first half-swing. The energy equation is

[h] Loss of p.e. = Work against friction.

Call the required friction torque M_f. Refer to diagram (a).

[a, c]
$$mgh = M_f\theta.$$
$$\therefore\ mg \times OG \sin\theta = M_f\theta.$$

(If the left-hand side of this equation is not immediately clear, assume that the pendulum swings through an angle θ which is less than 90° as shown in diagram (b). The loss of height is clearly seen to be OG sin θ. This is valid for any value of θ.)

$$\therefore\ M_f = \frac{mg \times OG \sin\theta}{\theta}$$
$$= \frac{24 \times 9.81 \times 0.64 \sin 173°}{173 \times \dfrac{\pi}{180}}$$
$$= 6.082 \text{ N m}.$$

We now write an energy equation from the initial position to the lowest point of swing (G_2 in diagram (a)). Although we label this point G_2 the pendulum passes this point *before* reaching G_1.

[h] Initial p.e. − Work against friction = Final k.e.

[b] $$mg \times OG - M_f \times \theta = \tfrac{1}{2}I\omega^2.$$

$$24g \times 0.64 - 6.082 \times \frac{\pi}{2} = \tfrac{1}{2} \times 12.6\omega^2.$$

$$\therefore\ \omega^2 = \frac{2}{12.6}(150.7 - 9.553) = 22.40.$$

$$\therefore\ \omega = \underline{4.73 \text{ rad s}^{-1}}.$$

Part (b)

We now write an energy balance equation from the initial position G_0 to the second position of instantaneous rest G_3 (see diagram (a)):

Work against friction = Loss of p.e.

In calculating the work against friction we must recognise that the pendulum has now swung through two half-swings. The total angle is (180 – 7 = 173) degrees to the left plus (173 – θ_3) degrees to the right.

$$\text{Work against friction} = M_f(173 + 173 - \theta_3)\frac{\pi}{180}$$

$$= 6.082 \times \frac{\pi}{180}(346 - \theta_3)$$

$$= 36.73 - 0.1062\theta_3.$$

(θ_3 is specified in *degrees* in this equation.)

$$\text{Loss of p.e.} = mg \times OG \sin\theta_3.$$

Equating, $$36.73 - 0.1062\theta_3 = 24g \times 0.64 \sin\theta_3.$$

$$36.73 = 150.7 \sin\theta_3 + 0.1062\theta_3.$$

This equation cannot be solved analytically, of course; we resort to 'trial and error'. If this sounds alarming, remember that in the first half-swing the pendulum 'lost' 7°. It is reasonable to suppose that it will lose approximately the same on the return.

We start by assuming $\theta_3 = 14°$:

$$150.7 \sin 14° + 0.1062 \times 14 = 37.94.$$

The figure is too high. Since both terms on the left-hand side are positive, we try a smaller value: say 13°:

$$150.7 \sin 13° + 0.1062 \times 13 = 35.28.$$

Judging by the spacing of the two answers either side of the required value, a final assessment of 13.6° will suffice:

$$150.7 \sin 13.6° + 0.1062 \times 13.6 = 36.88.$$

$$\therefore\ \theta = \underline{13.6°},$$

which is near enough.

If you need a more accurate answer without continuing to guess (and assuming you have time to do so) then you may calculate the intermediate value by

assuming that the function varies linearly with θ over a small range of θ. Calling the required value θ, we may express this by

$$\frac{\theta - 13}{14 - 13} = \frac{36.73 - 35.28}{37.94 - 35.28}.$$

$$\therefore \theta = \frac{1.45}{2.66} + 13 = 13.55°.$$

Checking: 150.7 sin 13.55° + 0.1062 × 13.55 = 36.75.
We will use this method again (see Example 5.12).

Example 5.3

A steel spring of negligible mass lies on a rigid surface with its axis vertical. A test shows that it compresses 5 mm under a load of 50 N. A body of mass m = 5 kg is dropped on to the spring from a height h = 100 mm above the top of the spring. Calculate the resulting maximum force in the spring.

Solution 5.3

Let the maximum instantaneous compressive force in the spring be F and the corresponding compression be x. The diagrams illustrate the question.

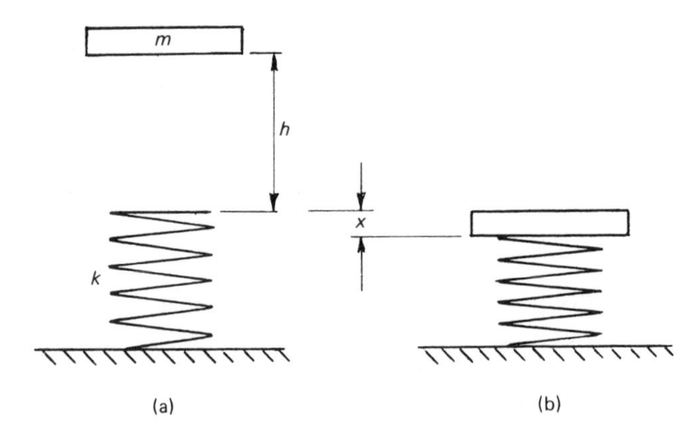

In diagram (a) the mass is stationary and the spring unloaded. In diagram (b) the mass has lost potential energy and the spring, now loaded, has gained strain energy. The mass is again momentarily at rest. We may write

[d] $$mg(h + x) = \tfrac{1}{2}Fx,$$

[f] but $$x = \frac{F}{k}.$$

$$\therefore mg\left(h + \frac{F}{k}\right) = \tfrac{1}{2}F\left(\frac{F}{k}\right).$$

Multiply all terms by $(2k)$:

$$2kmgh + 2mgF = F^2.$$

$$\therefore F^2 - F(2mg) - 2kmgh = 0.$$

$$\therefore F = \tfrac{1}{2}[2mg \pm \sqrt{(2mg)^2 + 4(2kmgh)}]$$
$$= mg \pm \sqrt{(mg)^2 + 2kmgh}.$$

You will note that the first term is the spring force resulting from the static weight of the body. Substituting numbers,

$$k = \frac{50}{5 \times 10^{-3}} = (10^4)\ \text{N m}^{-1}.$$

$$\therefore\ F = 5g + \sqrt{(5g)^2 + 2 \times 10^4 \times 5g \times 0.1}$$

$$= 49.05 + \sqrt{2406 + 98\,100}$$

$$= 49.05 + 317.0$$

$$= \underline{366.1\ \text{N}}.$$

It is interesting to note that this is nearly eight times the weight of the falling body.

The negative root was rejected; it would in any case have led to a smaller value of F. But the interpretation is of interest. Once attached to the spring, the mass would oscillate up and down, and the negative root indicates the maximum resulting tensile force in the spring as the mass is at the highest point of its oscillation.

Example 5.4

A spring is suspended from a fixed point and carries a body of mass m_1 at its lower end. A second body of mass m_2 is dropped onto the first body from a height h. Show that the increase of force in the spring resulting from the falling body is independent of the value of m_1.

Solution 5.4

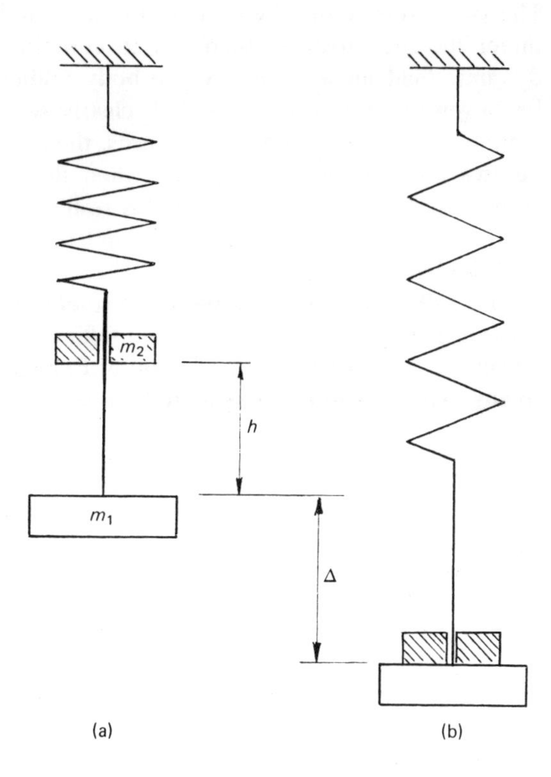

In diagram (a) both m_1 and m_2 have some potential energy (if we reckon from the lowest point, in (b)). Also, the spring has some strain energy. In (b), potential

energy is lost by both m_1 and m_2 and the spring has gained strain energy. The energy equation is

Total loss of potential energy = Gain of strain energy.

Let the spring force at (b) be F. The spring force in (a) is $(m_1 g)$. The additional stretch is Δ.

[d] $$m_2 g(h + \Delta) + m_1 g\Delta = \frac{F^2}{2k} - \frac{(m_1 g)^2}{2k}. \quad (1)$$

Δ is the additional extension resulting from the spring force increasing from $(m_1 g)$ to F. Therefore

$$\Delta = \frac{F - m_1 g}{k}.$$

Expanding equation 1 and substituting for Δ:

$$m_2 gh + m_2 g\left(\frac{F}{k}\right) - \frac{m_2 g m_1 g}{k} + m_1 g\left(\frac{F}{k}\right) - \frac{(m_1 g)^2}{k} = \frac{F^2}{2k} - \frac{(m_1 g)^2}{2k}.$$

Multiply throughout by $(2k)$ and arrange as a quadratic in F:

$$F^2 - F(2m_1 g + 2m_2 g) + (m_1 g)^2 + 2m_1 m_2 g^2 - 2km_2 gh = 0.$$

$$\therefore\ F = \tfrac{1}{2}\left[2g(m_1 + m_2) \pm \sqrt{4g^2(m_1 + m_2)^2 - 4((m_1 g)^2 + 2m_1 m_2 g^2 - 2km_2 gh)}\right]$$

$$= m_1 g + m_2 g \pm \sqrt{m_1^2 g^2 + m_2^2 g^2 + 2m_1 m_2 g^2 - (m_1 g)^2 - 2m_1 m_2 g^2 + 2km_2 gh}$$

$$= m_1 g + m_2 g + \sqrt{(m_2 g)^2 + 2km_2 gh}.$$

The two terms before the root are the static weights of the two masses. The terms under the root are now seen not to contain m_1. This root term represents the dynamic load increase due to the body falling on to the spring as distinct from being gently lowered on to it. It is clearly seen that even if $h = 0$, there is a dynamic load increase of $(m_2 g)$. In words, the increase of spring force due to m_2 will be twice the weight of m_2 even when the mass is dropped from zero height. (Example 5.6 illustrates this.) If this result is compared with the quadratic solution in Example 5.3 then it is seen that the dynamic part, i.e. the part under the root, is identical.

This rather tedious example is nevertheless useful, because it means that henceforward the dynamic effect of a load falling onto an elastic spring or structure can be simply calculated by the method of Example 5.3, the load simply being added to any existing load in the system.

Example 5.5

A steel bar has a diameter of 10 mm and a length of 1 m. When tested, the bar is observed to extend by 0.0637 mm under a tensile load of 1 kN. The maximum safe load on the bar is 4500 N. The bar is vertical, with its upper end attached to a fixed point. The lower end carries a plate on which a load can fall.

(a) Calculate from what height above the plate a load of mass 10 kg can fall without the maximum load in the bar being exceeded.

(b) Calculate the maximum instantaneous load in the bar if the 10 kg load falls from a height of 4 mm above the plate.

Solution 5.5

Part (a)

The problem may be solved by the method of Example 5.3. It must be appreciated that a steel bar behaves as a spring, in that a tensile load produces a proportionate extension (provided the load is not great enough to exceed the limit of proportionality).

If the height of fall is h and the maximum extension of the bar is x, we may write as before

[c, d]
$$mg(h + x) = \tfrac{1}{2}Fx.$$

$$\therefore\ h = \frac{Fx}{2mg} - x. \qquad (1)$$

$F = 4500$ N as stated,

$$\therefore\ x = \frac{0.0637}{1000} \times 4500 = 0.2867 \text{ mm}.$$

Substituting in equation 1,

$$h = \frac{4500 \times 0.2867 \times 10^{-3}}{2 \times 10 \times 9.81} - x$$

$$= (6.576 - 0.2867) \text{ mm}$$

$$= \underline{6.289 \text{ mm.}}$$

Part (b)

Exactly the same equation may be used as in Example 5.3. This is repeated here.

$$F = mg \pm \sqrt{(mg)^2 + 2kmgh.}$$

$$k = \frac{1000}{0.0637 \times 10^{-3}}$$

$$= 15.7 \times 10^6 \text{ N m}^{-1}.$$

$$\therefore\ F = 10g \pm \sqrt{(10g)^2 + 2(15.7 \times 10^6) \times 10g \times 0.004}$$

$$= \underline{3609.6 \text{ N.}}$$

As with Example 5.3 we reject the negative root. This answer is seen to be more than 36 times the actual static weight of the falling load.

Example 5.6

A helical spring hangs vertically from a fixed point. It carries a light plate at its lower end. A body of mass m is lowered on to the plate and is released from rest in this position. Obtain expressions for (a) the maximum force in the spring, and (b) the maximum velocity of the mass.

Solution 5.6

Both parts of the question may be approached by writing a general energy equation for a displacement x from the initial position. The diagram illustrates this: the mass rests on the plate at level *a–a* and is released at this point. When it has

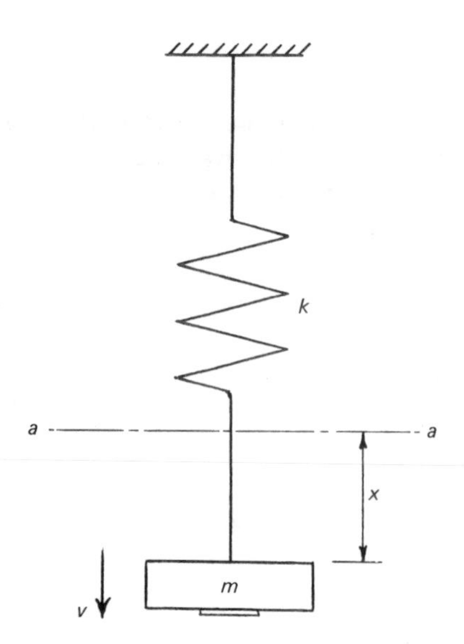

moved a distance x downwards its velocity is v. It has lost potential energy and gained kinetic energy. Also the spring has gained strain energy.

The energy equation is

$$\text{Loss of p.e.} = \text{Gain of k.e.} + \text{Gain of s.e.}$$

[b, c, d] $$\therefore\ mgx = \tfrac{1}{2}mv^2 + \tfrac{1}{2}Fx$$

[d] $$= \tfrac{1}{2}mv^2 + \tfrac{1}{2}kx^2.$$

Rearranging: $$v^2 = 2gx - \frac{k}{m}x^2. \qquad (1)$$

Part (a)

It is clear that at the lowest point of travel m, v, and therefore v^2, must be zero.

$$2gx - \frac{k}{m}x^2 = 0.$$

$$\therefore\ x_{\max} = 2\left(\frac{mg}{k}\right).$$

$$\therefore\ F_{\max} = 2mg.$$

This is exactly twice the load the spring would carry if it supported only the static weight of m.

Part (b)

If v is to be a maximum, then (v^2) also will be maximum. We obtain the differential $\dfrac{\mathrm{d}(v^2)}{\mathrm{d}x}$:

$$\frac{\mathrm{d}(v^2)}{\mathrm{d}x} = 0 = 2g - \frac{k}{m}(2x).$$

$$\therefore\ x = \frac{mg}{k}.$$

We might have deduced this result without differentiating. This is the amount the spring would stretch under the static weight of m. If x is less than this value then the weight exceeds the upward spring force and the body accelerates downwards. When this value of x is exceeded then the spring force becomes more than the weight and the body begins to slow down.

Substituting this value of x in equation 1 gives the maximum velocity.

$$v_{max}^2 = 2g\left(\frac{mg}{k}\right) - \frac{k}{m}\left(\frac{mg}{k}\right)^2$$

$$= \frac{2g^2 m}{k} - \frac{g^2 m}{k}$$

$$= \frac{g^2 m}{k} .$$

$$\therefore \; v_{max} = g\sqrt{\frac{m}{k}} .$$

Example 5.7

A body of mass m hangs from a helical spring of stiffness k. The upper end of the spring is moving vertically downwards with a constant velocity u when it is instantaneously brought to rest. Develop an expression for the maximum tensile force F in the spring due to this change of motion. If m = 20 kg and u = 4 m s^{-1}, calculate the maximum permissible value for the stiffness k if the maximum tensile force is not to exceed six times the weight of the body.

Solution 5.7

While the system is moving downwards, the body will possess kinetic energy. When the upper end of the spring is stopped then the body will move an additional distance downwards, Δ, thus losing potential energy, and the spring will stretch additionally by this amount, thus gaining additional strain energy. Thus:

Initial k.e. + p.e. + s.e. = Final s.e.

A simpler way to express this is:

k.e. + p.e. = Gain of s.e.

$$\tfrac{1}{2}mu^2 + mg\Delta = E_{s2} - E_{s1} . \qquad (1)$$

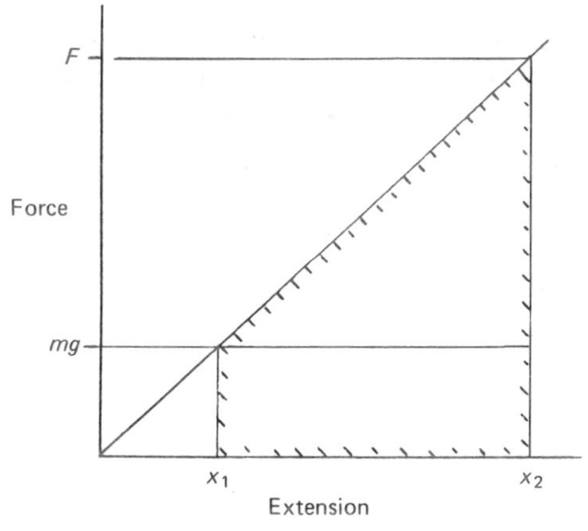

The diagram (p. 203) is a sketch of the force–extension graph for the spring. While the system is moving at a constant speed, the force in the spring is the weight of the body, mg. When a spring is stretched, the increase of strain energy arises out of the work down by the stretching force. Thus, in this case:

$$U = \Sigma\ (F\ \delta x)$$

which is the area under the force–extension graph. In this particular example, the change of strain energy is seen to be the shaded area of the graph.

This area is conveniently divided into a rectangle and a triangle.

$$\therefore\ E_{s2} - E_{s1} = mg\,(x_2 - x_1) + \tfrac{1}{2}(x_2 - x_1)\,(F - mg)$$

$$= mg\Delta + \tfrac{1}{2}\Delta\,(F - mg).$$

But the *increase* of tensile force, $(F - mg)$, and the increase of extension, Δ, are related by

[f]
$$\Delta = \frac{F - mg}{k}.$$

$$E_{s2} - E_{s1} = mg\Delta + \tfrac{1}{2}\,\frac{(F - mg)^2}{k}.$$

Substituting in the energy equation, equation 1,

$$\tfrac{1}{2}mu^2 + \cancel{mg\Delta} = \cancel{mg\Delta} + \tfrac{1}{2}\,\frac{(F - mg)^2}{k}.$$

$$\therefore\ F - mg = \sqrt{kmu^2}$$

$$= u\sqrt{km}.$$

$$\therefore\ \underline{F = mg + u\sqrt{km}.}$$

The second term on the right-hand side is seen to be the increase of tensile force due to stopping the downward motion.

Substituting the given data,

$$6mg = mg + 4\sqrt{k \times 20}.$$

$$\therefore\ 5mg = 4\sqrt{20k}.$$

$$\therefore\ k = \frac{(5 \times 20g)^2}{4^2 \times 20}$$

$$= \underline{3007\ \text{N m}^{-1}},$$

and from the expression for F it is seen that this must be a maximum, as the higher the value of k the greater the corresponding value of F. It will be noted in several of the examples in this chapter that forces arising from suddenly applied loads can be reduced by introducing flexibility into the system, which is the opposite of stiffness.

The most likely source of error in a problem of this type is to fail to take account of the initial strain energy in the spring that is due to the weight of the body.

Example 5.8

(a) A body of mass m = 2 kg rests on a horizontal frictionless table. It is attached to a spring of stiffness k = 650 N m^{-1}, the other end of the spring being secured to a fixed mounting. The body is pulled to one side, extending the spring by a distance $x = 50$ mm, and it is released from rest in this position. Calculate the maximum velocity of the resulting oscillation.

(b) The same mass connected to the same spring now rests on a horizontal table having a coefficient of friction of $\mu = 0.1$. It is again moved 50 mm and released. Calculate the maximum velocity of the mass in the subsequent motion. Also find how far it moves before again coming to rest.

Solution 5.8

Part (a)

The maximum velocity will occur when all the initial strain energy of the stretched spring is transferred to the kinetic energy of the body.

$$\text{s.e. of spring} = \text{k.e. of body.}$$

$$\tfrac{1}{2}kX^2 = \tfrac{1}{2}mv^2 .$$

$$\therefore\ v_{\text{max}} = \sqrt{\frac{k}{m}}\,X$$

$$= \sqrt{\frac{650}{2}} \times 0.05$$

$$= \underline{0.901 \text{ m s}^{-1}}.$$

Part (b)

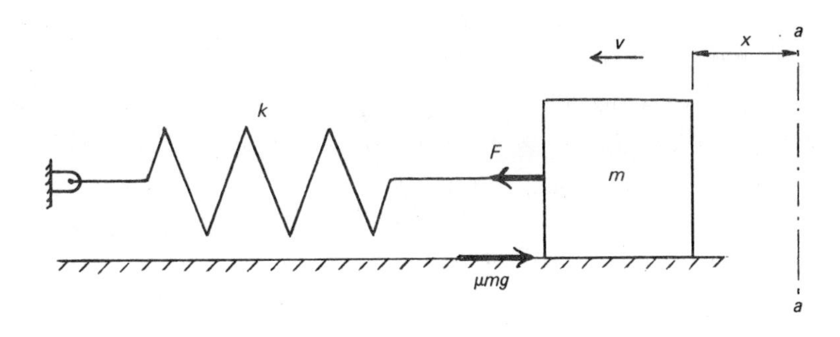

Let the displacement from the point of release (line *a–a* in the diagram), be x. Let the velocity of the body at this point be v, and let the force in the spring be F. The stretch of the spring will be $(X - x)$. The friction force will be (μmg). The energy equation is

[h] Initial spring s.e. – Work (friction) = Final spring s.e. + body k.e.

$$\tfrac{1}{2}kX^2 - (\mu mg)\,x = \tfrac{1}{2}k\,(X - x)^2 + \tfrac{1}{2}mv^2 .$$

$$\therefore\ v^2 = \frac{kX^2}{m} - \frac{k}{m}\,(X - x)^2 - 2\mu gx. \qquad (1)$$

We may differentiate straight away:

$$\frac{\mathrm{d}(v^2)}{\mathrm{d}x} = 0 - \frac{k}{m}\,2\,(X - x)\,(-1) - 2\mu g$$

$$= 0 \text{ (for max.).}$$

$$\therefore\ X - x = 2\mu g\ \frac{m}{2k}.$$

$$\therefore\ x = X - \frac{\mu g m}{k}.$$

Substituting, $$x = 0.05 - \frac{0.1 \times 9.81 \times 2}{650} = 0.046\,98\ \text{m}$$

and $$(X - x) = 0.003\,02\ \text{m}.$$

Substituting these values in equation 1,

$$(v_{max})^2 = \frac{650 \times (0.05)^2}{2} - \frac{650 \times (0.003\,02)^2}{2} - 2 \times 0.1 \times 9.81 \times 0.046\,98$$

$$= 0.7174.$$

$$\therefore\ v_{max} = \underline{0.847\ \text{m s}^{-1}}.$$

When the body again comes to rest, $v = 0$. Equating the left-hand side of equation 1 to zero,

$$0 = \frac{kX^2}{m} - \frac{k}{m}(X - x)^2 - 2\mu g x.$$

Expanding,

$$0 = \frac{kX^2}{m} - \frac{k}{m}(X^2 + x^2 - 2Xx) - 2\mu g x$$

$$= \frac{kX^2}{m} - \frac{kX^2}{m} - \frac{kx^2}{m} + \frac{2kXx}{m} - 2\mu g x.$$

$$\therefore\ \frac{kx^2}{m} = \frac{2kXx}{m} - 2\mu g x.$$

$$\therefore\ x = 2X - \frac{2\mu g m}{k}.$$

Substituting,

$$x = 2 \times 0.05 - \frac{2 \times 0.1 \times 9.81 \times 2}{650}$$

$$= (100 - 6.04)\ \text{mm}$$

$$= \underline{93.96\ \text{mm}}.$$

Example 5.9

A ship's tow-rope has a breaking strength of 250 kN. A test on the rope shows that a length of 1 m extends 0.8 mm per kN of tensile force. The rope is 150 m long and connects two vessels, one of mass 30 000 kg, moving at 4 m s^{-1} directly away from the other, which is stationary and has a mass of 40 000 kg. The rope tightens, causing the second vessel to start moving. At the instant both vessels are moving at the same speed, calculate the force in the tow-rope. What minimum length of rope is required if the load under these circumstances is not to exceed one-fifth of its breaking load?

Solution 5.9

We make use of the momentum principle to find the common speed of the two vessels. See Chapter 3. You are warned, in that chapter, not to use the method of energy to analyse problems of colliding bodies. But the reason is that energy is

absorbed by the bodies on collision which is not always recovered. Here it may be assumed that the rope absorbs the energy of collision, so that energy 'lost' by the vessels is gained by the rope.

The momentum equation is

$$m_A u_A + m_B u_B = m_A v_A + m_B v_B .$$

$$40\,000 \times 0 + 30\,000 \times 4 = (40\,000 + 30\,000)\, v .$$

$$\therefore\ v = \frac{120\,000}{70\,000} = 1.714 \text{ m s}^{-1}.$$

We now use an energy equation:

Initial k.e. of towing vessel = k.e. of both vessels + Strain energy of rope.

[b, d] $\quad \frac{1}{2} \times 30\,000 \times 4^2 = \frac{1}{2} \times 70\,000 \times (1.714)^2 + E_s$

$$\therefore\ E_s = 240\,000 - 102\,823$$

$$= 137\,177 \text{ J}.$$

For rope, $\quad E_s = \frac{1}{2}kx^2 .$

From data, $\quad k = \dfrac{1000}{0.8 \times 10^{-3} \times 150} = 8333 \text{ N m}^{-1}.$

$$\therefore\ 137\,177 = \frac{1}{2} \times 8333 \times x^2 .$$

$$\therefore\ x = \sqrt{\frac{2 \times 137\,177}{8333}} = 5.738 \text{ m}.$$

The force F is obtained from

[f] $\quad F = kx$

$$= 8333 \times 5.738 = 47.81 \text{ kN}.$$

Let the minimum length of rope be L. The energy absorbed is the same.

$$E_s = \frac{1}{2}Fx = \frac{1}{2}F\left(\frac{F}{k}\right).$$

F is required to be $\dfrac{250}{5} = 50$ kN.

$$\therefore\ k = \frac{F^2}{2 \times E_s}$$

$$= \frac{50^2 \times 10^6}{2 \times 137\,177}$$

$$= 9.112 \text{ kN m}^{-1}.$$

Since 1 m extends 0.8 mm per kN load, a length L extends

$(0.8L \times 10^{-3})$ m per kN load.

$$\therefore\ \frac{1000}{0.8L \times 10^{-3}} = 9112.$$

$$\therefore\ L = \frac{1000}{0.8 \times 10^{-3} \times 9112} = \underline{137.2 \text{ m}},$$

and if the rope was shorter than this the load would be greater than 50 kN. It may not be immediately obvious that the longer a rope is, the more energy it can absorb for a given maximum load. For this reason, tow-ropes are required by authorities to be of a certain minimum length which is determined by the purpose for which they are used.

Example 5.10

An aircraft arrester comprises a pair of springs joined together, the two outer ends being attached to two fixed points 50 m apart. Each spring is 25 m long, so that when joined they are just unstrained. An aircraft of mass 35 000 kg travelling at 50 m s^{-1} catches the springs on its arrester-hook at the mid-point. Determine the required stiffness of the springs if the aircraft is to be brought to rest in 24 m.

Solution 5.10

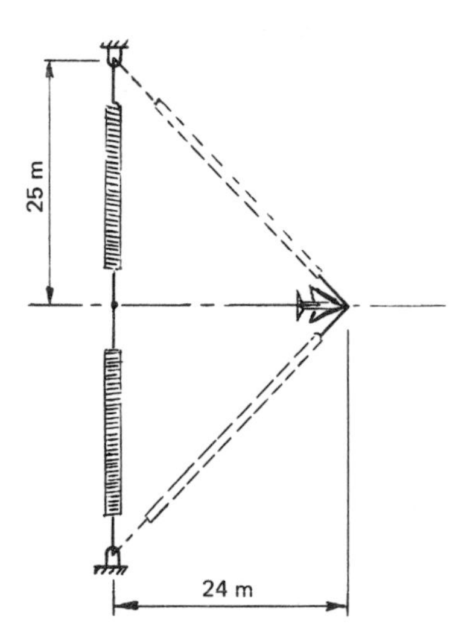

The simple equation of energy is

$$\text{Loss of k.e. of aircraft} = \text{Gain of s.e. of spring.}$$

From the diagram the length of the stretched spring is $\sqrt{25^2 + 24^2} = 34.66$ m.

$$\text{Spring stretch} = 34.66 - 25.0 = 9.66 \text{ m.}$$

[d] $$\tfrac{1}{2}mv^2 = 2 \times \tfrac{1}{2}kx^2 \qquad \text{(for two springs).}$$

$$\therefore\ k = \frac{mv^2}{2x^2}$$

$$= \frac{35\,000 \times 50^2}{2 \times (9.66)^2}$$

$$= \underline{468.8 \text{ kN m}^{-1}}.$$

Example 5.11

A winding-drum and rope comprise part of a mine hoist. The ends of the rope hang vertically from opposite sides of the drum and support a load of 2600 kg and a counterweight of 2000 kg, the counterweight descending as the load is raised. The motion of the load and that of the counterweight are both opposed by a friction force of 100 N. The drum is driven by an electric motor through gearing, and the efficiency of the transmission is 80 per cent. If the load is to be raised in 45 seconds up a vertical shaft of depth 320 m, calculate the required average power of the driving motor. Neglect any variation of power due to acceleration and retardation of the system. If the motor speed is 1440 rev min^{-1}, what will be the average shaft torque of the motor?

Solution 5.11

Work done by the drum = Gain of p.e. of load − Loss of p.e. of counterweight + Work done against friction.

$$\text{w.d.}_{\text{drum}} = m_L gh - m_C gh + Fh \times 2$$

$$= 9.81 \times 320(2600 - 2000) + 100 \times 320 \times 2$$

$$= 1\,947\,520 \text{ J.}$$

[g] $$\therefore \text{ Motor work} = 1\,947\,520 \times \frac{100}{80}.$$

[e] $$\text{Power} = \text{Work s}^{-1}$$

$$= \frac{1\,947\,520}{0.8 \times 45}$$

$$= \underline{54.10 \text{ kW}}.$$

For the motor,

$$W_{\text{motor}} = T\omega.$$

$$\therefore T = \frac{W}{\omega} = \frac{54\,100 \times 60}{2\pi \times 1440} = \underline{358.8 \text{ N m}}.$$

Example 5.12

A car of mass 980 kg has an engine capable of a power output of 60 kW. The efficiency of the transmission system is 85 per cent. Resistance to motion due to wind may be assumed to be $(0.9v^2)$ N where v is the velocity in m s^{-1}. Estimate the maximum speed at which the car can travel (a) on a straight horizontal road; (b) up a slope of 5°; (c) down a slope of 5°.

Solution 5.12

When the car is travelling at its maximum speed, the tractive force must be exactly equal to the total resistance to motion. Let the wind resistance be R.

Part (a)

[e] $$\text{Wheel power} = 60\,000 \times 0.85 = R \times v = 0.9v^3.$$

$$\therefore v = \sqrt[3]{\frac{60\,000 \times 0.85}{0.9}}$$

$$= \underline{38.41 \text{ m s}^{-1}}.$$

Part (b)

In one second, the car will travel a distance v, and will thereby gain a vertical height of $(v \sin 5°)$.

$$\text{Wheel power} = \text{Work per second}$$

$$= R \times v + mg(v \sin 5°).$$

$$\therefore\ 60\,000 \times 0.85 = 0.9v^3 + 980g \times v \sin 5°.$$

$$\therefore\ v^3 + 931v = 56\,667.$$

An analytical solution is not practical. A programmable calculator, providing your examiners allow it, is useful in solving such an equation. But trial-and-error is quite feasible. The required answer is obviously less than 38.41. Try 36, 34 and so on.

v	36	34	32	30
l.h. side of equation	80 172	70 958	62 560	54 930

If we assume a linear variation between 32 and 30,

$$\frac{v - 30}{32 - 30} = \frac{56\,667 - 54\,930}{62\,560 - 54\,930}.$$

$$\therefore\ v = \frac{2 \times 1737}{7630} + 30 = 30.46 \text{ m s}^{-1}.$$

Checking,

$$(30.46)^3 + 931 \times 30.46 = 56\,619$$

and this solution is close enough to the slightly more accurate value of 30.473 to satisfy the question requirement.

Hence $\qquad v = \underline{30.46 \text{ m s}^{-1}}.$

Part (c)

The argument is the same as for part (b) but the potential energy must now be subtracted.

$$60\,000 \times 0.85 = 0.9v^3 - 980g \times v \sin 5°.$$

$$\therefore\ v^3 - 931v = 56\,667.$$

The answer for (b) was approximately 8 m s^{-1} below that for (a); let us try 8 m s above, say 46 m s^{-1}:

$$(46)^3 - 931 \times 46 = 54\,510.$$

Already very close. Try 47:

$$(47)^3 - 931 \times 47 = 60\,066.$$

Assuming 'linearity' as before,

$$\frac{v - 46}{47 - 46} = \frac{56\,667 - 54\,510}{60\,066 - 54\,510}.$$

$$\therefore\ v = \frac{2157}{5556} + 46 = 46.39.$$

$$(46.39)^3 - 931 \times 46.39 = 56\,644,$$

which is quite accurate enough. (The more accurate answer would be 46.394.)

$$v = \underline{46.39 \text{ m s}^{-1}}.$$

(You would not, of course, be expected to go to this sort of trouble in solving an examination question.)

Example 5.13

A motor vehicle has a mass of 1400 kg and the wheel radius is 0.3 m. It is equipped with four disc brakes, each comprising a circular disc attached to the wheel axle, upon which friction pads press from both sides at a mean radius of 0.11 m as shown in the diagram. The 'active' pad is actuated by a pressure cylinder of diameter 20 mm, the maximum pressure being 80 MN m^{-2}. The friction coefficient between disc and pads is 0.1. Assuming equal effective braking on all four wheels and that the wheels do not skid, calculate the least distance in which the car can be brought to rest from a speed of 30 m s^{-1} (a) on a straight level road; (b) down an incline of slope $\sin^{-1} 0.1$. An average wind resistance of 650 N may be assumed in each case.

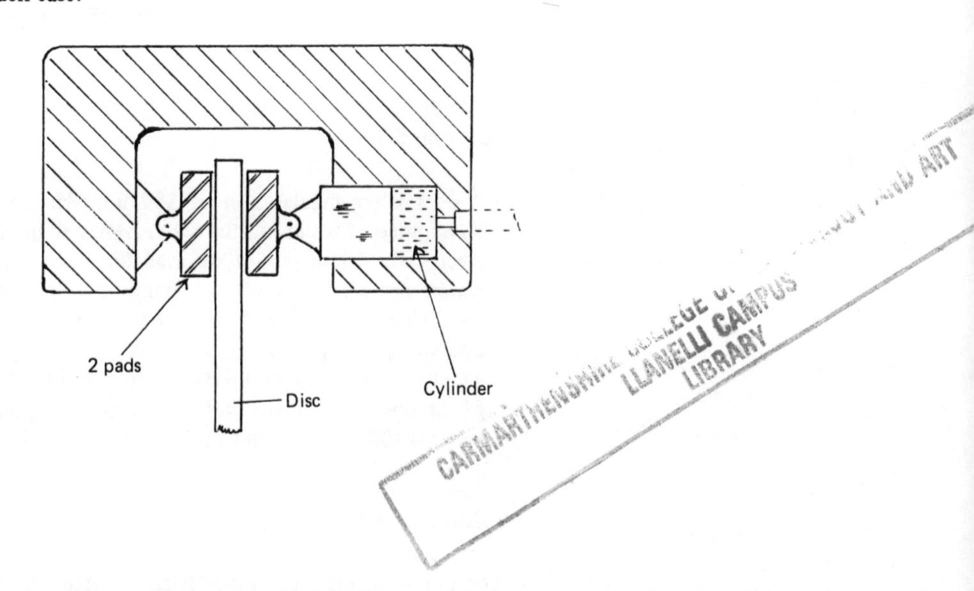

Solution 5.13

Part (a)

The initial kinetic energy of the vehicle is expended, partly in doing work against wind resistance and partly in doing work against the brakes.

$$\text{Load on brake pad} = \text{Pressure} \times \text{Area}$$

$$= (80 \times 10^6) \times \frac{\pi}{4} (20 \times 10^{-3})^2$$

$$= (8000\pi) \text{ N.}$$

$$\text{Friction force per pad} = (8000\pi)\, 0.1 = (800\pi) \text{ N.}$$

[a] $\therefore$ Work done per rev per pad $= (800\pi)(2\pi \times 0.11) = 1737$ J rev^{-1}.

$$\therefore \text{ Total brake work per rev} = 1737 \times 8 = 13\,896 \text{ J rev}^{-1}.$$

$$\text{Distance travelled per rev} = (2\pi \times 0.3) \text{ m.}$$

$$\therefore \text{ Brake work per metre} = \frac{13\,896}{2\pi \times 0.3} = 7372 \text{ J m}^{-1}.$$

The energy equation is

[h] Initial k.e. = Work done (brakes) + Work done (wind).

$$\therefore \tfrac{1}{2}mv^2 = 7372x + 650x = 8022x.$$

$$\therefore x = \frac{1400 \times (30)^2}{2 \times 8022} = \underline{78.53 \text{ m}}.$$

Part (b)

In this case, the vehicle loses potential energy as well as kinetic. The energy equation now becomes

[h] Initial (k.e. + p.e.) = Work done (brakes + wind).

In a distance x, the loss of p.e. is $(mgx \sin \theta)$.

$$\therefore \tfrac{1}{2}mv^2 + mgx \sin \theta = 8022x.$$

$$\therefore x = \frac{\tfrac{1}{2}mv^2}{8022 - mg \sin \theta}$$

$$= \frac{1400 \times (30)^2}{2(8022 - 1400 \times 9.81 \times 0.1)}$$

$$= \underline{94.76 \text{ m.}}$$

Example 5 14

A dock is rectangular in shape, 180 m by 35 m. It contains water to a depth of 12 m. It is to be emptied of water, all the water having to be raised 13 m above the floor of the dock. Six electrically driven pumps are available, each capable of delivering 12 000 m^3 per hour of water. The efficiency of each pump is 86 per cent, and that of the driving motors 84 per cent. Calculate the electrical power input required to each motor (a) at the start of the operation, and (b) at the end, and the corresponding current taken by each motor, the supply voltage being 400 volts. Determine the total work done by all six pumps, and the cost of the operation, if one unit of electricity (1 kilowatt hour) costs 4.5 pence. The density of water is 1000 kg m^{-3} and 1 kW h is 3.6 x 10^6 J.

Solution 5.14

At the beginning, the water is to be lifted only 1 m. For each pump,

$$\dot{m} \text{ (mass per second)} = \frac{12\,000}{60 \times 60} \times 1000 = 3333 \text{ kg s}^{-1}$$

$$\text{Power output} = \dot{m}gh = 3333 \times 9.81 \times 1 = 32.70 \text{ kW.}$$

$$\text{Power input} = \frac{32.70}{0.86} = 38.02 \text{ kW.}$$

$$\text{Motor input} = \frac{38.02}{0.84} = \underline{45.27 \text{ kW.}}$$

[e] $$I = \frac{W}{V} = \frac{45.27 \times 10^3}{400} = \underline{113.2 \text{ A.}}$$

At the end, the water has to be raised from the dock floor, a height of 13 m. Otherwise the calculations are the same.

$$\text{Motor input} = (45.27 \times 13) = \underline{588.5 \text{ kW.}}$$

$$I = 113.2 \times 13 = \underline{1472 \text{ A.}}$$

$$\text{Total work} = mgh = (180 \times 35 \times 12) \times 1000g \times 7$$

$$= 5191 \times 10^6 \text{ J.}$$

$$\text{Total work by motors} = \frac{5191}{0.84 \times 0.86}$$

$$= 7186 \text{ MJ}$$

$$= \frac{7186}{3.6} = \underline{1996 \text{ kWh.}}$$

$$\text{Cost} = 1996 \times 4.5 = 8982 \text{ pence} = \underline{£89.82.}$$

Example 5.15

A fire tender is required to operate five hoses simultaneously, each capable of delivering a jet of water 75 mm in diameter which can ascend a vertical height of 20 m. Neglecting energy losses, calculate the power at which the tender pump must operate. The density of water is 1000 kg per cubic metre.

Solution 5.15

To calculate the rate of flow of water we equate the kinetic energy of the jet with the potential energy of the water at its highest point. For a mass m of water,

$$\tfrac{1}{2}mv^2 = mgh$$

$$\therefore\ v = \sqrt{2gh} = \sqrt{2g \times 20} = 19.81 \text{ m s}^{-1}.$$

For one pump,

$$\text{Flow rate} = \text{Area of jet} \times \text{Velocity}$$

$$= \frac{\pi}{4}(75 \times 10^{-3})^2 \times 19.81 = 0.0875 \text{ m}^3 \text{ s}^{-1}.$$

$$\text{Total flow rate} = 0.0875 \times 5 = 0.4375 \text{ m}^3 \text{ s}^{-1}.$$

Calling the mass of water per second $\dot{m}$,

$$\text{Power} = \dot{m}gh = (0.4375 \times 1000) \times 9.81 \times 20$$

$$= \underline{85.8 \text{ kW.}}$$

Example 5.16

A load of weight 200 N is placed at the end of a horizontal cantilever, and causes a vertical deflection at that point of 12 mm. The same load is then raised a height of 15 mm above the unloaded cantilever and allowed to fall on to the cantilever from this height. Calculate the resulting maximum deflection, and determine also the magnitude of the equivalent static load, W', which, if applied gradually at the same point, would produce the same deflection. It may be assumed that the cantilever is not stressed at any point beyond the limit of proportionality.

PCL

Solution 5.16

It must first be appreciated that a horizontal cantilever can be treated as a spring. (See Example 5.5.) It is a member which deforms proportionally to the load acting on it; in other words, it obeys Hooke's law. (At least, this is true within the limit of proportionality of the material of the cantilever, but the question states that this may be assumed.) The information relating to the deflection resulting from the load of 200 N enables the stiffness to be calculated. Recall that stiffness *k* is the force per unit deflection. (Fact Sheet (f).)

When the load is dropped on to the spring from a height *h* the resulting maximum deflection Δ will be greater than if the load were merely lowered gently on to the beam. The diagram shows the two states: (a) before releasing the load, and (b) the point of maximum deflection of the beam after release.

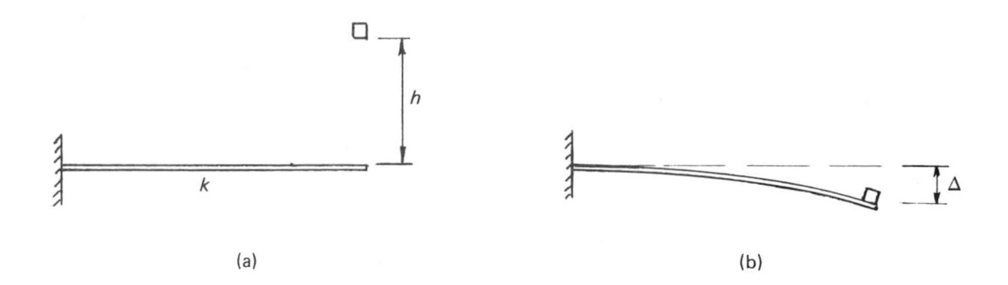

At the instant of maximum deflection, the load will be instantaneously at rest and the beam will contain the maximum strain energy. The energy equation is therefore:

Loss of potential energy (a) = Gain of strain energy (b).

Note that the load falls a distance $(h + \Delta)$.

To determine the strain energy of the deformed beam, we use the concept of the equivalent static load. This is the load which, applied directly and gradually to the beam, would result in the same deformation as that produced by the falling load. When a spring is deformed by a load, the strain energy is given by:

[d] $$U = \tfrac{1}{2}W'\Delta.$$

The energy equation is therefore

$$mg(h + \Delta) = \tfrac{1}{2}W'\Delta. \quad (1)$$

(Actually, this is not strictly accurate, as it assumes that the strain energy in the beam is the same for a given end deflection whether it is produced by a falling load or by a gradually applied static load. In fact, the beam would adopt a slightly different shape for a falling load. But the difference of strain energy would be very small.)

Equation 1 contains the two unknown terms, W' and Δ. But Δ can be expressed in terms of W' and stiffness k.

$$\Delta = \frac{W'}{k}. \quad (2)$$

Substituting this in equation 1,

$$mg\left(h + \frac{W'}{k}\right) = \tfrac{1}{2}W'\left(\frac{W'}{k}\right).$$

Expanding, and multiplying throughout by $2k$:

$$2kmgh + 2mgW' = (W')^2$$

which is seen to be a quadratic in W'. Rearranging,

$$(W')^2 - W'(2mg) - 2kmgh = 0.$$

$$\therefore\ W' = \frac{2mg \pm \sqrt{4(mg)^2 + 4 \times 2kmgh}}{2}$$

$$= mg \pm \sqrt{(mg)^2 + 2kmgh}, \quad (3)$$

and it is seen that the equivalent static load is greater than the purely static weight of the falling load by the quantity under the square root.

Evaluating stiffness k,

[f] $$k = \frac{F}{x} = \frac{200}{12 \times 10^{-3}} \text{ N m}^{-1}.$$

$$\therefore\ W' = 200 \pm \sqrt{(200)^2 + 2\left(\frac{200}{0.012}\right) \times 200 \times 0.015}$$

$$= 200 \pm \sqrt{40\,000 + 100\,000.}$$

We choose the larger positive solution. The negative answer is explained by the circumstance that the beam would begin to vibrate when the load falls on it.

$$\therefore\ W' = (200 + 374.2) = \underline{574.2 \text{ N.}}$$

Substituting this in equation 2,

$$\Delta = \frac{W'}{k}$$

$$= \frac{574.2}{200} \times 0.012$$

$$= (34.45 \times 10^{-3}) \text{ m}$$

$$= \underline{34.45 \text{ mm.}}$$

Before leaving this example, it is interesting to examine equation 3 in its algebraic form. First, observe the effect of the stiffness k on the value of W'. A large value will give a correspondingly large value of W'. Putting this another way, a very stiff spring or member, subjected to a suddenly applied load, will experience greater stress than a less stiff one. Thus, by holding a length of light string in your two hands, and jerking them apart, you can break the string, whereas if you hold a piece of elastic you cannot break it in the same way.

Second, note the effect of substituting a value $h = 0$ in the expression. This is equivalent to dropping a load from a height of zero. You can see that in this case W' has a value of $2mg$, which is twice the weight of the load. So, in general, when a load acts suddenly on a member or a structure, even though it is not actually dropped on, the resulting stresses and deformations are twice the values they will have if the same load is applied gradually.

Example 5.17

Part of a switching device is shown in the diagram. A thin uniform rod of mass 0.02 kg and length 100 mm is pivoted to a fixed point 25 mm from one end of the rod. The spring attached to the upper end of the rod has a free length of 50 mm and is initially extended 25 mm in the position shown. A constant moment of 5 N mm is applied in a clockwise direction to the rod causing it to rotate about the fixed point. It is required that the rod will hit the stop at A with a velocity of 0.5 m s^{-1}. Determine the necessary stiffness of the spring.

The stop at A is removed. Show that in this case the total rotation of the rod from the vertical position would be approximately 98° before it came to rest, the moment on the rod being maintained.

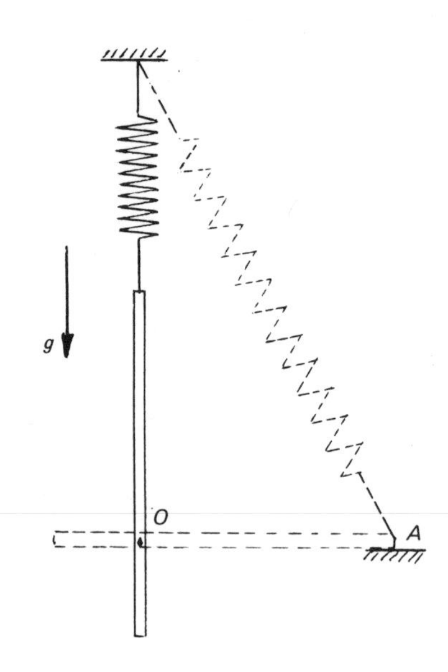

U. Lond. K.C.

Solution 5.17

The application of the clockwise moment to the rod causes:
(a) a reduction of potential energy of the rod.
(b) an increase of strain energy of the spring.
(c) an increase of kinetic energy of the rod.

The corresponding equation of energy is therefore:

$$\text{Initial p.e. (rod)} + \text{Initial s.e. (spring)} + \text{Work done} = \text{Final k.e. (rod)} + \text{Final s.e. (spring)}. \quad (1)$$

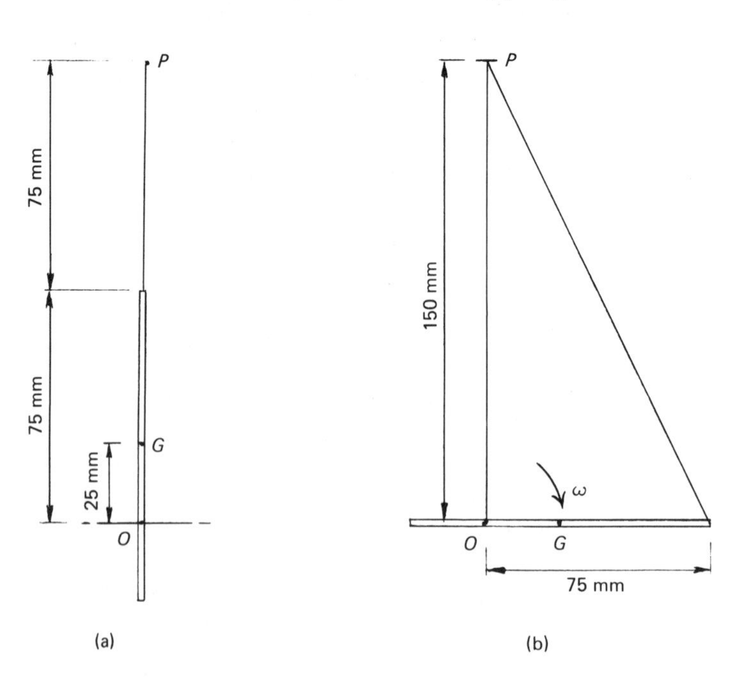

Diagrams (a) and (b) illustrate the two conditions. Taking the terms in order: We will reckon p.e. relative to a horizontal datum through O.

p.e. of rod in (a):

$$mgh = 0.02g(25 \times 10^{-3}) = 0.004\,905 \text{ J}.$$

s.e. of spring in (a):

[d] $$\tfrac{1}{2}kx^2 = \tfrac{1}{2}k(25 \times 10^{-3})^2 = (0.000\,312\,5k) \text{ J}.$$

Work done by moment:

[a] $$M\theta = (5 \times 10^{-3})\,\frac{\pi}{2} = 0.007\,854 \text{ J}.$$

k.e. of rod in (b):

We first require the moment of inertia of the rod about an axis through O, I_o.

Using the parallel-axis theorem,

$$\begin{aligned} I_o &= I_g + mh^2 \\ &= \tfrac{1}{12}mL^2 + mh^2 \\ &= \tfrac{1}{12} \times 0.02(0.1)^2 + 0.02(25 \times 10^{-3})^2 \\ &= (2.917 \times 10^{-5}) \text{ kg m}^2. \end{aligned}$$

$$\omega = \frac{v}{R} = \frac{0.5}{75 \times 10^{-3}}.$$

[b] $$\begin{aligned} \text{k.e.} &= \tfrac{1}{2}I\omega^2 \\ &= \tfrac{1}{2}(2.917 \times 10^{-5})\left(\frac{0.5}{75 \times 10^{-3}}\right)^2 \\ &= 0.000\,648\,2 \text{ J}. \end{aligned}$$

s.e. of spring in (b):
The diagrams show that the extended length of the spring is

$$\sqrt{(75 + 75)^2 + (75)^2} = 75\sqrt{5} \text{ mm}.$$

$$\text{Initial length} = 50 \text{ mm}.$$

$$\text{Extension} = 75\sqrt{5} - 50 = 117.71 \text{ mm}.$$

[d] s.e. of spring in (b): $\tfrac{1}{2}kx^2 = \tfrac{1}{2}k(0.117\,71)^2 = 0.006\,927k$ J.

Substituting these terms in equation 1,

$$0.004\,905 + 0.000\,312\,5k + 0.007\,854 = 0.000\,648\,2 + 0.006\,927k.$$

$$\therefore\ 0.012\,11 = 0.006\,614\,5k.$$

$$\therefore\ \underline{k = 1.8308 \text{ N m}^{-1}}.$$

The wording of the final part of the question should be carefully noted. We are not required to *prove* that the total rotation is 98° but merely to *show* it. We shall therefore assume this value in our second energy equation. This equation is:

Initial p.e. + Initial s.e. + Work done = Final p.e. + Final s.e.

(Note that there will now be no final k.e.)

Initial p.e. = 0.004 905 J as before.

Initial s.e. = $0.000\,312\,5k$ J (as before)

$$= 0.000\,312\,5 \times 1.8308$$

$$= 0.000\,572\,1 \text{ J.}$$

Work done by moment:

$$M\theta = (5 \times 10^{-3})\left(98° \times \frac{\pi}{180}\right)$$

$$= 0.008\,552 \text{ J.}$$

Final p.e. will be negative, as we are still reckoning to a horizontal datum through O. The vertical depth of G below O is given by

$$h = 25 \sin 8° = 3.479 \text{ mm.}$$

$$\text{Final p.e.} = -mgh$$

$$= -0.02g \times 3.479 \times 10^{-3}$$

$$= -0.000\,682\,6 \text{ J.}$$

Final s.e.: we require the final stretched length of the spring. The diagram illustrates.

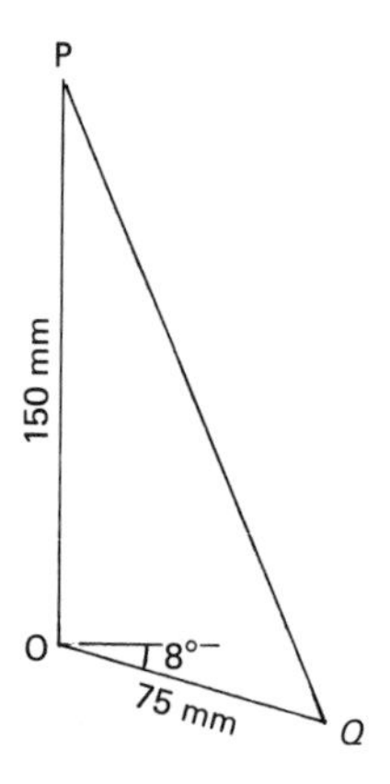

Using the cosine rule,

$$PQ = (150^2 + 75^2 - 2 \times 150 \times 75 \cos 98°)^{1/2}$$

$$= 176.79 \text{ mm.}$$

$$\therefore \text{ Final extension} = 176.79 - 50 = 126.79 \text{ mm.}$$

[d] $$\text{Final s.e.} = \tfrac{1}{2}kx^2$$

$$= \tfrac{1}{2} \times 1.8308\,(0.126\,79)^2$$

$$= 0.014\,72 \text{ J.}$$

Substituting terms in the left-hand side of the energy equation:

$$\text{Initial energy} = 0.004\,905 + 0.000\,672\,1 + 0.008\,552$$

$$= 0.014\,03 \text{ J.}$$

And, substituting in the right-hand side,

$$\text{Final energy} = -0.000\,682\,6 + 0.014\,72$$

$$= 0.014\,04 \text{ J,}$$

which is a satisfactory close agreement.

Example 5.18

The diagram shows a crate resting on a slowly elevating ramp. If the critical angle of the ramp is 20° such that the crate just begins to slip at this elevation, find the mass of the crate which will cause a 50 mm compression of the spring. Assume the coefficient of kinetic friction is 50 per cent of the static coefficient and solve by an energy method.

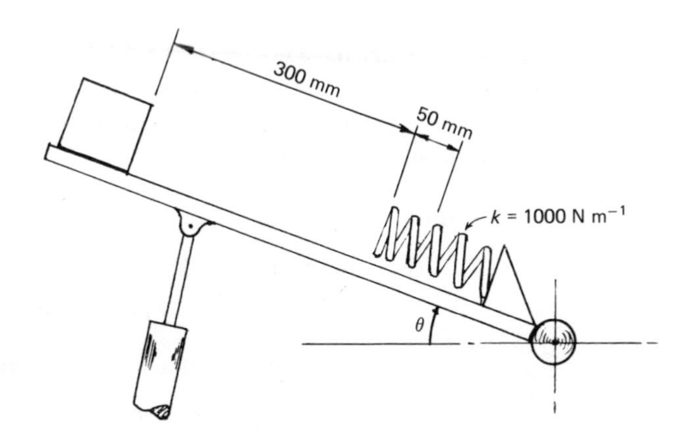

KP

Solution 5.18

The sliding mass will lose potential energy; work will be done against friction, and the spring will gain strain energy.

The energy equation is:

$$E_p - \text{Work done} = E_s.$$

Potential energy:

$$E_p = mgh$$

$$= mg(350 \times 10^{-3}) \sin 20°$$

$$= (1.174m) \text{ J}.$$

Work done:

$$\text{Work} = \text{Force} \times \text{Displacement}.$$

The mass just begins to slide when the plane is inclined at 20°. The free-body diagram is as shown.

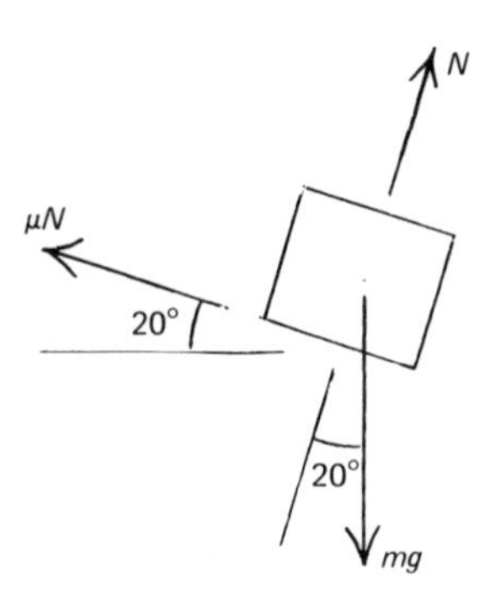

At the critical point, the mass is about to slide; the force down the plane is exactly equal to the force up the plane.

$$mg \sin 20° = \mu_s N.$$

Across the plane:

$$mg \cos 20° = N.$$

$$mg \sin 20° = \mu_s mg \cos 20°.$$

$$\therefore \ \mu_s = \tan 20°.$$

(You will probably have encountered this result earlier, in work on friction.)

μ_s is the static friction coefficient. It is found from experiment that the friction coefficient is less for a sliding body than for a stationary one. We shall call this sliding friction coefficient μ_k.

We are told that μ_k is $\frac{1}{2}\mu_s$. Hence:

$$\text{Friction force} = \mu_k N$$

$$= \tfrac{1}{2}\mu_s \times mg \cos 20°.$$

$$\text{Work done} = \text{Force} \times \text{Displacement}$$

$$= \tfrac{1}{2}\mu_s \times mg \cos 20° \times 350 \times 10^{-3}$$

$$= \tfrac{1}{2} \tan 20° \times mg \cos 20° \times 350 \times 10^{-3}$$

$$= (0.5872m) \text{ J}.$$

Strain energy:

[d]

$$E_s = \tfrac{1}{2}kx^2$$

$$= \tfrac{1}{2} \times 1000 (50 \times 10^{-3})^2$$

$$= 1.25 \text{ J}.$$

Substituting the terms in the energy equation,

$$1.174m - 0.5872m = 1.25.$$

$$\therefore \ m = \frac{1.25}{0.5868}$$

$$= \underline{2.13 \text{ kg}}.$$

Example 5.19

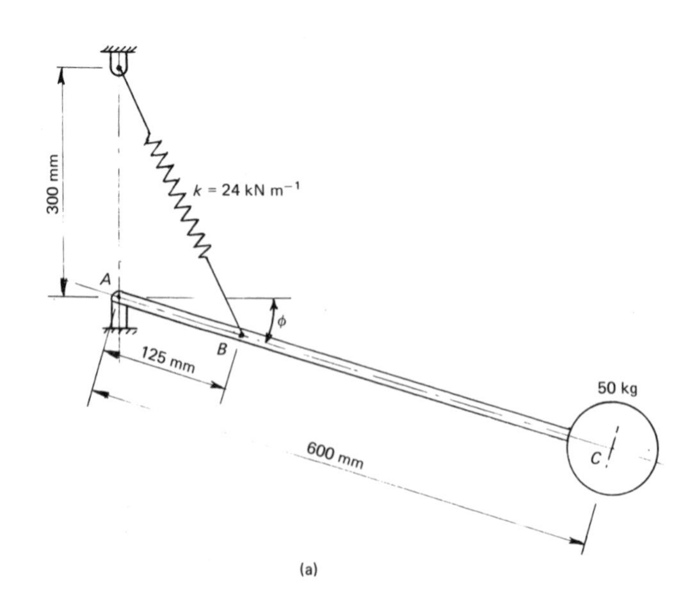

Diagram (a) shows a spring-mass system which is originally at rest when $\phi = 0$. If the speed of the block is 2.5 m s^{-1} when $\phi = 90°$ determine the required value of the initial tension in the spring assuming the bar is weightless and the block has a mass of 50 kg.

KP

Solution 5.19

The energy equation, in words, is:

Initial strain energy of spring + Initial potential energy of mass
= Final strain energy of spring + Final kinetic energy of mass.

$$E_{s1} + E_p = E_{s2} + E_k.$$

$$E_{s2} - E_{s1} = E_p - E_k. \quad (1)$$

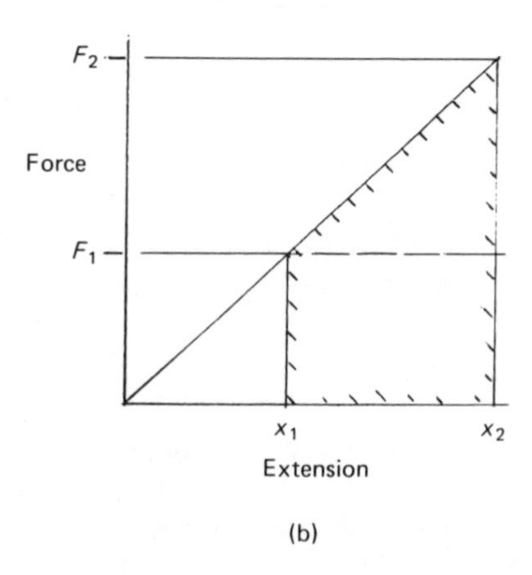

(b)

Diagram (b) shows a graph of force F against spring extension, x. F_1 is the initial force in the spring (when the bar is horizontal). F_2 is the final force when the bar is vertical and ϕ = 90°. The increase of strain energy is the area of the shaded portion. This comprises the sum of the rectangle and the triangle.

$$E_{s2} - E_{s1} = F_1(x_2 - x_1) + \tfrac{1}{2}(x_2 - x_1)(F_2 - F_1).$$

$$F_2 = F_1 + k(x_2 - x_1).$$

$$\therefore\ F_2 - F_1 = k(x_2 - x_1).$$

Substituting,

$$E_{s2} - E_{s1} = F_1(x_2 - x_1) + \tfrac{1}{2}(x_2 - x_1)\,k(x_2 - x_1)$$
$$= F_1(x_2 - x_1) + \tfrac{1}{2}k(x_2 - x_1)^2.$$

Spring length when AB is vertical = (300 + 125)

= 425 mm.

Spring length when AB is horizontal = $\sqrt{300^2 + 125^2}$

= 325 mm.

$$\therefore\ x_2 - x_1 = 425 - 325 = 100 \text{ mm} = 0.1 \text{ m}.$$

$$\therefore\ E_{s2} - E_{s1} = F_1 \times 0.1 + \tfrac{1}{2} \times 24\,000 \times (0.1)^2$$

$$= 0.1F_1 + 120.$$

Initial p.e.: $$E_p = mgh$$
$$= 50g(600 \times 10^{-3})$$
$$= 294.3 \text{ J}.$$

Final k.e.: $$E_k = \tfrac{1}{2}mv^2$$
$$= \tfrac{1}{2} \times 50(2.5)^2$$
$$= 156.25 \text{ J}.$$

Substituting in the energy equation (equation 1),

$$0.1F_1 + 120 = 294.3 - 156.25.$$
$$\therefore\ F_1 = \frac{18.05}{0.1} = \underline{180.5 \text{ N}.}$$

The potential mistake in this problem is to misuse the formula for strain energy of a spring:

[d] $$E_s = \tfrac{1}{2}kx^2,$$

substituting the length *increase* $(x_2 - x_1)$ for the true extension, x, to calculate the increase of strain energy. A look back through the working will show you that this gives an incorrect result; it is seen that it would give only the area of the triangle on the graph (diagram (b)).

Example 5.20

The diagram shows a block A, of mass 10 kg, which can move on a smooth plane inclined at 30° to the horizontal. A second block, B, of mass 50 kg, is connected to A by a light inextensible cord which passes through a frictionless guide at P. A tension spring of stiffness 200 N m^{-1} is also attached to A. The block A is held in the position shown with the spring stretched by 100 mm and then released.

Calculate the velocity of A when it reaches the position C, when the velocity of B is zero.

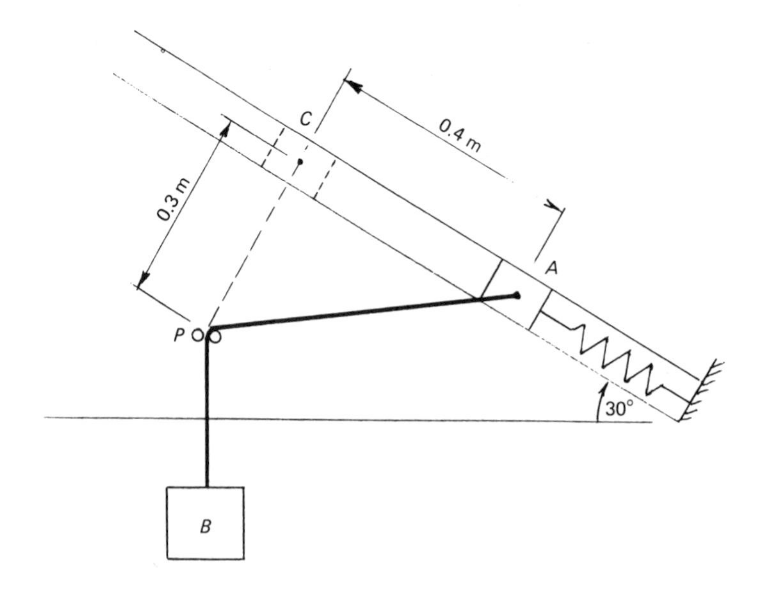

U. Lond. U.C.

Solution 5.20

The energy equation will be:

Initial (p.e. of B + s.e. of spring) = Final (p.e. of A + k.e. of A + s.e. of spring).

$$E_{pb} + E_{s1} = E_{pa} + E_{ka} + E_{s2}.$$

From the diagram, it can be seen from the triangle APC, which is a '3–4–5' triangle, that the diagonal AC is 0.5 m. (It is clear that CP is perpendicular to AC; this is confirmed by the information that when block A reaches C, block B will be stationary. Thus, C must be the nearest point on the plane to P, so that PC must be perpendicular to the plane.) So when A reaches point C, only 0.3 m of cord remains above P; thus, B must be lowered by 0.2 m.

Hence, loss of p.e. of B is

$$E_{pb} = mgh$$
$$= 50g \times 0.2$$
$$= 98.1 \text{ J.}$$

[d]

$$E_{s1} = \tfrac{1}{2}kx_1^2$$
$$= \tfrac{1}{2} \times 200(0.1)^2$$
$$= 1 \text{ J.}$$

Block A gains height by an amount (0.4 sin 30°).

$$\therefore\ E_{pa} = mgh$$
$$= 10g \times 0.4 \sin 30°$$
$$= 19.62 \text{ J.}$$

The final length of the spring is 100 mm + 0.4 m = 0.5 m.
Hence, final strain energy E_{s2} is

$$E_{s2} = \tfrac{1}{2}kx^2$$
$$= \tfrac{1}{2} \times 200 \times (0.5)^2$$
$$= 25 \text{ J.}$$

Kinetic energy of A:

$$E_{ka} = \tfrac{1}{2}mv^2$$
$$= \tfrac{1}{2} \times 10v^2 = 5v^2.$$

Substituting in equation 1,

$$98.1 + 1 = 19.625 + 5v^2 + 25.$$

$$\therefore\ v^2 = \frac{54.55}{5} = 10.891.$$

$$\therefore\ v = \underline{3.300 \text{ m s}^{-1}.}$$

Example 5.21

The elements of a simple hoist are shown in the diagram (p. 224). A truck of total mass m_1 = 1100 kg has four wheels, each of radius 0.3 m and moment of inertia 20 kg m^2. A rope attached to the truck passes round a winding drum such that there is no slip of the rope on the drum. The drum radius is 1.9 m and its moment of inertia is 280 kg m^2. The other end of the rope hangs vertically, and a counterweight mass m_2 = 260 kg hangs from it. The truck

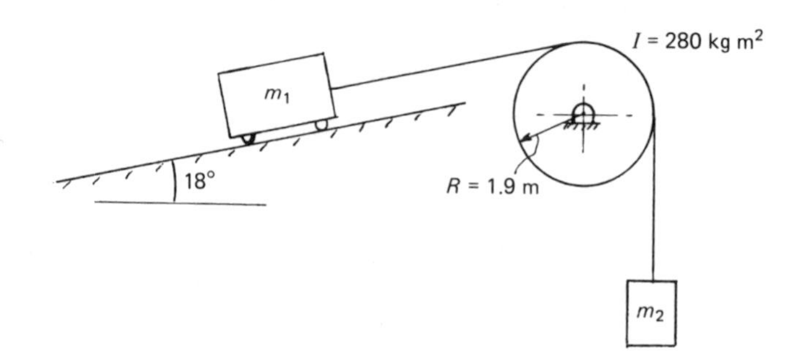

is released from rest at the top of a slope of 18°. When in motion, the truck is subjected to a frictional resistance F which may be assumed constant. Calculate the magnitude of this force, given that the truck attains a velocity of 0.5 m s^{-1} when the truck has travelled 120 m down the slope.

An energy equation is suggested.

PCL

An alternative answer to this question is given in Chapter 4, using force-mass-acceleration equations. See Example 4.18.

Solution 5.21

The energy equation may be stated thus:

$$\text{Initial p.e. of } m_1 - \text{Work done (friction)}$$
$$= \text{Final p.e. of } m_2 + \text{Final k.e. } (M_1 + m_2 + I). \quad (1)$$

Since energy is involved, we require to take into account the effect of the rolling wheels of the truck. See the examples in Chapter 4.

Effective mass of truck + wheels, m_e, is

$$m_e = m + \Sigma\left(\frac{I}{R^2}\right)$$
$$= 1100 + \frac{4 \times 20}{(0.3)^2}$$
$$= 1988.9 \text{ kg.}$$

p.e. of m_1: $E_p = mgh$

$= 1100g \times 120 \sin 18°$

$= 400\,152.3$ J.

Work done: $E = Fx$

$= (120F)$ J.

p.e. of m_2: $E_p = mgh$

$= 260g \times 120$

$= 306\,072$ J.

Kinetic energy: $E_{k1} = \frac{1}{2}m_e v^2$

$= \frac{1}{2} \times 1988.9\,(0.5)^2$

$= 248.6$ J.

$$E_{k2} = \tfrac{1}{2}I\omega^2$$

$$= \tfrac{1}{2} \times 280 \times \left(\frac{0.5}{1.9}\right)^2$$

$$= 9.7 \text{ J.}$$

$$E_{k3} = \tfrac{1}{2}m_2 v^2$$

$$= \tfrac{1}{2} \times 260(0.5)^2$$

$$= 32.5 \text{ J.}$$

Substituting in equation 1,

$$400\,152.3 - 120F = 306\,072 + 248.6 + 9.7 + 32.5$$

$$= 306\,362.8.$$

$$\therefore\ F = \frac{400\,152.3 - 306\,362.8}{120}$$

$$= \underline{781.6 \text{ N.}}$$

Example 5.22

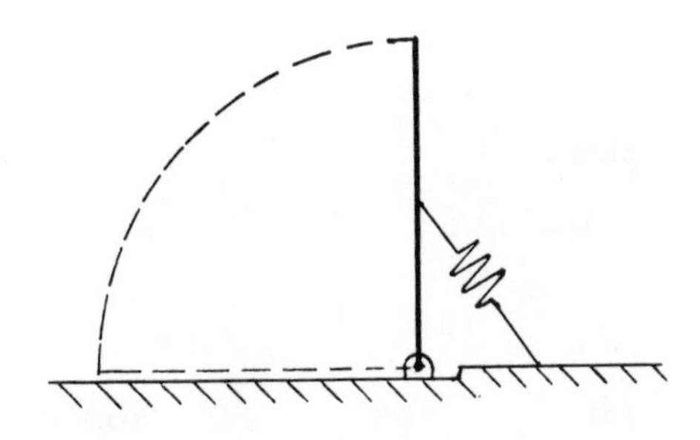

A slender uniform rod of mass m and length L is pivoted about its lower end as shown in the diagram. A light spring of stiffness K and unstretched length $\tfrac{1}{2}L$ has one end attached to the mid-point of the rod and the other end attached to a fixed point a distance $\tfrac{3}{8}L$ from the pivot measured in the horizontal plane.

Derive an expression for the angular velocity of the rod as it passes the vertical position in an anticlockwise direction if it is to reach the horizontal position with zero velocity.

Assume a constant frictional moment of $\dfrac{5KL^2}{24\pi}$ to be acting at the pivot.

U. Surrey

Solution 5.22

The energy equation is

Initial energy (p.e. + k.e. + s.e.) – Work done = Final s.e.

Taking the terms in order,

[c] Initial p.e. $= mg\tfrac{1}{2}L.$

[b] Initial k.e. $= \tfrac{1}{2}I\omega^2$

$$= \tfrac{1}{2}(\tfrac{1}{3}mL^2)\omega^2$$

$$= \frac{mL^2\omega^2}{6}.$$

($I = \frac{1}{3}mL^2$ for a uniform rod about an axis through one end. This expression and that for the moment of inertia of a thin rod about an axis through the centre ($\frac{1}{12}mL^2$) are met so frequently that one would normally be expected to know them.)

[d] $$\text{Initial s.e.} = \tfrac{1}{2}Kx_1^2,$$

where x_1 is the initial stretch of the spring.

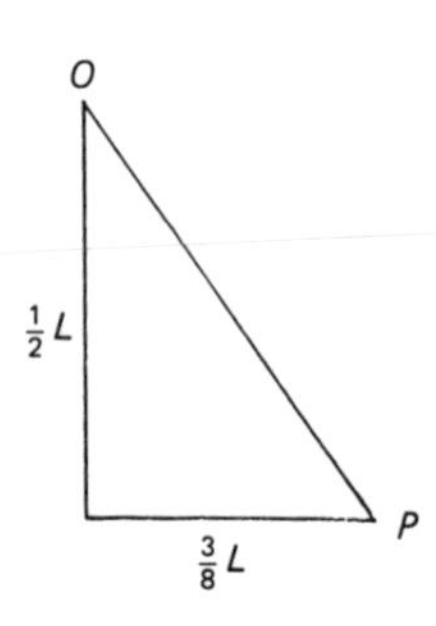

It is seen from the diagram that the stretched length of the spring, OP, is

$$OP = \sqrt{(\tfrac{3}{8}L)^2 + (\tfrac{1}{2}L)^2}$$

$$= \tfrac{5}{8}L,$$

giving $$x_1 = \tfrac{1}{8}L.$$

Initial s.e. $$= \tfrac{1}{2}K(\tfrac{1}{8}L)^2$$

$$= \frac{KL^2}{128}.$$

[a] $$\text{Work done} = M\theta$$

$$= \frac{5KL^2}{24\pi} \times \frac{\pi}{2}$$

$$= \frac{5}{48}KL^2.$$

[d] $$\text{Final s.e.} = \tfrac{1}{2}Kx_2^2.$$

A diagram should not be necessary to show that x_2 is $\frac{3}{8}L$.

$$\therefore \text{Final s.e.} = \tfrac{1}{2}K(\tfrac{3}{8}L)^2$$

$$= \frac{9}{128}KL^2.$$

Substituting the terms in the energy equation,

$$mg\tfrac{1}{2}L + \tfrac{1}{6}mL^2\omega^2 + \frac{KL^2}{128} - \frac{5}{48}KL^2 = \frac{9}{128}KL^2.$$

$$mg\tfrac{1}{2}L + \tfrac{1}{6}mL^2\omega^2 = KL^2\left(\frac{9}{128} + \frac{5}{48} - \frac{1}{128}\right)$$

$$= \tfrac{1}{6}KL^2.$$

$$\therefore \tfrac{1}{6}mL^2\omega^2 = \tfrac{1}{6}KL^2 - \tfrac{1}{2}mgL.$$

$$\therefore \omega^2 = \frac{K}{m} - \frac{3g}{L}.$$

Example 5.23

A door AB, as shown in the diagram, is initially at rest in the position shown, mounted on a horizontal hinge at O by a light frame. The horizontal spring CD is stretched 1.0 m at this position. A moment M = 400 N m is applied. Find the angular velocity of the door as it reaches its highest position with AB horizontal at $A'B'$.

Data: Mass of door 100 kg.
Stiffness of spring 1500 N m^{-1}.
The radius of gyration of the door about its mass centre is $\frac{\text{length}}{\sqrt{12}}$.

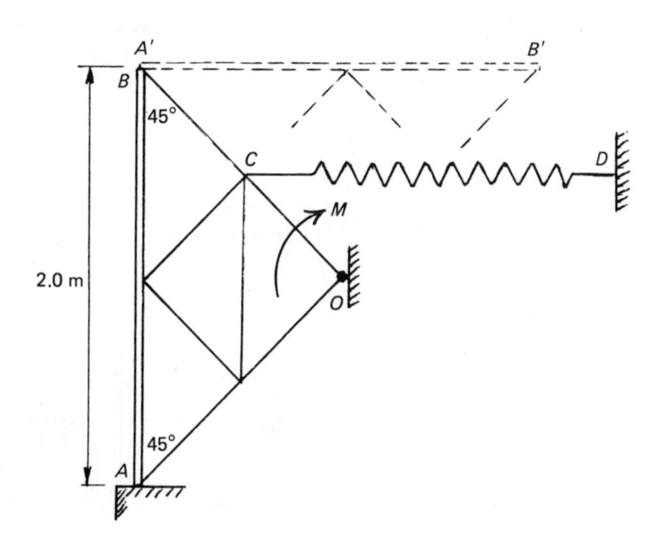

Thames Poly.

Solution 5.23

The diagram shows that as the door moves upwards to the highest position, the end C of the spring moves to the right a distance equal to the length of the originally vertical strut through C, i.e. a distance of 1 m. Thus, the spring just becomes completely unloaded as the door reaches the top position. The energy equation is:

Initial strain energy of spring + Work done by moment
= Final kinetic energy of door + Final potential energy of door.

The kinetic energy of the door may be considered as purely rotational if we calculate the moment of inertia of the door with respect to the axis of rotation, i.e. the axis through O. (The radius of gyration given with the data indicates that the door may be considered to be a uniform thin member.) The diagram again shows that the distance between the axis through O and the centre of the door is 1 m.

$$\begin{aligned} I_o &= I_g + mh^2 \\ &= \tfrac{1}{12} \times 100 \times (2)^2 + 100 \times (1)^2 \\ &= 133.3 \text{ kg m}^2. \end{aligned}$$

Writing the energy equation algebraically:

$$\tfrac{1}{2}kx^2 + m\theta = \tfrac{1}{2}I_o\omega^2 + mgh.$$

Substituting values,

$$\tfrac{1}{2} \times 1500 \times 1^2 + 400 \times \frac{\pi}{2} = \tfrac{1}{2} \times 133.3\omega^2 + 100g \times 1.$$

$$\therefore\ \omega^2 = \frac{750 + 628.3 - 981}{\tfrac{1}{2} \times 133.3}$$

$$\therefore\ \omega = \underline{2.442 \text{ rad s}^{-1}}.$$

Example 5.24

Make a formal statement of Newton's laws of motion.

In an experiment to determine ring friction, the piston in the diagram is released from rest when pressure $p = p_0$ and distance $x = x_0$. Atmospheric pressure on the other side of the piston is p_A ($< p_0$), and the cross-sectional area of the bore is A. If the total frictional force from the rings is F, a constant, and if the piston comes to rest again further along the bore with $x = x_1$, show that

$$F = A\left[p_0 \left(\frac{x_0}{x_1 - x_0}\right) \ln\left(\frac{x_1}{x_0}\right) - p_A\right].$$

You may assume that the gas over the piston satisfies pV = constant where V = volume.

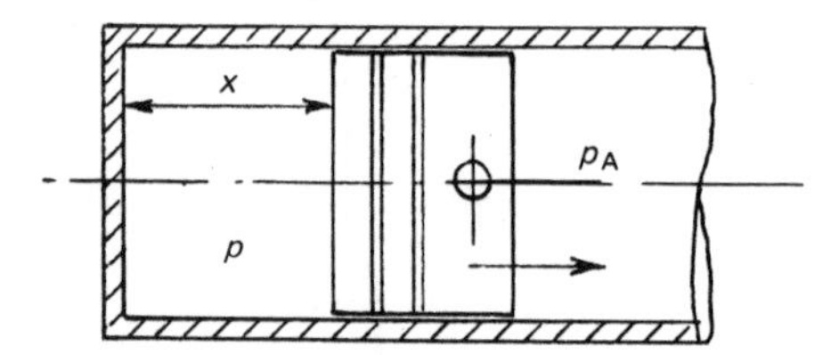

U. Manchester

Solution 5.24

For a displacement x we argue that the work done by the expanding gas is the same as the work done against friction and against the atmospheric pressure force. Since the gas force varies with x, we can write:

$$\Sigma\,(pA\ \delta x) = F(x_1 - x_0) + p_A A\,(x_1 - x_0).$$

From the gas equation,

$$pV = p_0 V_0.$$

$$\therefore\ p = p_0 \left(\frac{V_0}{V}\right) = p_0 \left(\frac{Ax_0}{Ax}\right).$$

$$\therefore\ Ap_0 \int \left(\frac{x_0}{x}\right) dx = (x_1 - x_0)(F + p_A A).$$

$$\therefore\ Ap_0 x_0 \Big[\ln(x)\Big]_{x_0}^{x_1} = (x_1 - x_0)(F + p_A A).$$

$$\therefore\ Ap_0 x_0 \ln\left(\frac{x_1}{x_0}\right) = (x_1 - x_0)(F + p_A A).$$

$$\therefore\ Ap_0 \left(\frac{x_0}{x_1 - x_0}\right) \ln\left(\frac{x_1}{x_0}\right) = F + p_A A.$$

$$\therefore\ \underline{F = A\left[p_0 \left(\frac{x_0}{x_1 - x_0}\right) \ln\left(\frac{x_1}{x_0}\right) - p_A\right]}.$$

A candidate might be forgiven for mistaking this question for one on variable acceleration (see Chapter 6). The key lies in the fact that the mass of the piston is not stated, and hence we cannot write an equation of motion for the piston. The question could actually be solved by writing the equation of motion, assuming a mass m for the piston; this would be found to 'disappear' and the correct answer obtained. But the energy approach is much neater. A further pointer is that information is given, or required, at the beginning and at the end of the process, and not in between, and that the time for the process is not asked for.

5.3 Problems

Problem 5.1

A wagon of mass 7200 kg is projected with an initial velocity of u along a rail which is horizontal for 100 m and then slopes upwards at an angle of $\sin^{-1} 0.01$. The rail exerts a constant frictional resistance to motion of 140 m.

(a) If $u = 3 \text{ m s}^{-1}$ how far up the slope will the wagon travel before coming to rest?
(b) If $u = 3 \text{ m s}^{-1}$ how far from its starting-point will the wagon finally come to rest?
(c) At what value of u should the wagon be propelled if it is to come finally to rest at its starting-point?

Problem 5.2

Firing a shell from a gun causes the barrel to recoil backwards with an initial velocity of 25 m s^{-1}. The recoil is absorbed partly by a spring and partly by a frictional force of 450 N. The mass of the barrel is 290 kg. Calculate the required stiffness of the spring to bring the barrel to rest in a distance of 0.9 m when the gun is fired with an elevation of 60°.

Problem 5.3

In a simple emergency fire-escape, a rope is connected to a drum of diameter 0.2 m. The drum is attached to the low-speed shaft of a gearbox of ratio 12 and a flywheel which has a moment of inertia of 0.41 kg m^2 is attached to the high-speed shaft. A person of weight 800 N hangs from the rope and descends a distance of 12 m to the ground. Calculate the velocity with which he strikes the ground.

Problem 5.4

The maximum safe load allowed at the centre of the horizontal beam of a travelling crane is 80 000 N and this load is sufficient to cause a vertical deflection of the beam at that point of 64 mm. The crane supports a load of mass 2000 kg. Calculate from what maximum height an additional load of 500 kg may be dropped on to the top of this load without exceeding the maximum permissible load on the beam. Neglect any deformation of the crane cables.

Problem 5.5

A loaded mine cage has a total mass of 2300 kg. It hangs from a cable which stretches 0.3 mm per kN of load for a length of 1 m. The load is descending the shaft at a speed of 3 m s^{-1} when the winding gear is suddenly shut off, causing the top end of the cable suddenly to stop. Calculate the resulting maximum force in the cable if the length of cable supporting the load is (a) 10 m; (b) 100 m. Neglect the mass of the cable itself.
Hint: see Example 5.7.

Problem 5.6

A wheel of radius R = 0.8 m is mounted on one end of a horizontal shaft which has a torsional stiffness of $k_t = 1500$ N m per radian of twist. The shaft is fixed at the other end, and arranged so that it can twist but not bend. A rope hangs from the wheel rim, and carries a plate on which a load can fall. If the load is 50 kg and falls from a height of h = 15 mm on to the plate, calculate the resulting angle of twist of the shaft, and the corresponding torque. The rope may be assumed to be inextensible.

Hints: See Fact Sheet (d). The vertical deflection of the falling load, x, is related to the angle of twist of the shaft, θ, by $x = \theta R$.

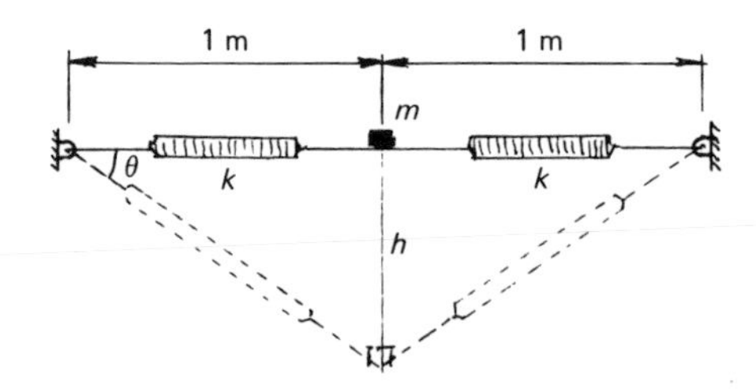

Problem 5.7

Two identical springs, each having a stiffness of 500 N m and unstrained length 1 m, are joined and attached to two fixed points at the same level as shown in the diagram. A body of mass m = 1 kg is secured to the centre point as shown, and is released from rest. Show that at the instant the springs make an angle θ with the horizontal, the velocity v of the mass will be given by the expression

$$v^2 = 19.62 \tan \theta - 1000 (\sec \theta - 1)^2.$$

By calculating the potential and strain energy changes for a value of θ of 23.81° show that this corresponds to the maximum deformation of the system.

Problem 5.8

In demolishing a building, a steel-wire cable is attached to a part of the building, the other end being secured to a tractor which drives away. The breaking load of the cable is 60 kN and a test-length of 1 m extends 0.7 mm per kilonewton of tensile load. The tractor has a mass of 6200 kg and is driven at 1.3 m s^{-1}. Assuming the total kinetic energy of the tractor to be absorbed by the cable, calculate the minimum length of cable if the maximum load is not to exceed half the breaking load.

Problem 5.9

A car has a mass of 1100 kg. It can travel along a level road at a maximum speed of 45.4 m s^{-1} and up a slope of $\sin^{-1}$ 0.1 at a maximum speed of 35.0 m s^{-1}. Assuming the engine works at a constant power, and that the only resistance to motion is due to the air, which may be assumed to be $R = kv^2$, calculate the maximum speed it can travel down a slope of $\sin^{-1}$ 0.1.

Hints: See Example 5.12. From the two items of information given, obtain two equations in W (power) and k. Solve for W and k. (Ans. 69 715 W and 0.745.) Use these to obtain a cubic equation for motion down the slope ($v^3 - 1448v = 93\,577$). 'Try' v = 55 and 56, and interpolate between.

Problem 5.10

Oil is contained in a rectangular tank 12 m by 24 m to a depth of 30 m. The density of the oil is 840 kg m^{-3}. The tank is to be emptied by a pump which delivers the oil as a jet through a delivery pipe of diameter 0.25 m at a height of 8 m above the surface of the full tank. If the tank is required to be emptied in 4 hours, estimate the maximum power required for the pump, assuming that it is 0.78 efficient and that it delivers at a constant rate.

5.4 Answers to Problems

5.1 (a) 21.74 m; (b) 12.05 m; (c) 3.115 m s^{-1}.
5.2 228 kN m^{-1}.
5.3 1.791 m s^{-1}.
5.4 249 mm.
5.5 (a) 105.6 kN; (b) 48.83 kN.
5.6 31.02°; 812 N m.
5.7 $E_p = E_s = 4.33$ J No E_k.
5.8 16.63 m.
5.9 55.89 m s^{-1}.
5.10 289.1 kW.

6 Non-linear Acceleration

Analysis of motion of bodies subjected to variable force. Force as a function of time. Force as a function of velocity. Force as a function of displacement. Approximate methods of solution from graphical and tabular data.

6.1 The Fact Sheet

(a) Simple Equations of Motion

When the resultant force acting on a mass is not constant, the simple kinematic equations of motion ($v = u + at$, etc.) *must not be used* as they are derived on the assumption of a constant acceleration.

(b) Differential Equations

In such cases, the kinetic equation of motion ($\Sigma(F) = ma$ for linear translation, $\Sigma(M) = I\alpha$ for rotation) will be a differential equation with time t as the independent variable.

(c) Force as a Function of Time

If the force F is a function of time, t, i.e.

$$\mathrm{f}(t) = ma = m\frac{\mathrm{d}^2 x}{\mathrm{d}t^2},$$

then the equation can be solved by separating the variables and integrating twice.

(d) Force as a Function of Velocity

If the force F is a function of velocity, v, i.e.

$$\mathrm{f}(v) = ma = m\frac{\mathrm{d}v}{\mathrm{d}t}$$

then the equation can be solved by separating the variables and integrating to give t as a function of v.

The equation may be written alternatively

$$\mathrm{f}(v) = ma = m\frac{\mathrm{d}v}{\mathrm{d}x}\frac{\mathrm{d}x}{\mathrm{d}t} = mv\frac{\mathrm{d}v}{\mathrm{d}x}$$

and this may be solved by separating variables and integrating to give x as a function of v.

(e) Force as a Function of Displacement

If the force F is a function of displacement, x, i.e.

$$\mathrm{f}(x) = ma = mv\,\frac{\mathrm{d}v}{\mathrm{d}x}$$

then the equation can be solved by separating variables and integrating to give v as a function of x, i.e.

$$\mathrm{f}(x) = v = \frac{\mathrm{d}x}{\mathrm{d}t}$$

and this can be integrated again to give t as a function of x.

(f) Alternative Method of Solution

The equation of motion may be written as a second-order equation in x and t, and the solution obtained directly as x as a function of t. v may then be obtained by direct differentiation of x.

Important note: In the *general* solution to a second-order equation, there will be *two* constants of integration. The values of these may be found by applying two conditions of motion; these are usually specifications of velocity and/or displacement at the initial state of the motion (i.e. when $t = 0$).

When the equation is integrated between specific limits then the need for the inclusion of constants of integration does not arise.

(g) Revision: Solutions of Differential Equations

Equation: $\dfrac{\mathrm{d}^2x}{\mathrm{d}t^2} = -Kx.$

Solution: $x = A\sin(\sqrt{K}t) + B\cos(\sqrt{K}t).$

Equation: $\dfrac{\mathrm{d}^2x}{\mathrm{d}t^2} = +Kx.$

Solution: $x = A\sinh(\sqrt{K}t) + B\cosh(\sqrt{K}t).$

(Recall $\sinh(y) = \frac{1}{2}(\mathrm{e}^y + e^{-y})$; $\cosh(y) = \frac{1}{2}(\mathrm{e}^y - \mathrm{e}^{-y})$.)

Equation: $\dfrac{\mathrm{d}^2x}{\mathrm{d}t^2} = -K\,\dfrac{\mathrm{d}x}{\mathrm{d}t}.$

Solution: $x = A\mathrm{e}^{-Kt} + B.$

6.2 Worked Examples

Example 6.1

In an experiment, a miniature rocket comprises a mass of 20 kg which is projected by a force F which decreases with time according to the expression

$$F = (315 - 15t - 0.4t^2) \text{ newtons}$$

where t is the time in seconds. Calculate how far the mass will travel, starting from rest, in 15 seconds, and its velocity after that time, if it is propelled along a straight frictionless track, (a) if the track is horizontal; (b) if it has an upward slope of 20°. In case (b) determine the maximum velocity of the mass.

Solution 6.1

Part (a)

A free-body diagram is not necessary; the propelling force F is the only relevant force, as weight and track reaction force both act at right angles to the line of motion.

[d]

$$F = ma = m\frac{dv}{dt}.$$

$$\therefore \int m\,dv = \int F\,dt = \int (315 - 15t - 0.4t^2)\,dt.$$

$$\therefore mv = 315t - \tfrac{1}{2} \times 15t^2 - \tfrac{1}{3} \times 0.4t^3 + A.$$

The 'initial condition' $v = 0$ when $t = 0$ shows that the constant $A = 0$.

$$\therefore v = \frac{1}{m}(315t - \tfrac{1}{2} \times 15t^2 - \tfrac{1}{3} \times 0.4t^3). \quad (1)$$

Writing $v = \dfrac{dx}{dt}$,

$$\frac{dx}{dt} = \frac{1}{m}(315t - \tfrac{1}{2} \times 15t^2 - \tfrac{1}{3} \times 0.4t^3).$$

$$\therefore \int dx = \frac{1}{m}\int (315t - \tfrac{1}{2} \times 15t^2 - \tfrac{1}{3} \times 0.4t^3)\,dt.$$

$$\therefore x = \frac{1}{m}(\tfrac{1}{2} \times 315t^2 - \tfrac{1}{6} \times 15t^3 - \tfrac{1}{12} \times 0.4t^4) \quad (2)$$

(The second constant of integration, like the first, will be zero, since displacement $x = 0$ when $t = 0$.)

Substituting in equation 1,

$$v = \tfrac{1}{20}(315 \times 15 - \tfrac{1}{2} \times 15^3 - \tfrac{1}{3} \times 0.4 \times 15^3)$$

$$= \underline{129.4 \text{ m s}^{-1}}.$$

and in equation 2,

$$x = \tfrac{1}{20}(\tfrac{1}{2} \times 315 \times 15^2 - \tfrac{1}{6} \times 15^4 - \tfrac{1}{12} \times 0.4 \times 15^4)$$

$$= \underline{1265.6 \text{ m}}.$$

Part (b)

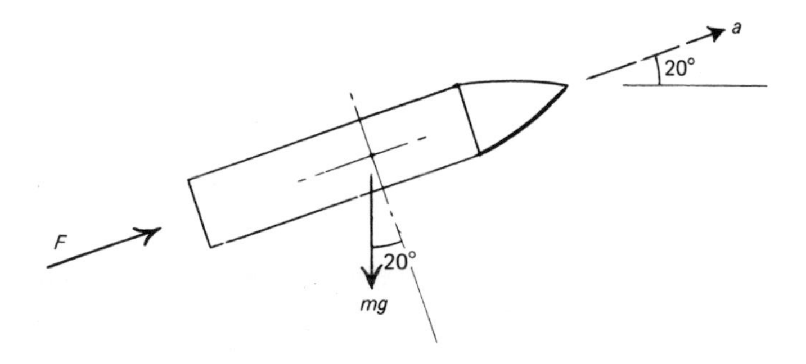

The free-body diagram now shows that a component of the weight acts down the track. The equation of motion is

[c] $$F - mg \sin 20° = ma = m\frac{\mathrm{d}v}{\mathrm{d}t}.$$

$$\therefore \int m\,\mathrm{d}v = \int F\,\mathrm{d}t$$

$$= \int (315 - 15t - 0.4t^2 - mg \sin 20°)\,\mathrm{d}t$$

(substituting values) $$= \int (315 - 15t - 0.4t^2 - 67.1)\,\mathrm{d}t$$

$$= \int (247.9 - 15t - 0.4t^2)\,\mathrm{d}t.$$

$$\therefore v = \frac{1}{m}(247.9t - \tfrac{1}{2} \times 15t^2 - \tfrac{1}{3} \times 0.4t^3). \qquad (3)$$

$$\frac{\mathrm{d}x}{\mathrm{d}t} = \frac{1}{m}(247.9t - \tfrac{1}{2} \times 15t^2 - \tfrac{1}{8} \times 0.4t^3).$$

$$\therefore \int \mathrm{d}x = \frac{1}{m}\int (247.9t - \tfrac{1}{2} \times 15t^2 - \tfrac{1}{3} \times 0.4t^3)\,\mathrm{d}t.$$

$$\therefore x = \frac{1}{m}(\tfrac{1}{2} \times 247.9t^2 - \tfrac{1}{6} \times 15t^3 - \tfrac{1}{12} \times 0.4t^4). \qquad (4)$$

(The integration constants again are zero in the two integrations.) Substituting in equation 3,

$$v = \tfrac{1}{20}(247.9 \times 15 - \tfrac{1}{2} \times 15^3 - \tfrac{1}{3} \times 0.4 \times 15^3)$$
$$= \underline{79.05 \text{ m s}^{-1}},$$

and in equation 4,

$$x = \tfrac{1}{20}(\tfrac{1}{2} \times 247.9 \times 15^2 - \tfrac{1}{6} \times 15^4 - \tfrac{1}{12} \times 0.4 \times 15^4)$$
$$= \underline{888.2 \text{ m.}}$$

Maximum velocity will occur when the differential of velocity, i.e. the acceleration, is zero. Since acceleration is proportional to the net force, we calculate the value of t for zero force. Therefore

$$247.9 - 15t - 0.4t^2 = 0.$$

Rearranging, and dividing by the coefficient of t^2,

$$t^2 + 37.5t - 619.8 = 0.$$

Solving the quadratic,

$$t = \frac{-37.5 \pm \sqrt{37.5^2 + 4 \times 619.8}}{2}$$

$$= -18.75 \pm 31.17$$

$$= 12.42 \text{ s},$$

and this value must be substituted into equation 3:

$$v_{max} = \tfrac{1}{20}\,[247.9 \times 12.42 - \tfrac{1}{2} \times 15 \times (12.42)^2 - \tfrac{1}{3} \times 0.4\,(12.42)^3\,]$$

$$= \underline{83.32 \text{ m s}^{-1}}.$$

Example 6.2

A body is projected horizontally through water. The resistance R to its motion is assumed to have the form

$$R = kv^2$$

where R is in newtons, and v is the velocity at any instant in metres per second. k is a constant. The initial velocity of the body is 400 m s^{-1} and it is reduced to one-tenth of this value in a distance of 160 m. Calculate the time taken to cover this distance. Find also the additional distance and time for the velocity to fall to one-hundredth of its initial value. Consider the horizontal motion only, and neglect any vertical motion due to the weight of the body.

Solution 6.2

As the resistance R is the only force, the equation of motion is

$$-kv^2 = ma = mv\,\frac{dv}{dx}.$$

$$\therefore \int dx = \int \frac{mv\,dv}{-kv^2}.$$

$$\therefore x = -\frac{m}{k}\Big[\ln(v)\Big]_{v_1}^{v_2}$$

$$= -\frac{m}{k}\ln\left(\frac{v_2}{v_1}\right).$$

$$\therefore x = +\frac{m}{k}\ln\left(\frac{v_1}{v_2}\right) \qquad (1)$$

(Note that integration constants are not required when integrating between limits.)

Rewriting the equation of motion,

[d]

$$-kv^2 = m\,\frac{dv}{dt}.$$

$$\therefore \int dt = \int \frac{m\,dv}{-kv^2}.$$

$$= +\frac{m}{k}\left[\frac{1}{v}\right]_{v_1}^{v_2}.$$

$$t = +\frac{m}{k}\left(\frac{1}{v_2} - \frac{1}{v_1}\right). \qquad (2)$$

Substitute the given values in equation 1:

$$160 = \frac{m}{k} \ln(10).$$

$$\therefore \quad \frac{m}{k} = \frac{160}{\ln(10)}.$$

Substitute this value of $\frac{m}{k}$ in equation 2:

$$t = \frac{160}{\ln(10)} \left(\frac{1}{40} - \frac{1}{400}\right)$$

$$= \underline{1.563 \text{ s.}}$$

For additional distance required, substitute in equation 1:

$$x = \left(\frac{160}{\ln(10)}\right) \ln(100) = 160 \times 2.$$

$$\therefore \text{ Additional distance} = \underline{160 \text{ m.}}$$

Substitute in equation 2:

$$t = \frac{160}{\ln(10)} \left(\frac{1}{4} - \frac{1}{400}\right).$$

$$\therefore \text{ Additional time} = \frac{160}{\ln(10)} \left[\left(\frac{1}{4} - \frac{1}{400}\right) - \left(\frac{1}{40} - \frac{1}{400}\right)\right]$$

$$= \frac{160}{\ln(10)} \left(\frac{1}{4} - \frac{1}{40}\right)$$

$$= \underline{15.63 \text{ s.}}$$

Example 6.3

A body having a mass of 8 kg is projected vertically upwards from the ground with an initial velocity of 70 m s^{-1}. The resistance R to its motion through the air, in newtons, is given by

$$R = 0.014v^2$$

where v is the velocity of the body at any instant, in m s^{-1}. Calculate the maximum height reached by the body, and the time to reach this height. Also calculate what height it would reach if there were no air resistance, and the corresponding time.

Solution 6.3

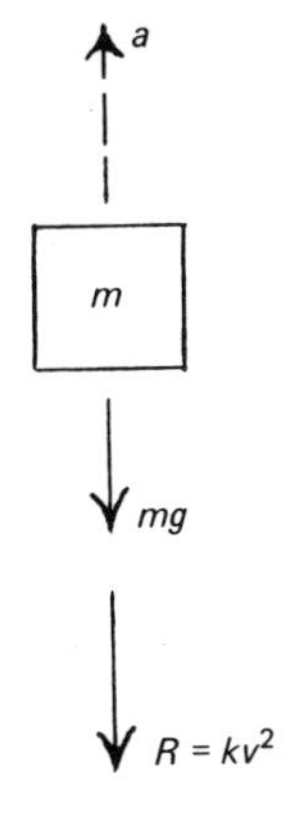

The free-body diagram shows the weight acting downwards, and also the resistance force R, which will also act downwards if the body is moving upwards. An arbitrary positive-upwards direction has been chosen, and the equation of motion must recognise this.

The equation therefore is

[d] $$-mg - 0.014v^2 = ma = m\frac{dv}{dt} = mv\frac{dv}{dx}.$$

The working is slightly simplified if we replace 0.014 by k temporarily.

$$\int dx = (-1)\int\left(\frac{mv}{mg + kv^2}\right) dv.$$

Let $(mg + kv^2)$ be u. Then

$$\frac{du}{dv} = 2kv.$$

$$\therefore\ v\,dv = \frac{du}{2k}.$$

$$\therefore\ \int dx = (-1)\int\left(\frac{m}{2ku}\right) du.$$

$$\therefore\ x = -\frac{m}{2k}\Big[\ln(u)\Big]_{v_1}^{v_2},$$

and replacing u $$\therefore\ x = -\frac{m}{2k}\left\{\ln(mg + kv_2^2) - \ln(mg + kv_1^2)\right\}$$

$$= +\frac{m}{2k}\ln\left(\frac{mg + kv_1^2}{mg + kv_2^2}\right).$$

The initial velocity (70 m s^{-1}) is v_1. The final velocity, v_2, will be zero at the maximum height. Substituting all values,

$$x = \frac{8}{2 \times 0.014}\ln\left(\frac{8 \times 9.81 + 0.014 \times (70)^2}{8 \times 9.81 + 0}\right)$$

$$= 285.7\ln(1.874)$$

$$= \underline{179.5\text{ m.}}$$

To determine the time, the equation of motion is written

[d] $$-mg - 0.014v^2 = m\frac{dv}{dt}.$$

$$\therefore\ \int dt = (-1)\int\left(\frac{m}{mg + kv^2}\right)dv$$

(again replacing 0.014 by k). The form of the denominator suggests a 'tan' substitution.

$$\int dt = -\frac{1}{g}\int\frac{dv}{\left(1 + \dfrac{kv^2}{mg}\right)}.$$

Let $\left(\dfrac{kv^2}{mg}\right)$ be $\tan^2\theta$. Then $v = \sqrt{\dfrac{mg}{k}}\tan\theta$.

$$\therefore\ \frac{dv}{d\theta} = \sqrt{\frac{mg}{k}}\sec^2\theta.$$

$$\therefore \int dt = -\frac{1}{g}\sqrt{\frac{mg}{k}}\int\left(\frac{\sec^2\theta}{(1+\tan^2\theta)}\right)d\theta$$

$$= -\sqrt{\frac{m}{gk}}\int\left(\frac{\sec^2\theta}{\sec^2\theta}\right)d\theta.$$

$$\therefore t = -\sqrt{\frac{m}{gk}}\,[\theta]$$

$$= -\sqrt{\frac{m}{gk}}\left[\tan^{-1}\left(\sqrt{\frac{k}{mg}}\,v\right)\right]_{v_1}^{v_2}$$

$$= +\sqrt{\frac{m}{gk}}\left\{\tan^{-1}\left(\sqrt{\frac{k}{mg}}\,v_1\right) - \tan^{-1}\left(\sqrt{\frac{k}{mg}}\,v_2\right)\right\}.$$

Substituting values,

$$\tan^{-1}\left(\sqrt{\frac{k}{mg}}\,v_1\right) = \tan^{-1}\left[\left(\sqrt{\frac{0.014}{8\times 9.81}}\right)70\right]$$

$$= \tan^{-1}(0.9349) = 0.7518 \text{ rad.}$$

(Note that the tan function must be in radians.)

Since v_2 is zero, $\tan^{-1}\sqrt{\frac{k}{mg}}\,v_2 = 0$

$$t = \left(\sqrt{\frac{8}{9.81\times 0.014}}\right)0.7518 = \underline{5.738 \text{ s.}}$$

With no air resistance, the acceleration has a constant value of $(-g)$. The elementary kinematic equations may be used.

$$v^2 = u^2 + 2ax$$

$$0 = (70)^2 + 2(-9.81)h.$$

$$\therefore h = \frac{4900}{2\times 9.81} = \underline{249.75 \text{ m.}}$$

For the time, $v = u + at$.

$$0 = 70 + (-9.81)t.$$

$$\therefore t = \frac{70}{9.81} = \underline{7.136 \text{ s.}}$$

Example 6.4

A body having a mass of 8 kg is released from rest at a vertical height above ground of 179.5 m. The resistance R to its motion through the air, in newtons, is given by

$$R = 0.014v^2$$

where v is its velocity at any instant, in m s^{-1}. Calculate its velocity on reaching the ground, and the time it takes to fall. Also calculate the approximate time of fall by assuming a constant value of resisting force equal to half the maximum value.

Solution 6.4

This problem is seen to follow on from Example 6.3. The free-body diagram shows that as the body is now falling, the resisting force acts upwards. Replacing the

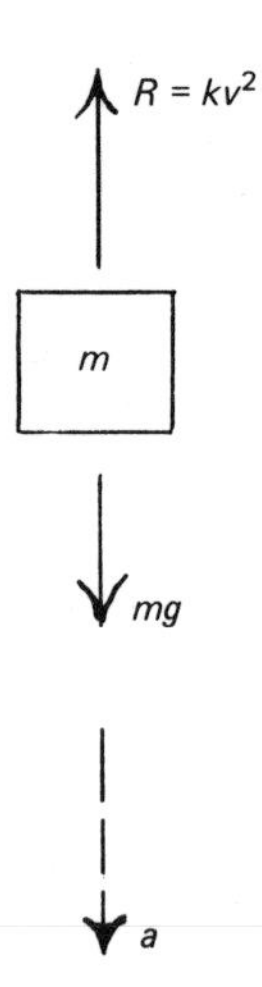

figure 0.014 by k, and assuming an arbitrary positive-downwards sign convention, the equation of motion now is

[d]

$$mg - kv^2 = ma = m\frac{dv}{dt} = mv\frac{dv}{dx}.$$

$$\therefore \int dx = \int \frac{mv}{mg - kv^2}\, dv.$$

Let $(mg - kv^2)$ be u. Then

$$\frac{du}{dv} = -2kv.$$

$$\therefore v\, dv = -\frac{du}{2k}.$$

$$\therefore \int dx = -\frac{m}{2k}\int \frac{du}{u}$$

$$= -\frac{m}{2k}\Big[\ln(u)\Big]_{v_1}^{v_2}$$

$$= +\frac{m}{2k}\left\{\ln(mg - kv_1^2) - \ln(mg - kv_2^2)\right\}$$

$$\therefore x = \frac{m}{2k}\ln\left(\frac{mg - kv_1^2}{mg - kv_2^2}\right).$$

The initial velocity v_1 is now zero, and x is 179.5 m.

$$179.5 = \frac{8}{2 \times 0.014}\ln\left(\frac{8g - 0}{8g - kv_2^2}\right).$$

$$\therefore \ln\left(\frac{8g}{8g - kv_2^2}\right) = \frac{179.5 \times 2 \times 0.014}{8} = 0.628\,25.$$

$$\therefore \frac{8g}{8g - kv_2^2} = 1.8743.$$

$$\therefore 8g = 8g \times 1.8743 - kv_2^2 \times 1.8743.$$

$$\therefore v_2^2 = \frac{8g\,(1.8743 - 1)}{k \times 1.8743}.$$

$$\therefore v_2 = \sqrt{\frac{8 \times 9.81 \times 0.8743}{0.014 \times 1.8743}} = \underline{51.14 \text{ m s}^{-1}}.$$

For the time of fall, we write the equation of motion:

[d]
$$mg - kv^2 = m\frac{dv}{dt}.$$

$$\therefore \int dt = \int \frac{m\,dv}{mg - kv^2}.$$

This time, the negative sign in the denominator suggests a 'difference of two squares' and a division into partial fractions. Rearranging

$$\int dt = \frac{1}{g}\,\frac{dv}{1 - (Av)^2} \qquad \left(\text{where } A = \sqrt{\frac{k}{mg}}\right)$$

$$= \frac{1}{g}\int \frac{dv}{(1 + Av)(1 - Av)}$$

$$= \frac{1}{g}\int \left(\frac{\frac{1}{2}}{1 + Av} + \frac{\frac{1}{2}}{1 - Av}\right)dv.$$

$$\therefore\ t = \frac{1}{g}\left[\frac{\frac{1}{2}}{A}\ln(1 + Av) + \frac{\frac{1}{2}}{(-A)}\ln(1 - Av)\right]_{v_1}^{v_2}$$

$$= \frac{1}{2Ag}\left[\ln\left(\frac{1 + Av}{1 - Av}\right)\right]_{v_1}^{v_2}.$$

Noting that again v_1 is zero,

$$t = \frac{1}{2Ag}\ln\left(\frac{1 + Av_2}{1 - Av_2}\right).$$

$$A = \sqrt{\frac{k}{mg}} = \frac{0.014}{8 \times 9.81} = 0.013\,36.$$

$$\therefore\ t = \frac{1}{2Ag}\ln\left(\frac{1 + 0.013\,36 \times 51.14}{1 - 0.013\,36 \times 51.14}\right)$$

$$= \frac{1}{2Ag}\ln(5.3137)$$

$$= \frac{1}{2 \times 0.01336 \times 9.81} \times 1.6703$$

$$= \underline{6.372\text{ s.}}$$

This answer shows that the time to fall is greater than the time to ascend, as we calculated in Example 6.3. This is to be expected, as when the body was ascending, both its weight and the resistance of the air acted against the motion. This must result in a greater retardation than when the air resistance acts upwards and gravity acts downwards.

The maximum resistance of the air will occur at the maximum speed of 51.14 m s^{-1}.

$$\text{Max. resistance} = 0.014 \times (51.14)^2 = 36.61\text{ N}.$$

Assuming a constant resistance of half this,

$$mg - 18.31 = ma.$$

This time, because the forces are constant, the acceleration a is also constant.

$$\therefore\ a = g - \frac{18.31}{8} = 7.521\ \text{m s}^{-2}$$

and again because the acceleration is constant, the simple kinematic equations of motion may be used.

$$x = ut + \tfrac{1}{2}at^2 = 0 + \tfrac{1}{2}at^2$$

$$\therefore\ t = \sqrt{\frac{2x}{a}} = \sqrt{\frac{2 \times 179.5}{7.521}} = \underline{6.909\ \text{s.}}$$

Example 6.5

A body having a mass m is moving with an initial velocity v_0. It is subjected to a retarding force R which is proportional to the speed at any instant, i.e. $R = kv$ where k is a constant. Show that the displacement x after time t is given by

$$x = \frac{m}{k} v_0 (1 - e^{-(k/m)t}).$$

If v_0 is 20 m s^{-1} and the velocity is reduced to 2 m s^{-1} after 15 seconds, how far does it travel in this time? What will be the velocity after a further 15 seconds?

Solution 6.5

As only a single force acts, we may dispense with a free-body diagram. The equation of motion is

[d]

$$-kv = ma = mv\frac{dv}{dx}.$$

$$\therefore \int dx = -\frac{m}{k}\int \frac{v\,dv}{v} = -\frac{m}{k}\int dv.$$

$$\therefore\ x = -\frac{m}{k}v + A. \qquad (1)$$

The constant A is determined from the initial condition that when $x = 0$, $v = v_0$, the initial velocity.

$$\therefore\ 0 = -\frac{m}{k}v_0 + A.$$

Substituting for A in equation 1,

$$x = -\frac{m}{k}v + \frac{m}{k}v_0. \qquad (2)$$

Rewriting the equation of motion,

[d]

$$-kv = ma = m\frac{dv}{dt}.$$

$$\therefore \int dt = -\frac{m}{k}\int \frac{dv}{v}.$$

$$\therefore\ t = -\frac{m}{k}\ln(v) + B. \qquad (3)$$

Substitute the second initial condition, i.e. when $t = 0$, $v = v_0$:

$$0 = -\frac{m}{k}\ln(v_0) + B$$

and substituting for B in equation 3

$$t = -\frac{m}{k}\ln(v) + \frac{m}{k}\ln(v_0).$$

$$\therefore\ t = \frac{m}{k}\ln\left(\frac{v_0}{v}\right).$$

$$\therefore\ \ln\left(\frac{v_0}{v}\right) = \frac{k}{m}t.$$

$$\therefore\ \frac{v_0}{v} = e^{(k/m)t}$$

$$\therefore\ \frac{v}{v_0} = e^{-(k/m)t}.$$

$$\therefore\ v = v_0 e^{-(k/m)t}. \qquad (4)$$

Substitute this value of v in equation 2:

$$x = -\frac{m}{k}v_0 e^{-(k/m)t} + \frac{m}{k}v_0.$$

$$\therefore\ x = \underline{\frac{m}{k}v_0(1 - e^{-(k/m)t})}. \qquad (5)$$

Equation 5 should be recognised as an example of 'exponential decay', which applies, as here, to the motion of a body against a resistance proportional to speed, and in modified forms, to radioactive decay, to the discharge of an electrical capacitance, to the pressure drop in a leaking container of gas, and to the growth of a bacterial culture.

Substituting the stated value in equation 4,

$$2 = 20e^{-(k/m)15}.$$

$$\therefore\ e^{(k/m)15} = 10.$$

$$\therefore\ \frac{k}{m} = \frac{\ln(10)}{15} = 0.1535.$$

Substituting this in equation 5,

$$x = \left(\frac{1}{0.1535}\right) \times 20\,(1 - e^{-0.1535 \times 15})$$

$$= \frac{20}{0.1535}(1 - 0.1)$$

$$= \underline{117.26 \text{ m.}}$$

For the final part of the question, we substitute $t = 30$ in equation 4:

$$v = 20e^{-0.1535 \times 30} = \underline{0.2 \text{ m s}^{-1}},$$

showing that in equal increments of time, the speed is reduced by a constant factor; in this example, the factor is 10. This is a characteristic of exponential decay.

An alternative solution of this problem can be found by writing the equation of motion thus:

$$\frac{d^2x}{dt^2} = -\frac{k}{m}\frac{dx}{dt}$$

and solving the equation directly. (See Fact Sheet (g).)

Example 6.6

A small body of mass 0.05 kg is allowed to fall vertically through a fluid. The resistance, R, in newtons, to the motion is given by

$$R = 0.4v$$

where v is the velocity at any instant in metres per second. If it is released from rest, calculate the time, and the distance fallen, when the velocity has increased to 1 m s^{-1}.

Solution 6.6

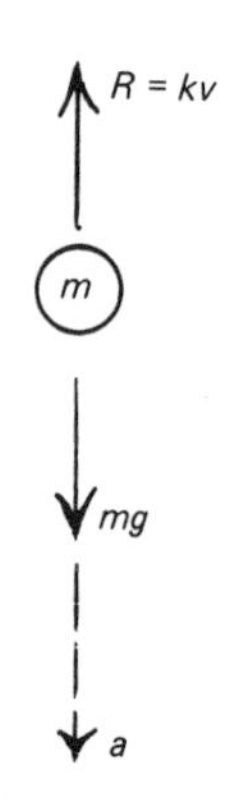

The free-body diagram shows the weight, acting downwards, and the upward-directed resistance, opposing the downward motion. The equation of motion therefore is

$$mg - kv = ma = m\frac{dv}{dt}.$$

(For the present, we replace 0.4 by k, and work in pure algebra.)
Rearranging,

$$\frac{m}{k}g - v = \frac{m}{k}\frac{dv}{dt}.$$

$$\therefore \int dt = \frac{m}{k}\int_{v_1}^{v_2} \frac{dv}{\frac{gm}{k} - v}.$$

$$\therefore t = \frac{m}{k}\left[(-1)\ln\left(\frac{gm}{k} - v\right)\right]_{v_1}^{v_2}.$$

$$\therefore t = \frac{m}{k}\left\{\ln\left(\frac{gm}{k} - v_1\right) - \ln\left(\frac{gm}{k} - v_2\right)\right\}.$$

$$\therefore \; t = \frac{m}{k} \ln \left(\frac{\dfrac{gm}{k} - v_1}{\dfrac{gm}{k} - v_2} \right). \tag{1}$$

We may now substitute numbers.

$$\frac{gm}{k} = 9.81 \times \frac{0.05}{0.4} = 1.226\,25$$

and substituting the two values for velocity, 0 and 1 m s^{-1} for v_1 and v_2 in equation 1,

$$t = \frac{0.05}{0.4} \ln \left(\frac{1.226\,25 - 0}{1.226\,25 - 1} \right) = \underline{0.2113 \text{ s.}}$$

Rewriting the equation of motion,

[d]
$$mg - kv = m \frac{dv}{dx} \frac{dx}{dt} = mv \frac{dv}{dx}.$$

$$\therefore \; \frac{gm}{k} - v = \frac{m}{k} v \frac{dv}{dx}.$$

$$\therefore \; \int dx = \frac{m}{k} \int_{v_1}^{v_2} \frac{v \, dv}{\dfrac{gm}{k} - v} = \frac{m}{k} \int_{v_1}^{v_2} \frac{v \, dv}{A - v}.$$

(temporarily replacing (gm/k) by the constant A).

$$\frac{v}{A - v} = \frac{A}{A - v} - 1.$$

$$\therefore \; x = \frac{m}{k} \int \left(\frac{A}{A - v} - 1 \right) dv$$

$$= \frac{m}{k} \Big[(-1) A \ln (A - v) - v \Big]_{v_1}^{v_2}$$

$$= \frac{m}{k} \Big\{ A \ln (A - v_1) - A \ln (A - v_2) - (v_2 - v_1) \Big\}.$$

$$\therefore \; x = \frac{m}{k} \left\{ (v_1 - v_2) + A \ln \left(\frac{A - v_1}{A - v_2} \right) \right\}.$$

Substituting values ($A = (gm)/k$ which has the same value as before),

$$x = \frac{0.05}{0.4} \left\{ (0 - 1) + 1.226\,25 \ln \left(\frac{1.226\,25 - 0}{1.226\,25 - 1} \right) \right\}$$

$$= 0.1341 \text{ m.}$$

Example 6.7

A uniform flexible chain of total length 1.5 m is laid in a straight line on the surface of a horizontal frictionless table, at right-angles to an edge, with a portion hanging over the edge. It is released from rest and allowed to slide.

(a) If the initial overhang is 0.1 m, calculate the time for all the chain to slide off the table.
(b) Calculate the velocity at the instant the end slides off the table.
(c) Show that if the time for all the chain to slide off is to be at least 2 seconds then the initial overhang must not be greater than 18 mm.

You may make use of the following integral:

$$\int \frac{dx}{\sqrt{x^2 - A^2}} = \ln (x + \sqrt{x^2 - A^2}).$$

Solution 6.7

As with many problems, the two parts to this question suggest that it would be an advantage to work purely in algebra, and insert numbers at the end.

Assume a total length of chain L. Assume that at any instant the length of overhang is x. Let the mass per unit length of chain be m.

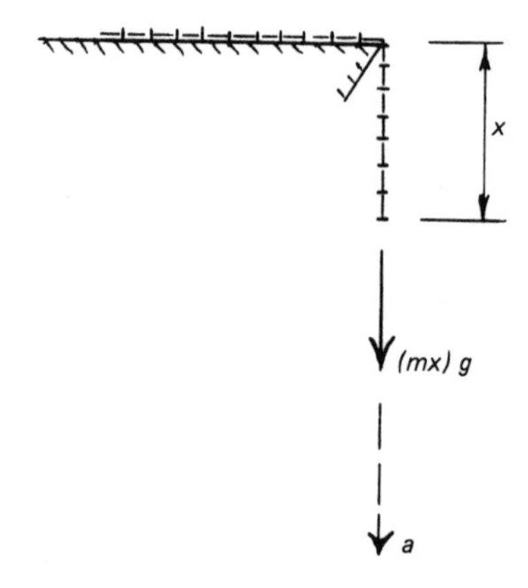

The accelerating force on the chain will be the weight of the overhanging portion only (see diagram). The remaining weight on the table is balanced by the table reaction force. But notice that the *total* mass of the chain has to be accelerated. The equation of motion is

[e] $$(mx)g = (mL)a = (mL)v\frac{dv}{dx}.$$

$$\therefore \int x \, dx = \frac{L}{g}\int v \, dv.$$

$$\therefore \tfrac{1}{2}x^2 = \frac{L}{g}(\tfrac{1}{2}v^2) + A. \qquad (1)$$

The constant of integration A may be found by substituting an 'initial condition' of the motion.

If we call the initial amount of overhang x_0 we know that when $x = x_0$, the velocity $v = 0$

(because the chain is released from rest).

$$\tfrac{1}{2}x_0^2 = 0 + A.$$

Substituting for A in equation 1,

$$\tfrac{1}{2}x^2 = \frac{L}{g}\tfrac{1}{2}v^2 + \tfrac{1}{2}x_0^2.$$

$$\therefore\ v^2 = \frac{g}{L}(x^2 - x_0^2). \qquad (2)$$

[e]
$$\frac{dx}{dt} = v = \sqrt{\frac{g}{L}(x^2 - x_0^2)}.$$

$$\therefore \int dt = \sqrt{\frac{L}{g}} \int \frac{dx}{\sqrt{x^2 - x_0^2}}$$

The significance of the integral given in the question is now seen.

$$\therefore\ t = \sqrt{\frac{L}{g}}\ \ln\left(x + \sqrt{(x^2 - x_0^2)}\right) + B. \qquad (3)$$

The constant B may be found from the condition $t = 0$ when $x = x_0$.

$$0 = \sqrt{\frac{L}{g}}\ \ln(x_0 + 0) + B.$$

Substitute for B in equation 3:

$$t = \sqrt{\frac{L}{g}}\ \ln\left[x + \sqrt{(x^2 - x_0^2)}\right] - \sqrt{\frac{L}{g}}\ \ln(x_0)$$

$$= \sqrt{\frac{L}{g}}\ \ln\left(\frac{x + \sqrt{x^2 - x_0^2}}{x_0}\right). \qquad (4)$$

The solution to part (a) of the question is obtained by substituting $L = 1.5$ m, $x_0 = 0.1$ m and $x = 1.5$ m in equation 4:

Part (a)

$$t = \sqrt{\frac{1.5}{9.81}}\ \ln\left(\frac{1.5 + \sqrt{1.5^2 - 0.1^2}}{0.1}\right)$$

$$= 0.3910 \times \ln(29.97)$$

$$= 0.3910 \times 3.400$$

$$= \underline{1.329 \text{ s.}}$$

Part (b)

We calculate velocity from equation 2. Again, $x = L = 1.5$ m and $x_0 = 0.1$ m.

$$v^2 = \frac{9.81}{1.5}(1.5^2 - 0.1^2)$$

$$= 14.65.$$

$$\therefore v = \underline{3.827 \text{ m s}^{-1}}.$$

Part (c)

It is tempting to repeat the procedure we adopted in part (a), substituting 0 for t, and calculating x_0. This would be found to lead to a tedious calculation. This is an example of the importance of reading the question carefully. If the question had stated, 'Calculate the amount of overhang corresponding to a time of 2 seconds for the chain to slide off,' then you would have no choice but to go through this calculation. But the explicit statement is, 'Show that . . .'. This is an invitation merely to substitute x_0 = 18 mm in equation 4 to verify that the answer, t, is at least 2 seconds. Thus:

$$x = 1.5 \text{ m}; x_0 = 0.018 \text{ m}.$$

$$t = 0.3910 \ln \left(\frac{1.5 + \sqrt{1.5^2 - 0.018^2}}{0.018} \right)$$

$$= 0.3910 \ln (166.66)$$

$$= 0.3910 \times 5.116$$

$$= \underline{2.000 \text{ s}}$$

and the question is correctly answered.

Example 6.8

An aircraft has a mass of 20 000 kg. It lands on the deck of a carrier with an initial speed of 60 m s^{-1} and is brought to rest by an arrester gear which exerts a resistance proportional to its extension, and which exerts a force of 2.5 MN for an extension of 30 m. Calculate the time for the aircraft to be brought to rest, and the maximum stretch of the arrester.

Solution 6.8

Let the force per unit extension of the arrester be k N m^{-1}. For an extension x, the equation of motion will be

$$-kx = ma = m \frac{d^2 x}{dt^2}.$$

Rearranging, $\dfrac{d^2 x}{dt^2} = -\dfrac{k}{m} x.$

The general solution to this equation is

[g] $$x = A \sin\left(\sqrt{\frac{k}{m}}\, t\right) + B \cos\left(\sqrt{\frac{k}{m}}\, t\right). \qquad (1)$$

Some examiners would definitely expect a candidate to know this solution to the differential equation. You should note the general form of the solution; because it is a second-order equation, the solution contains two constants of integration, A and B. These must be determined from a knowledge of the initial conditions of the motion. In this case, at the initial stage, when $t = 0$, the displacement $x = 0$ and the velocity $v = 60$ m s^{-1}. In order to apply this second initial condition, we need an expression for the velocity, v. Differentiate to determine v:

$$v = \frac{dx}{dt}$$

$$= A \sqrt{\frac{k}{m}} \cos\left(\sqrt{\frac{k}{m}}\, t\right) - B \sqrt{\frac{k}{m}} \sin\left(\sqrt{\frac{k}{m}}\, t\right). \qquad (2)$$

Condition $t = 0, x = 0$:

$$0 = A \sin 0 + B \cos 0.$$

$$\therefore B = 0.$$

Condition $t = 0$, $v = 60$ (m s^{-1}):

$$60 = A \sqrt{\frac{k}{m}} \cos 0 - 0.$$

$$\therefore \; A = 60 \sqrt{\frac{m}{k}} .$$

Rewriting equation 2, $\quad v = 60 \cos \left(\sqrt{\frac{k}{m}}\, t \right) .$

When brought to rest, $v = 0$. The first possible solution is

$$\sqrt{\frac{k}{m}}\, t = \frac{\pi}{2} .$$

$$\therefore \; t = \sqrt{\frac{m}{k}}\, \frac{\pi}{2} .$$

From given data, $k = \dfrac{2.5 \times 10^6}{30}$ N m^{-1}.

$$\therefore \; t = \frac{\pi}{2} \sqrt{\frac{20 \times 10^3 \times 30}{2.5 \times 10^6}}$$

$$= 0.7695 \text{ s.}$$

Substitute in equation 1 to find x:

$$x = A \sin\left(\frac{\pi}{2}\right)$$

$$= 60 \sqrt{\frac{m}{k}} \times 1$$

$$= 60 \sqrt{\frac{20 \times 10^3 \times 30}{2.5 \times 10^6}}$$

$$= \underline{29.30 \text{ m.}}$$

You may have met this type of problem in work on vibrations, and you will find more examples in Chapter 7. Needless to say, the aircraft executes only one quarter of a vibration; in an actual gear, some device is included to ensure that the arrester does not throw the aircraft forward again, after stopping it.

This problem may also be solved by writing the equation of motion

[e] $$-kx = m\,\frac{dv}{dt} = mv\,\frac{dv}{dx}$$

and integrating, as with other examples, to obtain $v = \dfrac{dx}{dt}$ **and then integrating a second time to obtain** x**. But the working is tedious, and the second integration requires a 'sine** θ**' substitution (as might be expected, in the light of the solution to the differential equation). The solution given here is the simpler one.**

Example 6.9

An aircraft of mass 5500 kg is propelled by a jet engine. The net thrust of the engine (i.e. the resultant forward thrust after overcoming air and ground resistance) is as shown in the graph of net thrust (kN against velocity (m s^{-1}). Values are also tabulated below in increments of velocity of 10 m s^{-1}.

The take-off speed of the aircraft is 80 m s^{-1}. By dividing the process into eight increments of 10 m s^{-1}, assuming a mean constant thrust during each increment, estimate the approximate distance required for take-off, and the approximate time taken.

Velocity (m s^{-1})	0	10	20	30	40	50	60	70	80
Net force (kN)	42.0	40.4	36.0	30.6	25.4	22.0	20.0	18.4	17.8

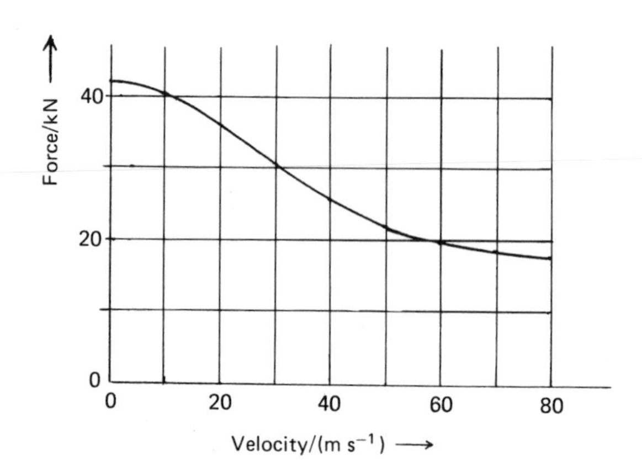

Solution 6.9

This is an example in which a purely mathematical solution cannot be obtained because the graph cannot be expressed as a mathematical function. We are told to divide the motion into eight stages, and to assume a mean constant force over each stage. If the force is constant, so will the acceleration be, and we can use the simple equations of kinematics.

For the first stage (0 to 10 m s^{-1}), the mean force, $\bar{F}$, is $\frac{1}{2}(42.0 + 40.4) =$ 41.2 kN.

$$\Sigma \bar{F} = ma.$$

$$\therefore\ a_1 = \frac{\bar{F}}{m} = \frac{41.2 \times 10^3}{5500} = 7.49 \text{ m s}^{-2}.$$

For the distance covered,

$$v^2 = u^2 + 2ax.$$

$$\therefore\ x = \frac{v^2 - u^2}{2a} = \frac{10^2 - 0}{2 \times 7.49} = 6.7 \text{ m (to one decimal place).}$$

For the time,

$$v = u + at.$$

$$\therefore\ t = \frac{v - u}{a} = \frac{10 - 0}{7.49} = 1.34 \text{ s.}$$

The rest of the calculation is not set out separately; the working is best presented in tabular form. Distances are given to one decimal place, and times to two places.

Velocity (m s^{-1})	0	10	20	30	40	50	60	70	80
Force (kN)	42.0	40.4	36.0	30.6	25.4	22.0	20.0	18.4	17.8

Increment No.	1	2	3	4	5	6	7	8
Mean force $\bar{F}$ (kN)	41.2	38.2	33.3	28.0	23.7	21.0	19.7	18.1
Acceleration a (m s^{-2})	7.49	6.95	6.05	5.09	4.31	3.82	3.49	3.29
x (m)	6.7	21.6	41.3	68.8	104.4	144.0	186.2	228.0
t (s)	1.34	1.44	1.65	1.96	2.32	2.62	2.87	3.04

The total distance covered is the sum of the x-row which is 801.0 m. The total time taken is the sum of the t-row which is 17.24 s.

This simple breaking down into increments can be done for most problems where the only data available are in graphical form. Of course, the smaller the increments, the greater the accuracy, but the calculation is proportionately more tedious.

Example 6.10

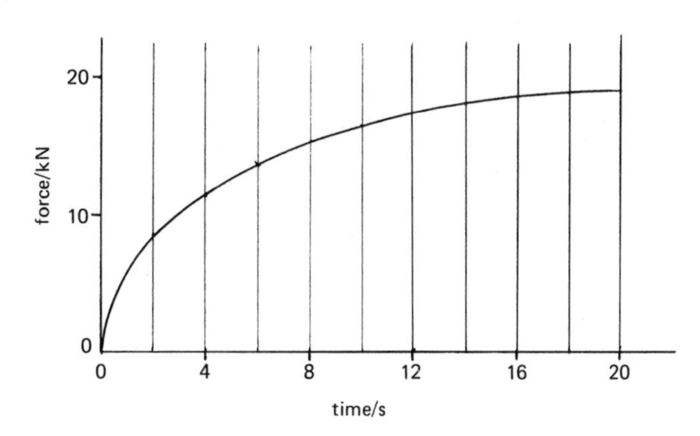

A missile is projected by a propulsive force which varies with time according to the graph and the table shown here; the table also gives incremental areas under the graph in time increments of 2 seconds.

Time (s)	0	2	4	6	8	10	12	14	16	18	20
Force (kN)	0	8.5	11.6	13.7	15.3	16.5	17.5	18.1	18.7	18.9	19.0

Area (kN s)	11.24	20.5	25.6	29.1	31.9	34.0	35.7	36.8	37.6	37.9

The mass of the missile is 3348 kg, and it travels along a straight frictionless track. Calculate its velocity after intervals of 10 and 20 seconds (a) if the track is horizontal; (b) if it slopes upwards at an angle of 15°.

Solution 6.10

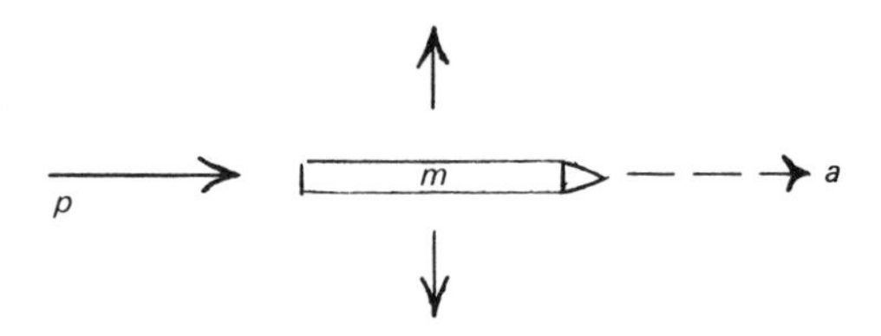

[a]

In the free-body diagram the only relevant force is seen to be the variable propulsive force, *p*. (Weight and track reaction will both be perpendicular to the line of acceleration.) The equation of motion is

$$p = ma = m\ \frac{dv}{dt}\ .$$

Rearranging, $\int p\ dt = \int m\ dv.$

The left-hand side is seen to be the area under the force–time graph.

$\therefore$ Area under graph $= m \left[v \right]_{v_1}^{v_2} = m\,(v_2 - v_1) = mv_2$ (if body starts from rest).

For 10 seconds, the appropriate area is the sum of the first five increments, which is 118.34 kN s.

$$\therefore\ mv_2 = 118.34 \times 10^3 .$$

$$\therefore\ v_2 = \frac{118.34 \times 10^3}{3348} = \underline{35.35 \text{ m s}^{-1}} .$$

For 20 seconds, the area is the sum of all ten increments, i.e. 300.34 kN s.

$$mv_2 = 300.34 \times 10^3 .$$

$$\therefore\ v_2 = \frac{300.34 \times 10^3}{3348} = \underline{89.71 \text{ m s}^{-1}} .$$

[b]

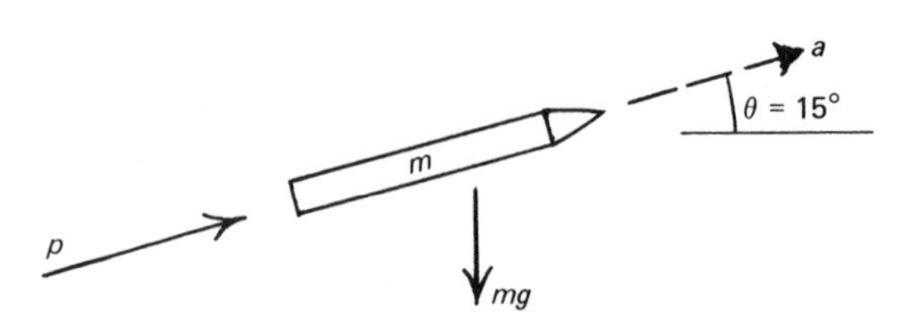

A free-body diagram now shows that the weight has a component along the line of acceleration. The equation of motion is

$$p - mg \sin\theta = ma = m\ \frac{dv}{dt}\ .$$

Rearranging: $\int p\ dt - \int mg \sin\theta\ dt = \int m\ dv.$

The first integral is again seen to be the area under the force–time graph. But it must be noted that this time we must not use the whole graph. The free-body diagram assumes the missile to be in motion. But it will not begin to move until the propulsive force attains a value equal to the weight component.

This initial value of p at which motion begins must be calculated.

$$p_0 = mg \sin 15^\circ$$
$$= 3348g \sin 15^\circ$$
$$= 8500 \text{ N.}$$

This is seen to coincide conveniently with 2 seconds on the graph. The integration must therefore start from this point, and not from $t = 0$. Hence, to determine the velocity after 10 seconds, $\int p \, dt$ will be the area from $t = 2$ to $t = 10$, i.e.

$$(20.5 + 25.6 + 29.1 + 31.9) = 107.1 \text{ kN s.}$$

$$\therefore \; 107.1 \times 10^3 - mg \sin\theta \left[t \right]_2^{10} = mv_2 \text{ (since } v_1 \text{ is again 0).}$$

$$\therefore \; v_2 = \frac{107\,100}{3348} - g \sin 15^\circ \times 8$$
$$= 31.99 - 20.31$$
$$= \underline{11.68 \text{ m s}^{-1}}.$$

And for a time of 20 seconds, the effective area will be the whole graph less the first increment up to 2 seconds, i.e. $(300.34 - 11.24) = 289.1$ kN s.

$$v_2 = \frac{289\,100}{3348} - g \sin 15^\circ \times 18$$
$$= 86.35 - 45.70$$
$$= \underline{40.65 \text{ m s}^{-1}}.$$

Graphical integration is not usually popular with students; they tend to prefer the clean-cut mathematical analysis. But graphical methods have considerable advantages. First, in a problem such as this, the force function cannot be expressed mathematically; it can only be found from a test. And second, even if the mathematical function is known then frequently the integration is difficult, or for all practical purposes impossible. But any student can plot a graph and count squares. An argument against such a method is the time required. But from the standpoint of an examination question, you may feel reasonably confident that the examiner will not give you more work than can be done in the allotted time. This example is typical, in that the examiner counted the squares for you, and left you to work out how to use the information.

Example 6.11

A projectile is fired into water, which exerts a resisting force proportional to the square of the velocity of the projectile. It is observed that the velocity is reduced from 40 m s^{-1} to 20 m s^{-1} in 2.2 s. Calculate (a) the distance travelled during this period; (b) the additional time required for the velocity to be reduced to 10 m s^{-1}.

PCL

Solution 6.11

The only force acting is the resistance to motion. We may dispense with the free-body diagram and write the equation of motion straight down.

$$-kv^2 = ma = m\,\frac{dv}{dt}\,. \qquad (1)$$

Note the negative sign, denoting a resistance as distinct from an accelerating force. We may leave the equation in this form, as the given data relate velocity to time.

$$\int dt = -\int \frac{m\,dv}{kv^2} = -\frac{m}{k}\int \frac{dv}{v^2}\,.$$

$$t = -\frac{m}{k}\left(\frac{(v^{-1})}{(-1)}\right) + A$$

$$= +\frac{m}{k}\left(\frac{1}{v}\right) + A.$$

When $t = 0$, $v = v_0$, the initial value.

$$0 = +\frac{m}{k}\left(\frac{1}{v_0}\right) + A.$$

$$\therefore\; A = -\frac{m}{k}\left(\frac{1}{v_0}\right).$$

Substitute A in the expression for t above.

$$t = \frac{m}{k}\left(\frac{1}{v}\right) - \frac{m}{k}\left(\frac{1}{v_0}\right)$$

$$= \frac{m}{k}\left(\frac{1}{v} - \frac{1}{v_0}\right).$$

Substituting the values given,

$$2.2 = \frac{m}{k}\left(\frac{1}{20} - \frac{1}{40}\right) = \frac{m}{k}\left(\frac{1}{40}\right).$$

$$\therefore\; \frac{m}{k} = 40 \times 2.2 = 88.$$

$$\therefore\; t = 88\left(\frac{1}{v} - \frac{1}{v_0}\right).$$

While the equation is in this form, we may as well substitute the data for the solution to part (b) of the question.

Part (b)

$$t = 88\left(\frac{1}{10} - \frac{1}{20}\right) = \underline{4.4\text{ s.}}$$

(Observe that we substitute 20 m s^{-1} for v_0 as the question asks for the *additional* time.)

Part (a)

To obtain the distance travelled, equation 1 is rearranged.

[d]

$$-kv^2 = m\frac{dv}{dt} = mv\frac{dv}{dx}.$$

$$\therefore \int dx = -\frac{m}{k}\int\left(\frac{v}{v^2}\right) dv = -\frac{m}{k}\int\frac{dv}{v}.$$

$$x = -\frac{m}{k}[\ln(v)] + B.$$

Where $x = 0$, $v = v_0$.

$$\therefore 0 = -\frac{m}{k}\ln(v_0) + B.$$

$$\therefore B = \frac{m}{k}\ln(v_0).$$

$$\therefore x = -\frac{m}{k}\ln(v) + \frac{m}{k}\ln(v_0)$$

$$= \frac{m}{k}\ln\left(\frac{v_0}{v}\right).$$

Substituting values:

$$x = 88\ln\left(\frac{40}{20}\right) = \underline{61.00\text{ m}}.$$

Example 6.12

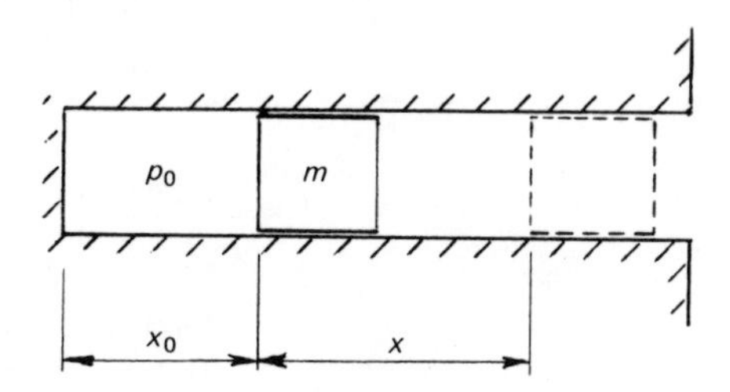

A cylindrical body of mass m is a perfect frictionless fit in a cylinder of cross-sectional area A as shown in the diagram. It is propelled along the cylinder by expansion of gas contained in the closed end. Initially, the mass is at rest at a distance x_0 from the closed end, and the pressure of the gas is p_0.

Assuming the gas expands according to the function

$$pv^{1.4} = K$$

where K is a constant, determine an expression for the velocity of the mass after it has travelled a distance x from the initial position.

PCL

Solution 6.12

The figure given with the question suggests deriving an equation of motion when the body has travelled a distance x from the initial position.

Let the gas pressure after displacement be p. The force acting on the body will be (pA). The equation of motion will therefore be

[d]

$$(pA) = ma = m\frac{dv}{dt} = mv\frac{dv}{dx}$$

(since we are required to relate v to x).
p is calculated from the function given.

$$p_0 V_0^{1.4} = p V^{1.4}.$$

$$\therefore\ p = p_0 \left(\frac{V_0}{V}\right)^{1.4}.$$

$$\therefore\ p_0 A \left(\frac{V_0}{V}\right)^{1.4} = mv\,\frac{\mathrm{d}v}{\mathrm{d}x}.$$

$$V_0 = A x_0\ :\ V = A(x_0 + x).$$

$$p_0 A \left(\frac{\cancel{A} x_0}{\cancel{A}(x_0 + x)}\right)^{1.4} = mv\,\frac{\mathrm{d}v}{\mathrm{d}x}.$$

$$\therefore \int v\,\mathrm{d}v = \frac{p_0 A}{m} \int \left(\frac{x_0}{x_0 + x}\right)^{1.4} \mathrm{d}x$$

$$= \frac{p_0 A}{m} \int \left(1 + \frac{x}{x_0}\right)^{-1.4} \mathrm{d}x.$$

$$\therefore\ \tfrac{1}{2} v^2 = \frac{p_0 A}{m} \left[\left(1 + \frac{x}{x_0}\right)^{-0.4} \left(\frac{1}{-0.4}\right)(x_0) + C\right]. \qquad (1)$$

When $x = x_0$, $v = 0$.

$$0 = \frac{p_0 A}{m} \left[(1 + 1)^{-0.4} \left(\frac{x_0}{-0.4}\right) + C\right].$$

$$\therefore\ C = \frac{1}{2^{0.4}} \left(\frac{x_0}{0.4}\right).$$

Substituting C in equation 1,

$$v^2 = \frac{2 p_0 A x_0}{-0.4 m} \left[\left(1 + \frac{x}{x_0}\right)^{-0.4} \left(-\frac{1}{2^{0.4}}\right)\right].$$

$$= \frac{2 p_0 A x_0}{0.4 m} \left[\frac{1}{2^{0.4}} - \left(1 + \frac{x}{x_0}\right)^{-0.4}\right].$$

A precaution with this type of question is to perform a dimension check. The bracketed terms are all ratios, and therefore dimensionless. You should be able to check that the coefficient outside the bracket has dimensions of (L^2/T^2) which is consistent with (v^2) on the left-hand side.

Example 6.13

A missile of mass m = 16 kg travels in a straight line with an initial velocity of 200 m s^{-1}. It is then retarded by a force f which is given by:

$$f = (16t + 0.4t^3) \text{ newtons}$$

where t is the time, in seconds, measuring from the instant of application of the force.

Calculate the time taken for the missile to come to rest, and the distance travelled in doing so.

PCL

Solution 6.13

The force f is the only force acting on the body. The equation of motion is therefore:

$$-(16t + 0.4t^3) = ma = m\frac{dv}{dt}. \qquad (1)$$

We can integrate directly, relating v to t:

$$\int m \, dv = -\int (16t + 0.4t^3)\, dt.$$

$$\therefore \; mv = (-16 \times \tfrac{1}{2}t^2 + 0.4 \times \tfrac{1}{4}t^4) + A.$$

When $t = 0$, $v = v_0$, the initial velocity.

$$\therefore \; mv_0 = 0 + A.$$

Substitute this value of A in the equation above.

$$mv = mv_0 - (8t^2 + 0.1t^4).$$

$$\therefore \; m(v_0 - v) = 8t^2 + 0.1t^4. \qquad (2)$$

To come to rest, $v = 0$. Substituting the values given:

$$16(200 - 0) = 8t^2 + 0.1t^4.$$

Rearranging,

$$t^4 + 80t^2 - 32\,000 = 0$$

and what threatened to be a formidable fourth-power equation in t is now seen to be a simple quadratic in (t^2).

Solving,

$$t^2 = -40 \pm \sqrt{1600 + 32\,000}$$

$$= 143.3.$$

$$\therefore \; t = \underline{11.97 \text{ s.}}$$

To determine the distance, we can rearrange equation 2 to make v the 'subject'.

$$m(v_0 - v) = 8t^2 + 0.1t^4.$$

$$\therefore \; v = v_0 - \frac{1}{m}(8t^2 + 0.1t^4).$$

v may now be written $\frac{dx}{dt}$:

$$\frac{dx}{dt} = v_0 - \frac{1}{m}(8t^2 + 0.1t^4).$$

$$\therefore \int dx = \int \left\{ v_0 - \frac{1}{m}(8t^2 + 0.1t^4) \right\} dt.$$

$$\therefore \; x = v_0 t - \frac{8t^3}{3m} - \frac{0.1t^5}{5m} + B.$$

When $x = 0$, $t = 0$, giving $B = 0$.

Substituting the value of t obtained in the first part of the solution,

$$x = 200 \times 11.97 - \frac{8 \times (11.97)^3}{3 \times 16} - \frac{0.1\,(11.97)^5}{5 \times 16}$$

$$= 2394 - 285.8 - 307.2.$$

$$x = \underline{1801 \text{ m.}}$$

Example 6.14

A rocket of initial total mass M, including fuel, is fired vertically from rest against a gravitational acceleration of g. The mass-time rate of burning of the fuel is μ and the exhaust gases have a velocity U relative to the rocket. Show, neglecting air resistance, that the acceleration a of the rocket at time t after take-off is

$$a = \frac{\mu U}{M - \mu t} - g.$$

Given that M is 500 kg, μ is 5 kg s^{-1}, U is 1500 m s^{-1} and g is 10 m s^{-2}, determine the acceleration of the rocket at take-off and the velocity of the rocket 20 seconds after take-off.

The total mass of fuel is 400 kg. What will be the velocity and acceleration of the rocket at the instant when the last of the fuel is being burnt?

U. Lond. K.C.

Solution 6.14

In this example, the acceleration is variable because the mass of the body is continually changing, although the thrust is constant. We first determine this thrust. This arises because the rocket thrusts effluent backwards at a constant rate, and thereby receives an equal forward thrust, in conformity with Newton's third law.

Considering the effluent:

$$\begin{aligned}\text{Thrust force} &= \text{Change of momentum per second}\\ &= \text{Mass per second} \times \text{Velocity change}\\ &= \mu U.\end{aligned}$$

After time t, the initial mass M is reduced by the amount (μt). The mass will thus be $(M - \mu t)$.

The rocket is subjected to (a) upward thrust, and (b) downward weight. The equation of motion will thus be:

$$\mu U - (M - \mu t)\, g = (M - \mu t)\, a.$$

$$\therefore\ a = \frac{\mu U}{M - \mu t} - g. \qquad (1)$$

$$\therefore\ \frac{dv}{dt} = \frac{\mu U}{M - \mu t} - g.$$

$$\therefore \int dv = \int \left(\frac{\mu U}{M - \mu t} - g\right) dt.$$

$$\therefore\ v = \mu U\left(-\frac{1}{\mu}\right)\ln(M - \mu t) - gt + A$$

$$= -U \ln(M - \mu t) - gt + A. \qquad (2)$$

A is determined from the 'initial condition' that when $t = 0$, $v = 0$.

$$0 = -U \ln(M) + A.$$

Substituting for A in equation 2,

$$\begin{aligned}v &= -U \ln(M - \mu t) - gt + U \ln(M)\\ &= U \ln(M) - U \ln(M - \mu t) - gt.\end{aligned}$$

$$v = U \ln\left(\frac{M}{M - \mu t}\right) - gt. \qquad (3)$$

We can now substitute given values. In equation 1, when $t = 0$ at take-off,

$$a_0 = \frac{\mu U}{M} - g$$

$$= \frac{5 \times 1500}{500} - 10$$

$$= \underline{5 \text{ m s}^{-2}}.$$

In equation 3, at $t = 20$ s,

$$v = 1500 \ln \left(\frac{500}{500 - 5 \times 20} \right) - 10 \times 20$$

$$= 1500 \ln (1.25) - 200$$

$$= \underline{134.7 \text{ m s}^{-1}}.$$

If there is 400 kg of fuel, then time t' for all fuel to be used is given by

$$\mu t' = 400.$$

$$\therefore \; t' = \frac{400}{5} = 80 \text{ s}.$$

Substitute in equation 3:

$$v = 1500 \ln \left(\frac{500}{500 - 400} \right) - 10 \times 80$$

$$= 2414.2 - 800$$

$$= \underline{1614.2 \text{ m s}^{-1}}$$

and the corresponding acceleration is found by substituting in equation 1:

$$a = \frac{5 \times 1500}{500 - 400} - 10$$

$$= \underline{65 \text{ m s}^{-2}}.$$

Example 6.15

When the fuel supply to the rocket engine of a space vehicle is cut off the thrust F of the engine does not cease instantly but decays according to the relationship

$$F = F_0 \mathrm{e}^{-kt}$$

where F is the thrust at a time t after cut-off and k is a constant.

If the total mass of the vehicle is assumed to remain constant, show that the increase in velocity after the fuel has been cut off is equal to that which would have resulted had the engine continued to run at full thrust F_0 for an additional time $1/k$.

(Neglect all forces other than engine thrust.)

U. Surrey

Solution 6.15

Since the forward thrust is the only active force, the equation of motion is:

$$F_0 \mathrm{e}^{-kt} = ma = m \frac{\mathrm{d}v}{\mathrm{d}t}$$

$$\therefore \int m \, \mathrm{d}v = \int (F_0 \mathrm{e}^{-kt}) \, \mathrm{d}t$$

$$\therefore\ mv = (F_0 e^{-kt})\left(-\frac{1}{k}\right) + A.$$

Let the velocity at the instant of cut-off be v_0.
When $t = 0$, $v = v_0$.

$$\therefore\ mv_0 = (F_0 e^0)\left(-\frac{1}{k}\right) + A.$$

$$\therefore\ A = mv_0 + \frac{F_0}{k}.$$

$$\therefore\ mv = (F_0 e^{-kt})\left(-\frac{1}{k}\right) + mv_0 + \frac{F_0}{k}.$$

$$\therefore\ m(v - v_0) = \frac{F_0}{k}(1 - e^{-kt}).$$

To find the total increase of velocity, we substitute a value of t = infinity. Thus, e^{-kt} becomes zero. Hence, the increase of velocity will be given by

$$v - v_0 = \frac{F_0}{mk}. \qquad (1)$$

For a constant thrust F_0 the acceleration a' is given by

$$a' = \frac{F_0}{m}$$

and the corresponding velocity increase is calculated from

$$v = u + a't.$$

For an additional time $1/k$,

$$v - u = a't = \frac{F_0}{m}\left(\frac{1}{k}\right),$$

which is the same result we obtained from equation 1.

Example 6.16

A particle is subjected to a force of $mk(c^2 + v^2)$, which causes it to decelerate in a straight line from an initial velocity u to a zero velocity in a displacement L. The mass of the particle is m, v is its instantaneous velocity, k and c are positive constants. Given that its speed is $u/3$ when it has moved a distance $L/2$, show that
(a) $u^2 = 63c^2$,

and (b) $63\frac{v^2}{u^2} = 64e^{-2ks} - 1$

when the particle has moved a distance s.

U. Surrey

Solution 6.16

The stated force causes a deceleration; it is therefore a negative force. The equation of motion will be

$$-mk(c^2 + v^2) = ma.$$

We are given a condition relating velocity and displacement. The equation must be written as a differential equation in x and v.

[d]
$$-mk(c^2 + v^2) = mv\frac{dv}{dx}.$$

$$\therefore \int dx = -\frac{1}{k}\int\frac{v\,dv}{c^2 + v^2}.$$

$$\therefore x = -\frac{1}{2k}\ln(c^2 + v^2) + A. \qquad (1)$$

(This integration should now be fairly familiar; we encountered it in Examples 6.3 and 6.4, where it was worked out in detail.)

We are given three conditions relating v and x.

(a) When $v = u$, $x = 0$.

(b) When $v = \frac{u}{3}$, $x = \frac{1}{2}L$.

(c) When $v = 0$, $x = L$.

By substituting the first of these in equation 1 we can determine a value for the integration constant A.

$$0 = -\frac{1}{2k}\ln(c^2 + u^2) + A,$$

giving
$$A = \frac{1}{2k}\ln(c^2 + u^2).$$

We insert this value of A into equation 1 to give the general equation relating v and x.

$$x = -\frac{1}{2k}\ln(c^2 + v^2) + \frac{1}{2k}\ln(c^2 + u^2).$$

$$x = \frac{1}{2k}\ln\left(\frac{c^2 + u^2}{c^2 + v^2}\right). \qquad (2)$$

Substituting the second and third conditions gives:

$$\tfrac{1}{2}L = \frac{1}{2k}\ln\left(\frac{c^2 + u^2}{c^2 + \frac{1}{9}u^2}\right).$$

$$L = \frac{1}{2k}\ln\left(\frac{c^2 + u^2}{c^2 + 0}\right).$$

Dividing the second of these by the first:

$$2 = \ln\left(\frac{c^2 + u^2}{c^2}\right)\Big/\ln\left(\frac{c^2 + u^2}{c^2 + \frac{1}{9}u^2}\right)$$

from which

$$\frac{c^2 + u^2}{c^2} = \left(\frac{c^2 + u^2}{c^2 + \frac{1}{9}u^2}\right)^2.$$

Letting the ratio $(u/c) = A$, this equation may be written more simply:

$$1 + A^2 = \left(\frac{1 + A^2}{1 + \frac{1}{9}A^2}\right)^2.$$

$$(1 + A^2)(1 + \tfrac{1}{9}A^2)^2 = (1 + A^2)^2.$$

Cancelling $(1 + A^2)$ and expanding,

$$\not{1} + \tfrac{2}{9}A^2 + \tfrac{1}{81}A^4 = \not{1} + A^2.$$

$$\tfrac{1}{81}A^4 = \tfrac{7}{9}A^2.$$

$$\therefore\ A^2 = 63.$$

$$\therefore\ \underline{u^2 = 63c^2}.$$

Now rewriting equation 2,

$$\ln\left(\frac{c^2 + u^2}{c^2 + v^2}\right) = 2kx.$$

$$\therefore\ \frac{c^2 + u^2}{c^2 + v^2} = e^{2kx}.$$

Substitute $x = s$ and $c^2 = \dfrac{u^2}{63}$:

$$\frac{(u^2/63 + u^2)}{(u^2/63 + v^2)} = e^{2ks}.$$

$$\frac{u^2 + 63u^2}{u^2 + 63v^2} = e^{2ks}.$$

Inverting both sides,

$$\frac{u^2 + 63v^2}{64u^2} = e^{-2ks}.$$

$$\therefore\ 63v^2 = 64u^2 \times e^{-2ks} - u^2.$$

$$\therefore\ \underline{63\frac{v^2}{u^2} = 64e^{-2ks} - 1.}$$

Example 6.17

An aircraft flying horizontally with a velocity V at a height h above the ground is to drop a projectile of mass m onto a train travelling with a constant velocity Z in the same direction as the aircraft and in the same vertical plane. In the vertical direction the projectile experiences no resistance to motion, whereas in the horizontal direction the resistance experienced is proportional to its instantaneous horizontal velocity, the constant of proportionality being K.

Derive an expression for the horizontal distance between the aircraft and the train at the instant the projectile is released if it is to score a direct hit.

U. Surrey

The vertical and horizontal components of motion of the projectile may be treated independently. Thus, the time between release of the projectile and it striking the target is the time for it to fall vertically through a height h, with no resistance to its vertical motion. We make use of the elementary formula

$$x = ut + \tfrac{1}{2}at^2.$$

$$h = 0 + \tfrac{1}{2}gt^2.$$

$$\therefore\ t = \sqrt{\frac{2h}{g}}.$$

The equation of horizontal motion is

$$-Kv = m\,\frac{\mathrm{d}v}{\mathrm{d}t} \tag{1}$$

(the negative sign recognising that the force is a resistance, not an accelerating force). Rearranging,

$$\int \mathrm{d}t = -\int \left(\frac{m}{Kv}\right) \mathrm{d}v.$$

$$\therefore\ t = -\frac{m}{K}\ln(v) + A.$$

When $t = 0$ (at the instant of release of the projectile) $v = V$, the velocity of the aircraft.

$$\therefore\ 0 = -\frac{m}{K}\ln(V) + A.$$

$$\therefore\ A = \frac{m}{K}\ln(V).$$

Substituting this value of A into the expression for t,

$$t = -\frac{m}{K}\ln(v) + \frac{m}{K}\ln(V).$$

$$t = \frac{m}{K}\ln\left(\frac{V}{v}\right). \tag{2}$$

We rewrite equation 1 to relate displacement x to velocity.

[d]

$$-Kv = mv\,\frac{\mathrm{d}v}{\mathrm{d}x}.$$

$$\therefore \int \mathrm{d}x = -\int \frac{m}{K}\,\frac{v\,\mathrm{d}v}{v} = -\frac{m}{K}\int \mathrm{d}v.$$

$$\therefore\ x = -\frac{m}{K}(v) + B.$$

When $x = 0$, $v = V$ as before.

$$\therefore\ 0 = -\frac{m}{K}(V) + B.$$

$$\therefore\ B = \frac{m}{K}(V).$$

Substituting in the expression for x,

$$x = -\frac{m}{K}(v) + \frac{m}{K}(V).$$

$$\therefore\ x = \frac{m}{K}(V - v). \tag{3}$$

Rearranging equation 2,

$$\ln\left(\frac{V}{v}\right) = \frac{tK}{m}.$$

$$\therefore \frac{V}{v} = e^{tK/m}.$$

$$\therefore v = Ve^{-tK/m}.$$

This can be substituted in equation 3:

$$x = \frac{m}{K}\left(V - ve^{-tK/m}\right)$$

$$= \frac{mV}{K}\left(1 - e^{-tK/m}\right).$$

Finally we substitute for t the value obtained at the beginning.

$$x = \frac{mV}{K}\left(1 - e^{-\sqrt{2gh}(k/m)}\right)$$

While the projectile is falling, the train moves forward a distance (Zt). Therefore, the required horizontal distance initially between aircraft and train is:

$$x' = \frac{mV}{K}\left(1 - e^{-\sqrt{2gh}(k/m)}\right) - Z\sqrt{\frac{2h}{g}}.$$

Example 6.18

The resistance to motion of a ship of mass 200 tonnes is directly proportional to the square of its speed within the speed range of 3 to 6 m s^{-1}, and the power required to overcome the resistance at 6 m s^{-1} is 1500 kW.

The ship is moving at 3 m s^{-1} and the propeller thrust is suddenly increased to 200 kN. Determine:

(a) the instantaneous acceleration of the ship;
(b) the time taken to increase speed to 5 m s^{-1};
(c) the distance travelled in this time.

U. Lond. U.C.

Solution 6.18

The data given relating to the power at a speed of 6 m s^{-1} permit the resistance function to be calculated; that is, we can derive a formula enabling the resisting force to be calculated for any speed. Resistance is given as being proportional to the square of the speed; therefore

$$R = Kv^2$$

and we begin by calculating the coefficient K. When the ship travels at a constant speed of 6 m s^{-1} the resisting force must be equal to the propelling force (because there is no acceleration).

$$\text{Power} = \text{Force} \times \text{Velocity}.$$

Hence:

$$R \times v = W.$$

$$\therefore (Kv^2)v = W.$$

$$\therefore K(6)^3 = 1\,500\,000.$$

$$\therefore\ K = \frac{1\,500\,000}{(6)^3} = 6944.4 \text{ kg m}^{-1}.$$

Part (a)

When the thrust is suddenly increased at 3 m s^{-1} the equation of motion is:

$$F - Kv^2 = ma.$$

$$\therefore\ 200\,000 - 6944.4 \times (3)^2 = 2000\,000a.$$

$$\therefore\ a = 1 - \frac{6944.4 \times 9}{200\,000}$$

$$= \underline{0.6875 \text{ m s}^{-2}}.$$

Part (b)

The general equation of motion, when propeller thrust is greater than resistance, is

$$F - Kv^2 = ma = m\,\frac{\mathrm{d}v}{\mathrm{d}t}. \qquad (1)$$

$$\therefore \int \mathrm{d}t = m \int \left(\frac{1}{F - Kv^2}\right) \mathrm{d}v$$

$$= \frac{m}{K} \int \left(\frac{1}{\frac{F}{K} - v^2}\right) \mathrm{d}v.$$

To simplify writing, let $\left(\frac{F}{K}\right) = b^2$. Then:

$$\int \mathrm{d}t = \frac{m}{K} \int \left(\frac{1}{b^2 - v^2}\right) \mathrm{d}v$$

$$= \frac{m}{K} \int \left(\frac{1}{(b+v)(b-v)}\right) \mathrm{d}v,$$

which can be simply divided into partial fractions:

$$\int \mathrm{d}t = \frac{m}{K} \int \left(\frac{\frac{1}{2}}{b+v} + \frac{\frac{1}{2}}{b-v}\right)\left(\frac{1}{b}\right) \mathrm{d}v.$$

$$\therefore\ t = \frac{m}{2bK}\,(\ln(b+v) - \ln(b-v) + A).$$

Observe that the integration constant A is within the bracket. It does not matter whether it is inside or outside the bracket, because the value of A will be determined by a specific condition relating t and v. You may verify this yourself by rewriting the last line with A outside the bracket, and calculating its value in accordance with the working that follows. You will have a different value for A than will be found here, but it will still give the correct expression for t.

t may be written in a simpler form:

$$t = \frac{m}{2bK}\left[\ln\left(\frac{b+v}{b-v}\right) + A\right].$$

Substituting the 'initial condition' that when $t = 0$, $v = v_0$, the initial velocity:

$$0 = \frac{m}{2bK}\left[\ln\left(\frac{b+v_0}{b-v_0}\right) + A\right].$$

$$\therefore\ A = -\ln\left(\frac{b + v_0}{b - v_0}\right).$$

Substituting this value in the equation for t:

$$t = \frac{m}{2bK}\left[\ln\left(\frac{b + v}{b - v}\right) - \ln\left(\frac{b + v_0}{b - v_0}\right)\right].$$

$$\therefore\ t = \frac{m}{2bK}\ \ln\left(\frac{b + v}{b - v}\right)\left(\frac{b - v_0}{b + v_0}\right).$$

$$b = \sqrt{\frac{F}{K}} = \sqrt{\frac{200\,000}{6944.4}} = 5.367.$$

$$\frac{m}{2bK} = \frac{200\,000}{2 \times 5.367 \times 6944.4} = 2.6831.$$

$$\therefore\ t = 2.6831\ \ln\left(\frac{5.367 + 5}{5.367 - 5}\right)\left(\frac{5.367 - 3}{5.367 + 3}\right)$$

$$= 2.6831\ \ln\,(7.9913)$$

$$= \underline{5.576\ \text{s}}.$$

Part (c)

To determine displacement, we rewrite equation 1:

$$F - Kv^2 = mv\,\frac{\mathrm{d}v}{\mathrm{d}x}.$$

$$\therefore\ \int \mathrm{d}x = m\int\left(\frac{v}{F - Kv^2}\right)\mathrm{d}v$$

$$= \frac{m}{K}\int\left(\frac{v}{\frac{F}{K} - v^2}\right)\mathrm{d}v$$

$$= \frac{m}{K}\int\left(\frac{v}{b^2 - v^2}\right)\mathrm{d}v \qquad \text{as before.}$$

This now familiar integral (see Examples 6.3, 6.4, 6.16) gives us

$$x = -\frac{m}{2K}\left[\ln\,(b^2 - v^2) + B\right].$$

When $x = 0$, $v = v_0$.

$$0 = -\frac{m}{2K}[\ln\,(b^2 - v_0^2) + B].$$

$$\therefore\ B = -\ln\,(b^2 - v_0^2).$$

$$\therefore\ x = -\frac{m}{2K}[\ln\,(b^2 - v^2) - \ln\,(b^2 - v_0^2)].$$

$$\therefore\ x = +\frac{m}{2K}\ln\left(\frac{b^2 - v_0^2}{b^2 - v^2}\right).$$

Substituting the values,

$$x = \frac{200\,000}{2 \times 6944.4} \ln\left(\frac{5.367^2 - 3^2}{5.367^2 - 5^2}\right)$$

$$= 14.4 \ln (5.206).$$

$$x = \underline{23.76 \text{ m.}}$$

Example 6.19

A free-fall parachutist drops from a stationary balloon at 2000 m and falls vertically downwards. If the resistance to motion is of the form mKv^2, where v is the parachutist's velocity and K is given as 5×10^{-3} m^{-1}, calculate his terminal velocity and the height at which he achieves 99 per cent of this velocity.

Outline a method for calculating the time to reach this height.

U. Lond. U.C.

Solution 6.19

As a body falls through air with a resistance which increases with speed, the greater the speed, the less the resultant downward force acting on the body. So the acceleration decreases, and (in theory at least) eventually becomes zero, when the resistance to motion is exactly equal to the weight of the body. This limiting speed is called the terminal velocity of the body. It is a theoretical concept to the extent that, assuming the resistance to vary as stated, it will be seen that the body only attains its terminal velocity after infinite time. In practice, falling bodies reach an approximately constant speed quite quickly.

At terminal velocity, weight is equal to resistance. Calling the required velocity $\hat{v}$,

$$mg = m \times K(\hat{v})^2.$$

$$\therefore \hat{v} = \sqrt{\frac{g}{K}} = \sqrt{\frac{9.81}{5 \times 10^{-3}}} = \underline{44.29 \text{ m s}^{-1}}.$$

The general equation of motion, taking down as positive, is:

[d] $$mg - mKv^2 = ma = mv\frac{dv}{dx}.$$

$$\therefore g - Kv^2 = v\frac{dv}{dx}.$$

$$\therefore \int dx = \int \frac{v}{g - Kv^2} dv.$$

$$\therefore \int dx = \frac{1}{K}\int\left(\frac{v}{\frac{g}{K} - v^2}\right)dv = \frac{1}{K}\int\left(\frac{v}{\hat{v}^2 - v^2}\right)dv.$$

Again we have the integration already met in Examples 6.3, 6.4, 6.16 and 6.18.

$$x = -\frac{1}{2K}[\ln(\hat{v}^2 - v^2) + A].$$

When $x = 0$, $v = 0$,

$$0 = -\frac{1}{2K}[\ln(\hat{v}^2) + A].$$

$$\therefore A = -\ln(\hat{v}^2).$$

$$\therefore x = -\frac{1}{2K}[\ln(\hat{v}^2 - v^2) - \ln(\hat{v}^2)].$$

$$\therefore x = \frac{1}{2K}\ln\left(\frac{\hat{v}^2}{\hat{v}^2 - v^2}\right).$$

Substituting values,

$$x = \frac{1}{2 \times 5 \times 10^{-3}}\ln\left(\frac{(\hat{v})^2}{(\hat{v})^2 - (0.99\hat{v})^2}\right)$$

$$= 100\ln\left(\frac{1}{1 - 0.9801}\right)$$

$$= 100\ln(50.25)$$

$$= 391.7 \text{ m}.$$

Since the body begins to fall from a height of 2000 m, the height above ground required is

$$h = 2000 - 391.7 = 1608.3 \text{ m}.$$

It is clear that, for practical purposes, we could say that the body had achieved terminal velocity at this height.

A more accurate assessment can be made merely by adopting a higher percentage of terminal velocity in the calculation. For example, you may show, by a similar calculation to the one above, that 99.9 per cent of terminal velocity will be reached for a value of x of 621.5 m.

To determine the time to reach the stated height, we write the equation of motion thus:

$$mg - mKv^2 = m\frac{dv}{dt}.$$

$$\therefore \int dt = \int\left(\frac{1}{g - Kv^2}\right)dv$$

$$= \frac{1}{K}\int\left(\frac{1}{\frac{g}{K} - v^2}\right)dv.$$

$$\therefore \int dt = \frac{1}{K}\int\left(\frac{1}{\hat{v}^2 - v^2}\right)dv,$$

and so far as the question is concerned this is as far as an examinee would be required to go; the question merely asks for an outline of the method. But we shall proceed here to complete the solution.

We met this integration before (see Example 6.18). Using the result obtained there,

$$t = \frac{1}{2\hat{v}K}\left[\ln\left(\frac{\hat{v} + v}{\hat{v} - v}\right) + B\right].$$

When $t = 0$, $v = 0$,

$$0 = \frac{1}{2\hat{v}K}\left[\ln\left(\frac{\hat{v}}{\hat{v}}\right) + B\right],$$

which gives $B = 0$ since ln (1) is 0.

$$\therefore\ t = \frac{1}{2\hat{v}K}\ \ln\left(\frac{\hat{v} + v}{\hat{v} - v}\right).$$

Substituting $v = 0.99\hat{v}$ and other values:

$$t = \frac{1}{2 \times 44.29 \times 5 \times 10^{-3}}\ \ln\left(\frac{1 + 0.99}{1 - 0.99}\right)$$

$$= 2.258 \ln (199)$$

$$= \underline{11.95 \text{ s.}}$$

A simple calculation shows that a body in free fall against zero resistance would take 8.9 seconds to fall the distance of 391.7 m calculated earlier. As expected, the time of fall is reduced.

Example 6.20

An aircraft lands on the deck of an aircraft carrier and is brought to rest in 4 s by the arrester gear. An accelerometer attached to the aircraft provides the acceleration record shown in the diagram. Estimate (a) the initial velocity of the aircraft relative to the deck, and (b) the distance the aircraft travels along the deck before coming to rest.

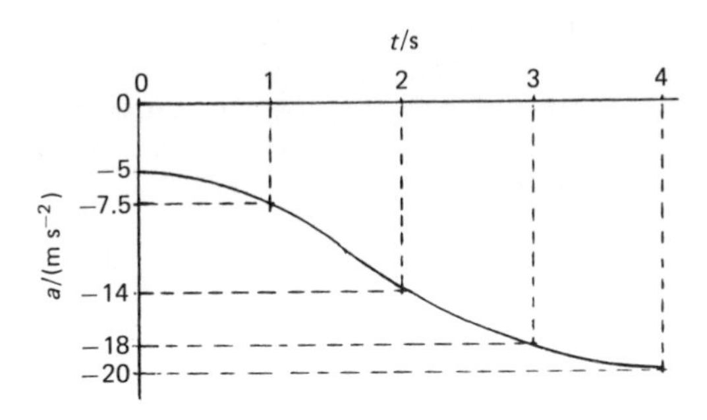

Thames Poly.

Solution 6.20

$$a = \frac{dv}{dt},$$

$$\int a \, dt = \int dv.$$

i.e. the area under an acceleration–time graph gives the change of velocity. We may calculate the area by assuming the graph to consists of four trapezia, each of width 1 second.

$$\Sigma\,(a\,\Delta t) = 1[\tfrac{1}{2}(5 + 7.5) + \tfrac{1}{2}(7.5 + 14) + \tfrac{1}{2}(14 + 18) + \tfrac{1}{2}(18 + 20)]$$

$$= 52 \text{ m s}^{-1}.$$

All accelerations are negative and the final speed is zero. Hence:

Initial velocity = $\underline{52 \text{ m s}^{-1}}$.

The velocity after each second can be calculated by subtracting each strip from the previous velocity. This is done in the table below. For each increment, we may then assume a mean velocity $\overline{v}$ being the average of the two velocities at the beginning and the end of that increment. Values of $\overline{v}$ are also given in the table.

Time t	0	1	2	3	4
Area	6.25	10.75	16	19	
v	52	45.75	35	19	0
$\overline{v}$	48.875	40.375	27	9.5	

Since each strip is 1 s wide, the corresponding distance travelled is numerically the same as $\overline{v}$.

Hence the total distance travelled is the sum of the $\overline{v}$ line:

$$\text{Distance} = \underline{125.75 \text{ m.}}$$

Remember that such a solution as the above is approximate only. Greater accuracy could be obtained by taking more and smaller increments. But it is clear from the diagram that the examiner intends four increments to be used.

Example 6.21

The kinematics of a particle are described by position x, velocity v and time t. If $v = u$ and $x = 0$ when $t = 0$, show that for constant acceleration f,
(a) $v = u + ft$;
(b) $x = ut + \frac{1}{2}ft^2$;
(c) $v^2 = u^2 + 2fx$.
A particle of mass m is projected vertically upwards with initial velocity u against gravitational acceleration g and air resistance kv^2 (k const.). Show that the maximum height attained is given by

$$x = \frac{m}{2k} \ln\left(1 + \frac{ku^2}{mg}\right).$$

U. Manchester

Solution 6.21

We will not present the solution to the first part of the question, as this can be found in any suitable textbook.

As the body ascends, weight and air resistance will both act downwards (i.e. negatively). The equation of motion is therefore

$$-kv^2 - mg = ma = mv\frac{dv}{dx}.$$

$$\therefore \int dx = -m\int \frac{v\,dv}{mg + kv^2}.$$

$$\text{Let } u = mg + kv^2.$$

$$\frac{du}{dv} = 2kv.$$

$$\therefore v\,dv = \frac{du}{2k}.$$

$$\therefore \int \mathrm{d}x = -\frac{m}{2k}\int \frac{\mathrm{d}u}{u}$$

$$= -\frac{m}{2k}\Big[\ln(u)\Big]$$

$$= -\frac{m}{2k}\Big[\ln(mg + kv^2)\Big]_{v_1}^{v_2}.$$

$$x = +\frac{m}{2k}\ln\left(\frac{mg + kv_1^2}{mg + kv_2^2}\right).$$

For maximum height, $v_2 = 0$; $v_1 = u$. Let max. height be $\hat{x}$.

$$\therefore \hat{x} = \frac{m}{2k}\ln\left(1 + \frac{ku^2}{mg}\right).$$

6.3 Problems

Problem 6.1

A body of mass m is subjected to a single accelerating force which is proportional to the velocity of the body at any instant. If, at a certain instant, the velocity of the body is 15 m s^{-1} and it increases to 150 m s^{-1} in 4 seconds, calculate how far the body will move during this time.

Problem 6.2

A missile having a mass of 60 kg is subjected to a propelling force along a horizontal frictionless track. The force F is calculated from

$$F = (1200 - 0.4t^3) \text{ newtons,}$$

where t is the time elapsed from the instant of firing. Determine (a) the maximum acceleration of the missile; (b) its maximum velocity; (c) the distance travelled in reaching maximum velocity. It is to be assumed that the force ceases to operate when its value becomes zero.

Problem 6.3

A uniform flexible chain of mass M and length L is wound around the periphery of a drum of radius R and of negligible moment of inertia. The drum is mounted on a frictionless horizontal axis. An initial length x_0 of chain is allowed to hang vertically from the drum, and the system is released from rest. Obtain an expression for the time elapsed before all of the chain has unwound from the drum. You may make use of the integral:

$$\int \frac{\mathrm{d}x}{\sqrt{x^2 - x_0^2}} = \ln\left(x + \sqrt{x^2 - x_0^2}\right).$$

If L is 5 m and it takes 2.79 seconds to leave the drum, calculate the initial overhang, and the final linear velocity as it leaves the drum.

Problem 6.4

A body having a mass of 25 kg moves along a horizontal track. It is subjected to a retarding force F given by the expression

$$F = (24 + 0.5v^2) \text{ newtons,}$$

where v is the velocity at any instant. Calculate the time taken for the body to come to rest, from an initial velocity of 10 m s^{-1}, and the distance travelled while coming to rest.

Problem 6.5

A vehicle of mass M moving along a straight horizontal track is subjected to a resisting force R which is given by the expression

$$R = a + bv^2$$

in which v is the velocity of the vehicle at any instant, and a and b are constants. If R is the only force affecting the motion, show that, if the initial velocity of the vehicle is V, it will come to rest in a distance of

$$\frac{M}{2b} \ln\left(1 + \frac{bV^2}{a}\right).$$

If M is 240 tonnes and $R = (75 + 0.65v^2)$ kN, calculate the distance travelled by the vehicle in slowing down from 100 km h^{-1} to 50 km h^{-1}.

Problem 6.6

A body moves along a straight horizontal path subjected to a single retarding force R which may be assumed to have the form

$$R = Kv^2$$

where v is the velocity at any instant, and K is a constant. The body initially has a velocity of 10 m s^{-1} and it is reduced to half this value while travelling a distance of 42 m. Calculate (a) the time taken to cover this distance, and (b) the time and distance for the velocity of the body to fall to 0.1 m s^{-1}.

Problem 6.7

A train has a total mass of 280 000 kg. The locomotive exerts a constant pull at all speeds. The resistance to motion varies as the square of the speed at any instant. On a straight level track, the train is capable of a maximum speed of 115 km h^{-1}, the power of the engine at this speed being 1350 kW. Determine how far the train has travelled from rest by the time its speed has reached 60 km h^{-1}, and the time for it to reach this speed.

Problem 6.8

A body having a mass of 5 kg is allowed to fall vertically from rest from a height of 2 m on to the top of a spring which exerts a retarding force on the body of ($400x$) newtons, where x is the amount of spring compression at any instant. The spring compresses and then extends, projecting the body upwards. Neglecting all forces other than the spring force and the weight of the body, show that the body will remain in contact with the spring for an approximate time of 0.39 seconds.

The general solution to the differential equation

$$\frac{d^2x}{dt^2} + \left(\frac{k}{m}\right)x = g$$

is

$$x = A \sin\left(\sqrt{\frac{k}{m}}\, t\right) + B \cos\left(\sqrt{\frac{k}{m}}\, t\right) + \frac{gm}{k}.$$

Problem 6.9

A body having a mass of 3 kg is projected vertically upwards with an initial velocity of 40 m s^{-1}. It reaches a maximum height of 31 m. Assuming the resistance R due to motion through the air to be of the form

$$R = kv^2,$$

show that this result is consistent with a value for k of 0.0822, and calculate the time for the body to reach maximum height.

Problem 6.10

A body having a mass of 220 kg is subjected to a propelling force which varies with time according to the table here.

Time, seconds	0	1	2	3	4	5	6	7	8	9	10	11	12	13
Force, kN	0	2.7	4.8	6.3	7.4	8.3	8.8	9.1	9.2	8.7	7.6	6.0	3.7	0

By dividing the motion into thirteen time-elements of 1 s each, determine the change of velocity of the body and the distance travelled during each increment, assuming it to start from rest, by assuming a mean constant force for each increment. Hence estimate the final velocity of the body, and the total distance travelled during the period of propulsion. Calculate velocity increments to two places of decimals only, and displacement increments to one place of decimals only.

6.4 Answers to Problems

6.1 234.5 m.

6.2 (a) 20 m s^{-2}; (b) 216.3 m s^{-1}; (c) 1872 m.

6.3 $t = \sqrt{\dfrac{L}{g}} \ln \left(\dfrac{L + \sqrt{L^2 - x_0^2}}{x_0} \right)$; 0.2 m; 6.998 m s^{-1}.

6.4 6.964 s; 28.15 m.

6.5 195.1 m.

6.6 (a) 6.059 s; (b) 600 s; 279 m.

6.7 1074 m; 122.5 s.

6.9 2.178 s.

6.10 Final velocity = 375.5 m s^{-1}; displacement = 2349 m.

7 Vibration

Newton's second law. Moment of inertia; radius of gyration. Simple harmonic motion. Period; frequency. Stress analysis; simple tension and torsion. Spring stiffness.

7.1 The Fact Sheet

(a) Newton's Second Law; Translation

$$\Sigma F = ma$$

where ΣF is the resultant force on a body in newtons (N),
a is the acceleration in metres per second per second (m s^{-2}), and
m is the mass in kilograms (kg).

Newton's Second Law: Rotation

$$\Sigma M = I\alpha$$

where ΣM is the resultant moment of all the forces on a body,
I is the moment of inertia in kilogram metres squared (kg m^2),
α is the angular acceleration in radians per second per second (rad s^{-2}).

(b) Moment of Inertia

This is defined by:

$$I = \Sigma\,(\delta m r^2)$$

$$= mk^2$$

where k is the radius of gyration of the body with respect to the spin axis.

Uniform disc of mass m, radius R:

$$I = \tfrac{1}{2}mR^2.$$

Uniform thin bar of mass m, length L:

$$I = \tfrac{1}{3}mL^2 \quad \text{(axis through one end);}$$

$$I = \tfrac{1}{12}mL^2 \quad \text{(axis through centre).}$$

Parallel-axis theorem:

$$I_o = I_G + mh^2.$$

(c) Simple Harmonic Motion, s.h.m.

$$x = A \sin(\omega t + \phi)$$

where x is the displacement, A is the amplitude, ω is the natural circular frequency and ϕ is a constant.

(This may be expressed in the alternative form:

$$x = A \sin(\omega t) + B \cos(\omega t).)$$

$$v = \dot{x} = A\omega \cos(\omega t + \phi).$$

$$a = \ddot{x} = -A\omega^2 \sin(\omega t + \phi)$$

$$= -\omega^2 x.$$

$$\tau = \frac{2\pi}{\omega}.$$

where τ is the period in seconds per cycle.

$$n = \frac{\omega}{2\pi}$$

where n is frequency in cycles per second or hertz (Hz).

Maximum velocity: $\hat{v} = A\omega$.
Maximum acceleration: $\hat{a} = A\omega^2$.

(d) Stress Analysis

For a uniform bar of length L and cross-section of area a, subject to a tensile or compressive force F within the limit of proportionality, extension or compression x is given by:

$$x = \frac{FL}{aE}$$

where E is Young's modulus of elasticity (newtons per square metre).

For a uniform bar of length L and circular cross-section subject to a torque M within the limit of proportionality, the angular twist θ (radians) is given by:

$$\theta = \frac{ML}{JG}$$

where G is the modulus of rigidity (newtons per square metre) and J is the polar second moment of area of cross-section (m^4).

For a bar of circular cross-section, outside and inside diameters D_o and D_i,

$$J = \frac{\pi}{32}(D_o^4 - D_i^4).$$

(e) Spring Stiffness

Spring stiffness k is the force per unit extension or compression:

$$k = \frac{F}{x}.$$

Torsional spring stiffness k_t is the torque per unit twist:

$$k_t = \frac{M}{\theta}.$$

(f) Springs in Series and in Parallel

Springs in series:

$$\frac{1}{k} = \frac{1}{k_1} + \frac{1}{k_2} + \text{etc.}$$

Springs in parallel:

$$k = k_1 + k_2 + \text{etc.}$$

(g) System Natural Frequency

When a system is given an initial disturbance, causing it to vibrate, and is not subsequently disturbed, the frequency of the vibration is called the system natural frequency.

(h) Procedure for Solution of Problems

(i) Ascribe +*ve* and −*ve* signs to directions of displacement.
(ii) Assume a small positive displacement.
(iii) Draw the resulting free-body diagram for the displaced body.
(iv) Write the equation of motion for the body ($\Sigma F = ma$ or $\Sigma M = I\alpha$ according to whether motion is translation or rotation). *Note sign.*
(v) Arrange the equation in the form

$$\ddot{x} = -Kx \qquad (\text{where } \ddot{x} = a)$$

or in the form

$$\ddot{\theta} = -K\theta \qquad (\text{where } \ddot{\theta} = \alpha)$$

where K is a constant in terms of the system parameters.
(vi) Equate constant K in (v) to ω^2; hence determine ω.
(vii) Thus evaluate frequency n ($= \omega/2\pi$) and/or period τ ($= 2\pi/\omega$).
(viii) If specific values of displacement, velocity and acceleration are required, determine the particular form of the displacement and velocity functions ((c) above) and determine the constants by substituting two initial conditions of motion.

7.2 Worked Examples

Example 7.1

A body of mass m is attached to one end of a helical spring of stiffness k. The other end of the spring is attached to a rigid mounting. The body is given an initial small displacement in the direction of the spring axis and then released. Show that, subject to certain assumptions, the subsequent motion will be simple harmonic (a) if the spring axis is horizontal and the body rests on a smooth horizontal surface; (b) if the spring axis is vertical and the body hangs freely. State in each case the assumptions made.

A body of mass 1.2 kg hangs from a helical spring. When disturbed and caused to oscillate, the body is observed to make 20 oscillations in 14.2 seconds. Caculate the stiffness of the spring.

Solution 7.1

Part (a)

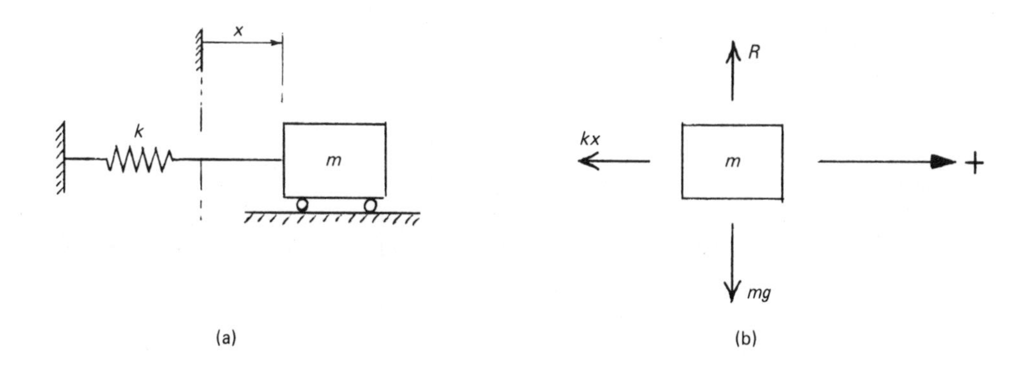

A sketch of the arrangement is shown in diagram (a). We assume that the body is actually in motion, and that we are looking at it at an instant when it has an instantaneous displacement x to the right of the statically neutral position, shown by the vertical chain-line. Our assumption that x is to the right is arbitrary; we could as easily have chosen displacement to the left. But the choice commits us from now on to the sign convention that 'right is positive'.

Diagram (b) is the free-body diagram. If all effects due to the presence of air are neglected, three forces are seen to act on the body. The weight mg acts vertically downwards and the surface must exert an upward force R on the body. Since motion is along a horizontal path, these two forces must be equal, and are not necessary to the solution. The third force is due to contact with the spring. Since displacement is to the right, the spring must exert a restoring force to the left, i.e. the spring must be in tension. (If we had chosen to assume a left-directed displacement, the spring would have been in compression and the force would have been to the right.) The magnitude of this force must be (kx).

The general equation of motion for translation is $\Sigma F = ma$. Having regard to sign,

[a] $$-kx = ma,$$

which can be rearranged:

$$\ddot{x} = -\left(\frac{k}{m}\right) x \qquad \text{(using the notation } a = \ddot{x}\text{).}$$

This is seen to have the same form as

[c] $$\ddot{x} = -(\omega^2)x,$$

proving that the motion is simple harmonic. The particular value of ω in this case is

$$\omega = \sqrt{\frac{k}{m}}.$$

The assumptions made are:

(i) The spring is linear (not strained beyond the limit of proportionality),
(ii) Resistance of air is negligible.
(iii) The mass of the spring is negligible.

Part (b)

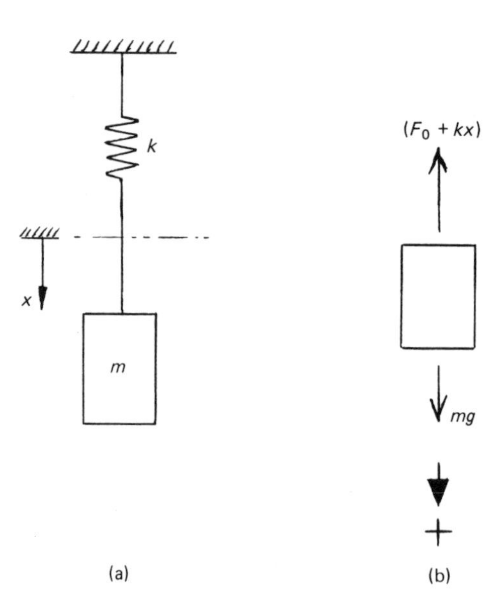

(a) (b)

Although this case looks simpler, extra care is needed. In diagram (a) we assume the body to have an instantaneous downward displacement x below the position of neutral equilibrium; thus, 'down is positive'. The spring is in tension and thus pulls upwards on the body. Unlike part (a), however, the spring force will *not* be zero in the position of static equilibrium but will have an initial value which we call F_0. So the spring force disclosed by the free-body diagram (b) will comprise this force plus the extra tension due to displacement x.

For translation, $\Sigma F = ma$. Having due regard to sign,

[a] $$-(F_0 + kx) + mg = ma.$$

It is clear that in the position of static equilibrium, $F_0 = mg$.
Thus

$$-(mg + kx) + mg = ma,$$

giving $$-kx = m\ddot{x}.$$

$$\therefore \ddot{x} = -\left(\frac{k}{m}\right)x,$$

which is the same equation as for part (a), proving that the frequency of motion is the same in both cases.

For the numerical problem we first calculate the frequency, n:

$$n = \left(\frac{20}{14.2}\right) \text{ cycles s}^{-1}\text{, or hertz (Hz).}$$

$$n = \frac{\omega}{2\pi}.$$

$$\therefore \ \omega = 2\pi n$$

$$= 2\pi \times \frac{20}{14.2}$$

$$= 8.85 \text{ s}^{-1}.$$

$$\omega^2 = \frac{k}{m}.$$

$$\therefore \ k = \omega^2 m$$

$$= (8.85)^2 \times 1.2.$$

$$\therefore \ k = \underline{93.98 \text{ N m}^{-1}}.$$

Oscillation of a body of known mass at the end of a spring is a useful practical way of determining the spring stiffness.

This particular case of a mass m on a spring of stiffness k arises so frequently that it is worth remembering the expression for the natural frequency,

$$n = \frac{1}{2\pi}\sqrt{\frac{k}{m}},$$

and we will refer to this result in several of the examples that follow.

Example 7.2

Show that a rotor having a moment of inertia I oscillating angularly at the end of a torsional spring of stiffness k_t will execute s.h.m. with a frequency n given by

$$n = \frac{1}{2\pi}\sqrt{\frac{k_t}{I}}.$$

A steel disc has a mass of 12.4 kg and a radius of gyration of 0.07 m. It is attached to the lower end of a vertical steel rod which has a solid circular cross-section 9 mm in diameter and a length of 1.75 m. The upper end of the rod is attached to a rigid mounting. The disc is displaced angularly by a small amount and then released. In the resulting oscillation it is observed that 20 cycles are completed in 5.71 seconds. Determine the value of G, the modulus of rigidity of the rod.

Solution 7.2

The problem is analogous to that of a mass on the end of a spring (see Example 7.1.) In diagram (a) here the rotor is assumed to have an instantaneous angular displacement θ clockwise viewed from above. Diagram (b) shows the rotor in plan view from above, as a free-body diagram. The effect of the spring is a single torque of magnitude ($k_t\theta$), anticlockwise, as the displacement is clockwise. The rotor will also be subjected to a downward weight and an equal upwards tension in the rod, but these two forces are not relevant to the motion and are not shown, being in any case perpendicular to the plane of the diagram. The direction of assumed displacement dictates a 'clockwise positive' sign convention. This is shown as an arrow in diagram (b).

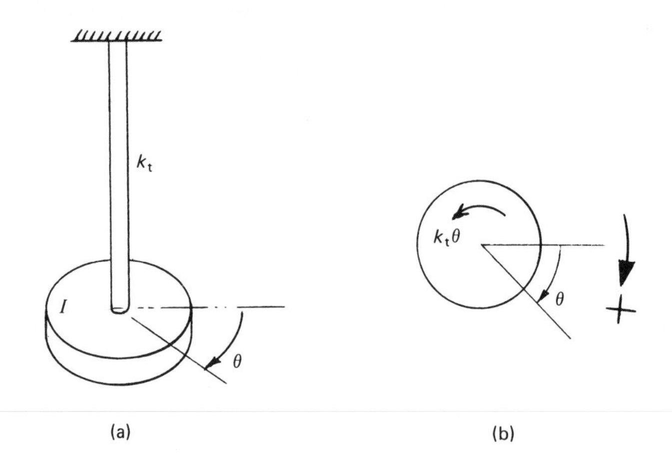

Motion is pure rotation: $\Sigma M = I\alpha$ applies. Noting the sign,

[a] $$-k_t\theta = I\ddot{\theta}.$$

Rearranging,

$$\ddot{\theta} = -\left(\frac{k_t}{I}\right)\theta.$$

This is seen to have the same form as $\ddot{x} = -(\omega^2)$, x and $\ddot{x}$ being replaced by θ and $\ddot{\theta}$. Thus s.h.m. is proved.

Equating the coefficient of θ to ω^2,

$$\omega = \sqrt{\frac{k_t}{I}}.$$

$$\therefore\ n = \frac{1}{2\pi}\sqrt{\frac{k_t}{I}}.$$

In solving the numerical problem, the value of I may first be calculated.

[b] $$I = mk^2$$
$$= 12.4 \times (0.07)^2$$
$$= 0.06076 \text{ kg m}^2.$$

Data concerning the number of oscillations permits evaluation of frequency n and hence ω.

$$n = \frac{20}{5.71} \text{ Hz.}$$

[c] $$\omega = 2\pi n$$
$$= 22.01 \text{ s}^{-1}.$$

From the proof in the first part of the problem,

$$k_t = \omega^2 I$$
$$= (22.01)^2 \times 0.06076$$
$$= 29.43 \text{ N m rad}^{-1}.$$

The rest of the solution requires stress-analysis theory.

[d] $$\theta = \frac{ML}{GJ},$$

giving $$\frac{M}{\theta} = \frac{GJ}{L}.$$

M/θ is the torsional stiffness k_t – the torque per unit twist.

$$\therefore\ G = k_t \frac{L}{J}.$$

[d] $$J = \frac{\pi}{32}(D_o^4 - D_i^4)$$

$$= \frac{\pi}{32}(9 \times 10^{-3})^4$$

$$= 6.441 \times 10^{-10}\ \text{m}^4.$$

$$\therefore\ G = 29.43 \times \frac{1.75}{(6.441 \times 10^{-10})}.$$

$$\therefore\ G = \underline{79.96 \times 10^9\ \text{N m}^{-2}}.$$

Example 7.3

A uniform solid steel disc has a mass of 15 kg and a radius of 0.14 m. It is mounted at the lower end of a vertical uniform steel shaft of diameter 8 mm and length 1.8 m, the upper end of the shaft being fixed.

(a) The mass is excited by a vibrator in an axial direction (i.e. along the direction of the shaft) and it is observed to 'resonate', i.e. vibrate at its own natural frequency, when the excitation frequency reaches a value of 97.8 Hz. Calculate the value of Young's modulus, E, for the shaft.

(b) The disc is then excited by an angular vibrator, causing it to resonate in angular mode at an excitation frequency of 1.732 Hz. Evaluate G, the modulus of rigidity for the shaft.

Solution 7.3

Part (a)

The steel rod may be treated as a spring. We may 'short-cut' the analysis by making use of the expression derived in Example 7.1:

$$n = \frac{1}{2\pi}\sqrt{\frac{k}{m}}.$$

The spring stiffness may be calculated. Stiffness is defined as force per unit extension, i.e. F/x.

[d] $$x = \frac{FL}{aE}.$$

$$k = \frac{F}{x} = \frac{aE}{L}.$$

$$\therefore\ n = \frac{1}{2\pi}\sqrt{\frac{aE}{Lm}}.$$

Rearranging, $$\therefore E = \frac{4\pi^2 n^2 Lm}{a}$$

$$= \frac{4\pi^2 (97.8)^2 \times 1.8 \times 15}{(\pi/4)(8 \times 10^{-3})^2}.$$

$$\therefore E = \underline{202.8 \times 10^9 \text{ N m}^{-2}}.$$

Part (b)

A rotor on the end of a shaft vibrating torsionally was analysed in Example 7.2. Using the result from that example,

$$n = \frac{1}{2\pi}\sqrt{\frac{k_t}{I}}.$$

We calculate k_t from the torsion formula:

[d] $$\theta = \frac{ML}{GJ}.$$

[e] $$k_t = \frac{M}{\theta} = \frac{GJ}{L}.$$

For a solid shaft,

[d] $$J = \frac{\pi}{32} D^4.$$

For a solid disc,

[b] $$I = \tfrac{1}{2}mR^2.$$

Substituting in n above,

$$n = \frac{1}{2\pi}\sqrt{\frac{GJ}{LI}}$$

$$= \frac{1}{2\pi}\sqrt{\frac{G(\pi/32)D^4}{L \times \frac{1}{2}mR^2}}.$$

Rearranging, $$G = \frac{4\pi^2 n^2 L(\frac{1}{2}mR^2) \times 32}{\pi(8 \times 10^{-3})^4}$$

$$= \frac{4\pi^2 (1.732)^2 \times 1.8 \times 0.5 \times 15 \times (0.14)^2 \times 32}{\pi(8 \times 10^{-3})^4}.$$

$$G = \underline{77.9 \times 10^9 \text{ N m}^{-2}}.$$

This is typical of many problems which are much simplified by working throughout in algebra, resisting the substitution of numerical values until the final stage. This results in only one operation with the calculator, and avoids possible errors arising out of manipulating and rewriting very long numbers and decimals.

Example 7.4

A simple pendulum consists of a body of mass m and of negligible size at the end of a massless string of length L, the upper end of the string being attached to a fixed support. The body is displaced sideways by a small amount and released from rest. Show that the resulting motion will be simple harmonic, and state any assumptions necessary.

A simple pendulum has a length of 6 m. It is displaced so that the string makes an angle of 5° with the vertical and is released from rest in this position.

(a) Calculate the angular displacement after 1 second.
(b) Calculate the least time for the pendulum to reach the lowest point of swing.
(c) Calculate the linear velocity of the body at this stage.

Solution 7.4

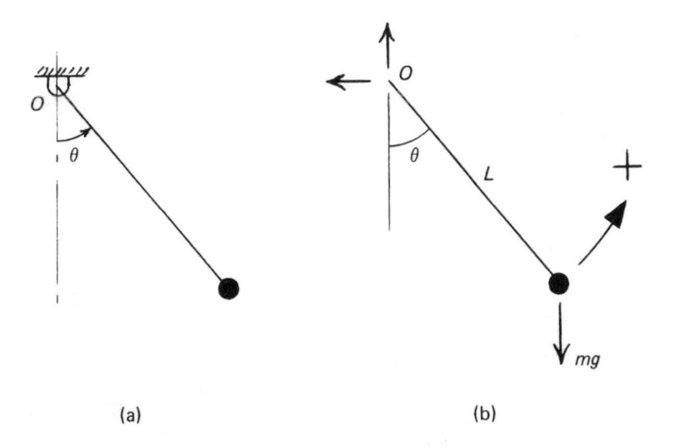

(a) (b)

Diagram (a) shows the pendulum arrangement; (b) is the free-body diagram, the 'body' including the massless string. Neglecting air resistance, the forces acting on the body comprise its weight *mg* and an unknown force exerted by the pivot at *O* on the string; this second force is shown as a pair of components. Displacement is assumed anticlockwise; this determines our sign convention.

Motion is seen to be pure rotation about *O*. $\Sigma M = I\alpha$ applies. Examining all forces, it is seen that (*mg*) is the only force exerting a moment about *O*. This moment is clockwise, and hence negative.

[a]
$$\Sigma M = I\alpha.$$
$$\therefore\ -(mgL \sin\theta) = I\ddot{\theta}.$$

[b]
$$I = \Sigma\,(\delta m r^2)$$
$$= mL^2$$

(since the size of the body is negligible).

$$\therefore\ -(mgL \sin\theta) = mL^2\ddot{\theta}.$$

Rearranging,

$$\ddot{\theta} = -\left(\frac{g}{L}\right)\sin\theta.$$

This equation does *not* conform with the equation which characterises s.h.m.: the variable on the left side is $\ddot{\theta}$ whereas we have $\sin\theta$ on the right hand. The motion will therefore not be s.h.m. But if the displacement is restricted to small values of θ then we can use the approximation

$$\theta \approx \sin\theta.$$

You can verify the truth of this. For example, use your calculator to determine sin 5°, and 5° in radian measure. Using this approximation,

$$\ddot{\theta} = -\left(\frac{g}{L}\right)\theta,$$

which is analogous to

$$\ddot{\theta} = -(\omega^2)\theta.$$

ω is seen to have the particular value $\sqrt{\frac{g}{L}}$.

[c] $$n = \frac{\omega}{2\pi} = \frac{1}{2\pi}\sqrt{\frac{g}{L}}.$$

Motion will therefore be s.h.m., provided displacement is limited to small angles – say 10° or less.

It is also assumed that resistance due to the air can be neglected.

The second part of the problem requires the angular form of the displacement function.

[c] $$x = A \sin(\omega t + \phi).$$

For angular motion,

$$\theta = \hat{\theta} \sin(\omega t + \phi).$$

Differentiating: $$\dot{\theta} = \hat{\theta}\omega \cos(\omega t + \phi).$$

$\hat{\theta}$ and ϕ can be found from information about the initial conditions of the motion. These are, (i) $t = 0$, $\theta = 5°$, and (ii) $t = 0$, $\dot{\theta} = 0$ (the pendulum is released from *rest*).

Substituting $t = 0$, $\theta = 5°$,

$$5 = \hat{\theta} \sin(0 + \phi). \tag{1}$$

Substituting $t = 0$, $\dot{\theta} = 0$,

$$0 = \hat{\theta}\omega \cos(0 + \phi). \tag{2}$$

Equation 2 may be solved directly:

$$\cos(0 + \phi) = 0$$

$$\therefore \phi = \frac{\pi}{2}.$$

Substituting in equation 1,

$$5 = \hat{\theta} \sin\left(\frac{\pi}{2}\right).$$

$$\therefore \hat{\theta} = 5°.$$

(Note that in this calculation we have an exception to the usual rule of converting displacement and velocity to radians and radians per second.)

From the first part of the problem,

$$\omega^2 = \frac{g}{L} = \frac{9.81}{6}.$$

$$\therefore \omega = 1.279 \text{ s}^{-1}.$$

Substituting in θ and $\dot{\theta}$,

$$\theta = 5 \sin\left(1.279t + \frac{\pi}{2}\right) \text{ deg.} \tag{3}$$

$$\dot{\theta} = (5 \times 1.279) \cos\left(1.279t + \frac{\pi}{2}\right) \text{ deg s}^{-1}. \tag{4}$$

Part (a)

Substitute $t = 1$ s in equation 3:

$$\theta = 5 \sin\left(1.279 \times 1 + \frac{\pi}{2}\right)$$

$$= 1.438°.$$

Part (b)

Substitute $\theta = 0$ in equation 3:

$$0 = 5 \sin\left(1.279t + \frac{\pi}{2}\right).$$

$$\therefore \quad \left(1.279t + \frac{\pi}{2}\right) = \pi \text{ (the lowest realistic solution).}$$

$$\therefore\ t = \frac{\pi - \frac{1}{2}\pi}{1.279}$$

$$= \underline{1.228 \text{ s.}}$$

Part (c)

The angular velocity of the pendulum is obtained by substituting this value of t in equation 4:

$$\dot{\theta} = (5 \times 1.279) \cos\left(1.279 \times 1.228 + \frac{\pi}{2}\right)$$

$$= -6.395 \text{ deg s}^{-1}.$$

You can check that the value of the cosine in this expression is (-1): the velocity is therefore the maximum value. This is to be expected at the lowest point of travel of the pendulum. Note also the negative sign: since the pendulum is returning to its lowest point, the direction of velocity must be in the opposite direction to the initial displacement.

The linear velocity of the pendulum bob is calculated from the simple kinematic formula $v = \Omega r$. Note that when using this formula, $\Omega\,(= \dot{\theta})$ must be in radians per second. We use Ω in preference to the more familiar ω to avoid confusion with the natural circular frequency of the motion.

$$v = \Omega r$$

$$= \dot{\theta} L$$

$$= \left(6.395 \times \frac{\pi}{180}\right) 6$$

$$= \underline{0.67 \text{ m s}^{-1}}.$$

Example 7.5

A body given an initial disturbance moves with s.h.m. with a natural circular frequency ω of 4 rad s^{-1}. It is initially displaced 1 m from its neutral position and then given an initial velocity, in the same direction, of 10 m s^{-1}. Calculate (a) its maximum displacement; (b) the least time for it to reach maximum displacement; (c) the least time for it to return to its neutral position; (d) its maximum velocity; (e) its maximum acceleration. Sketch a graph of displacement against time for approximately the first half-cycle of the motion.

Solution 7.5

This example is an exercise in the determination of the 'constants of integration' or 'initial condition constants' A and ϕ of Fact Sheet formula (c). When these are not stated, as in this example, they must be determined from some knowledge of initial conditions of the motion; hence the name 'initial conditions'. Since two constants are required, two items of information are needed; these are (i) the initial displacement (when $t = 0$) is +1 m, and (ii) the initial velocity (again when $t = 0$) is +10 m s^{-1}. The plus sign for displacement is chosen arbitrarily, but the plus sign for velocity is mandatory, as the velocity is stated as being 'in the same direction'.

The general expression for displacement for s.h.m. is:

[c] $$x = A \sin(\omega t + \phi).$$

Because we will need to substitute velocity when $t = 0$ we also need the velocity. Differentiating x,

[c] $$\dot{x} = A\omega \cos(\omega t + \phi).$$

Substituting the condition $x = 1$ m when $t = 0$,

$$+1 = A \sin(0 + \phi), \tag{1}$$

and the condition $\dot{x} = 10$ m s^{-1} when $t = 0$,

$$+10 = A \times 4 \cos(0 + \phi), \tag{2}$$

giving two simultaneous equations in A and ϕ.
Dividing equation 1 by equation 2,

$$\frac{1}{10} = \frac{A \sin \phi}{4A \cos \phi}.$$

$$\therefore\ 0.1 = \frac{\tan \phi}{4}.$$

$$\therefore\ \phi = \tan^{-1}(0.4) = 0.3805 \text{ rad}.$$

(Always work in radians, not degrees.)

Substituting in equation 1,

$$1 = A \sin(0.3805).$$

$$\therefore\ A = \frac{1}{0.3714} = 2.693 \text{ m}.$$

Displacement may now be rewritten in the particular form required for this problem:

$$x = 2.693 \sin(4t + 0.3805) \text{ m}. \tag{3}$$

Part (a)

Maximum displacement is $A = \underline{2.693 \text{ m}}.$

Part (b)

Substituting $x = 2.693$ in equation 3,

$$2.693 = 2.693 \sin(4t + 0.3805).$$

$$\therefore \quad 1 = \sin(4t + 0.3805).$$

$$\therefore \quad 4t + 0.3805 = \tfrac{1}{2}\pi,\ 1\tfrac{1}{2}\pi,\ 2\tfrac{1}{2}\pi, \text{ etc.}$$

Taking the lowest value,

$$t = \frac{1.5708 - 0.3805}{4} = \underline{0.2976 \text{ s}}.$$

Part (c)

Substituting $x = 0$ in (3),

$$0 = 2.693 \sin(4t + 3.805).$$

$$\therefore \quad 4t + 0.3805 = 0,\ \pi,\ 2\pi, \text{ etc.}$$

The first realistic solution must be π. (0 would result in a negative value for t.)

$$4t + 0.3805 = \pi.$$

$$\therefore \quad t = \frac{\pi - 0.3805}{4} = \underline{0.6902 \text{ s}}.$$

Part (d)

[c] $\hat{v} = A\omega = 2.693 \times 4 = \underline{10.772 \text{ m s}^{-1}}$.

Part (e)

[c] $\hat{a} = A\omega^2 = 2.693 \times 4^2 = \underline{43.088 \text{ m s}^{-2}}$.

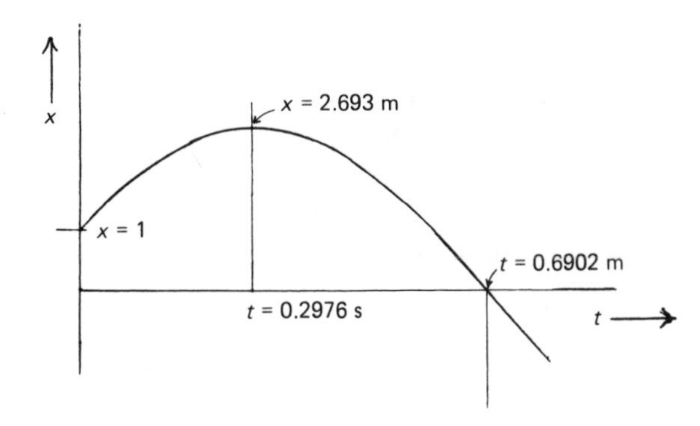

The sketch of displacement has been drawn accurately in the diagram, but in an examination answer it is sufficient merely to freehand sketch it and to indicate the significant values shown.

This example illustrates the importance of using the generalised form of the displacement function $x = A \sin(\omega t + \phi)$ for s.h.m. rather than the simpler and particular form $x = A \sin(\omega t)$ frequently used in elementary analysis of s.h.m.

Example 7.6

A compound pendulum comprises a rigid body pivoted at a point O to swing in a vertical plane, the mass-centre being a distance h from O. Show that if the pendulum swings with small amplitude and air resistance is ignored then the motion will be simple harmonic. Show also that if the position of O is varied, the frequency of oscillation will have a maximum value when h has the same value as the radius of gyration of the pendulum with respect to a transverse axis through G.

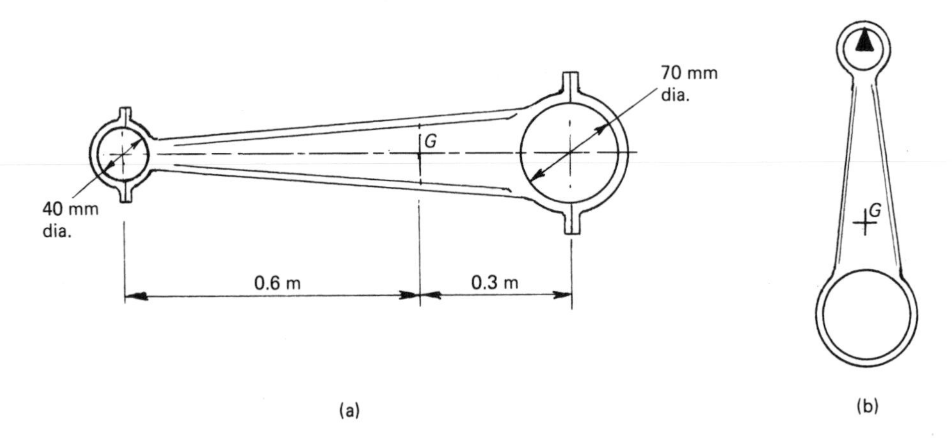

Some principal dimensions of an engine connecting-rod ('con-rod') are shown in diagram (a). G is the position of the mass centre and the mass is 15.7 kg. The rod is suspended by hanging it from a knife-edge passing through the small-end bore, as shown in diagram (b). In this position, the rod is given a small initial displacement causing it to swing with small amplitude. The time for 20 complete swings is 34.1 s. Determine I_g, the moment of inertia of the rod with respect to a transverse axis through G. Calculate how long 20 swings will take if the rod is swung from the knife-edge passing through the big-end bore.

Solution 7.6

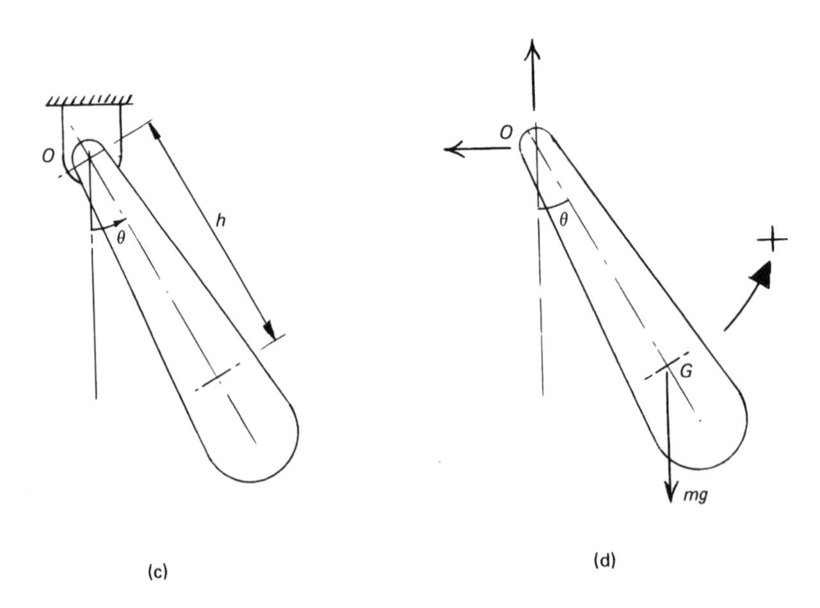

Any rigid body swinging in a vertical plane with the mass-centre below the point of support is called a compound pendulum if, in contrast to the simple pendulum, the body is not small in comparison with the distance between the mass-centre

and the point of support. The pendulum is represented in diagram (c) by a bar of random shape pivoted at O. We assume it is in motion with an instantaneous angular displacement θ arbitrarily assumed anticlockwise; thus the sign-convention is established. (d) is the free-body diagram. Motion consists of pure rotation about O; therefore $\Sigma M = I\alpha$ applies. Investigating moments about O, the pivot reaction force has zero moment; mg is the only relevant force.

Let the moment of inertia of the pendulum with respect to a transverse axis through O be I_o

[a] $$\Sigma M = I\alpha.$$

$$\therefore\ -mg\,(h \sin \theta) = I_o \ddot{\theta}.$$

As with the simple pendulum of Example 7.4, if a small displacement is prescribed then $\sin \theta \approx \theta$.

Rearranging, $$I_o \ddot{\theta} = -mgh\theta.$$

$$\therefore\ \ddot{\theta} = -\left(\frac{mgh}{I_o}\right)\theta,$$

thus proving s.h.m., ω having a value in this case of $\sqrt{\dfrac{mgh}{I_o}}$.

[b] $$I_o = I_g + mh^2$$

[b] $$= mk^2 + mh^2,$$

where k is the radius of gyration with respect to an axis through G.

$$\therefore\ \omega = \sqrt{\frac{mgh}{mk^2 + mh^2}}.$$

[c] $$\therefore\ n = \frac{1}{2\pi}\sqrt{\frac{gh}{k^2 + h^2}}.$$

The question now requires us to show that n will have a maximum value as h varies. To show this, we must differentiate the expression with respect to h. Treat the square root as power $\frac{1}{2}$; the function under the root is then differentiated using the 'quotient' formula $\mathrm{d}\,(u/v) = \dfrac{v\,\mathrm{d}u - u\,\mathrm{d}v}{v^2}$:

$$\frac{\mathrm{d}n}{\mathrm{d}h} = \frac{1}{2\pi}\,\frac{1}{2}\left(\frac{gh}{k^2 + h^2}\right)^{-1/2}\left(\frac{(k^2 + h^2)\,g - gh\,(2h)}{(k^2 + h^2)^2}\right).$$

The apparent complexity of this result is reduced if we see that for the expression to be zero (the mathematical condition for 'maximum') then the top line inside the right-hand bracket must be zero.

$$\therefore\ (k^2 + h^2)\,g - 2gh^2 = 0.$$

$$\therefore\ k^2 + h^2 = 2h^2.$$

$$\therefore\ k = h.$$

Referring to diagram (b), when the rod is suspended from the small-end bore, the corresponding value of h will be 0.6 m plus the bore radius, i.e. 0.62 m. We use the expression for n obtained in the above analysis.

$$n = \frac{1}{2\pi}\sqrt{\frac{gh}{k^2 + h^2}}.$$

$$\therefore \frac{20}{34.1} = \frac{1}{2\pi}\sqrt{\frac{9.81 \times 0.62}{k^2 + (0.62)^2}}.$$

$$4\pi^2 (0.5865)^2 = \frac{9.81 \times 0.62}{k^2 + (0.62)^2}.$$

$$k^2 = \frac{9.81 \times 0.62}{4\pi^2 (0.5865)^2} - (0.62)^2 = 0.0635.$$

$$\therefore k = 0.252 \text{ m}.$$

[b]

$$I_g = mk^2$$

$$= 15.7 \times (0.252)^2.$$

$$\therefore I_g = \underline{0.997 \text{ kg m}^2}.$$

Reversing the rod, and swinging from the big-end bore, give $h = (0.3 + 0.035) = 0.335$ m. k, of course, remains the same.

$$n' = \frac{1}{2\pi}\sqrt{\frac{9.81 \times 0.335}{(0.252)^2 + (0.335)^2}} = 0.6883 \text{ Hz}.$$

$$\therefore \text{Time for 20 swings} = \frac{20}{0.6883} = \underline{29.06 \text{ s.}}$$

This last calculation, although simple, is often a source of mistakes; students become confused as to 'what to divide by what'. Think of the units of each term. Frequency is 'cycles/second'. We are given 'cycles' (20) and we require 'seconds'. So we need to divide 'cycles' by 'cycles/second'. If we mistakenly divide 'cycles/second' by 'cycles' we finish up with '1/seconds', which is seen to be wrong.

Example 7.7

A mine cage hangs from a steel wire cable. The fully loaded cage has a total mass of 3400 kg. A laboratory test on a portion of the cable 1 m long shows that a force of 10 kN extends it by 0.228 mm. Estimate the natural frequency of vertical vibration of the suspended cage when the length of cable from the top support to the cage is: (a) 35 m (when the cage is at the top of the mine shaft); (b) 3200 m (when it is at the bottom of the shaft). In both cases, neglect the mass of the cable itself.

Solution 7.7

A mass hanging from a steel cable is the same as a mass hanging from a spring, as treated in Example 7.1. The 'spring' stiffness is not stated, but we have data about a tensile test on a 1-metre length. If 1 m extends 0.228 mm under a certain load then a length of 35 m must extend 35 times this amount for the same load; the stiffness can be calculated from the stress analysis formula.

Part (a)

[e]

$$k_a = \frac{F}{x} = \frac{10 \times 10^3}{(0.228 \times 10^{-3}) \times 35}.$$

From Example 7.1, $n_a = \dfrac{1}{2\pi}\sqrt{\dfrac{k_a}{m}}$

$$= \frac{1}{2\pi}\sqrt{\frac{10 \times 10^3}{(0.228 \times 10^{-3}) \times 35 \times 3400}} .$$

$$\therefore\ n_a = 3.055 \text{ Hz.}$$

Part (b)

The calculation will be exactly the same, except for the different length of cable.

$$n_b = \frac{1}{2\pi}\sqrt{\frac{10 \times 10^3}{(0.228 \times 10^{-3}) \times 3200 \times 3400}}$$

$$\therefore\ n_b = \underline{0.319 \text{ Hz.}}$$

You should note that neglecting the mass of the cable is a gross simplification, particularly in (b). It is probable that the mass of 3200 m of cable would actually be much greater than the mass of the cage itself.

Example 7.8

A large generator has a mass of 2570 kg. It is to be mounted on a horizontal steel girder. When it is placed in position, careful measurement shows that the girder deflects vertically 2.32 mm at the point where the machine is mounted. Estimate the natural frequency of transverse vibration of the machine-girder system, neglecting the mass of the girder itself. State whether your answer is likely to be higher or lower than the true value of frequency.

Solution 7.8

A steel girder will behave exactly as a spring. The data regarding the deflection due to weight is used to calculate spring stiffness. Recall that the deflecting force *F* is the weight, (*mg*), of the mass.

[e]

$$k = \frac{F}{x} = \frac{(2570 \times 9.81)}{2.32 \times 10^{-3}} .$$

From Example 7.1,

$$n = \frac{1}{2\pi}\sqrt{\frac{k}{m}} \qquad (1)$$

$$= \frac{1}{2\pi}\sqrt{\frac{2570 \times 9.81}{(2.32 \times 10^{-3}) \times 2570}} .$$

$$\therefore\ n = \underline{10.35 \text{ Hz.}}$$

It is of interest to observe that the value of the mass cancels out; the frequency can actually be calculated merely from the static deflection due to the weight of the load. For this particular type of problem, some books derive a special formula,

$$n = \frac{1}{2\pi}\sqrt{\frac{9.81}{\Delta}} ,$$

or even

$$n = \frac{15.76}{\sqrt{\Delta \text{ mm}}} ,$$

in which Δ is the 'self weight' deflection of the beam or structure, being stated in the second formula in millimetres. But the direct calculation is relatively simple, and hardly justifies remembering a special formula.

Since the mass of the beam is neglected, the *actual* vibrating mass must be greater than 2570 kg, as it must include a part of the supporting beam. Equation 1 above shows that an increase of the value of m would result in a reduction of frequency n. Therefore:

The estimate is higher than the true frequency.

Example 7.9

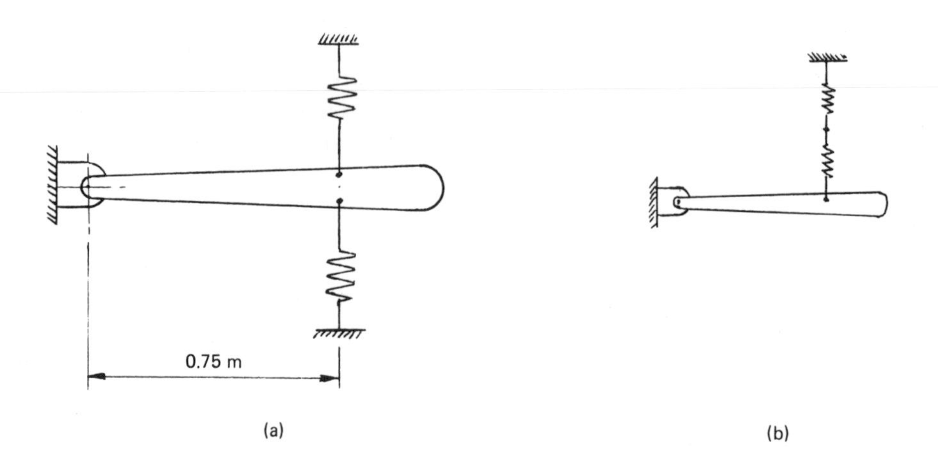

A bar, non-uniform in section, is pivoted at one end and supported by two springs, each of stiffness 650 N m^{-1} at a point 0.75 m from the pivot, as shown in diagram (a). The bar is given a small initial angular displacement and released. It is then observed to vibrate with a period of 0.87 s. Calculate the moment of inertia of the bar with respect to a transverse axis through the pivot. Calculate what the period of oscillation would be if both springs were on the same side of the bar, as shown in diagram (b).

Solution 7.9

There is a possible catch in this one: the arrangement of the two springs might lead you to think they are in series. They are actually in parallel. The test is that for springs in series the load is the same in each spring and the deformation is shared between them, whereas for springs in parallel the deformation is the same for each but the load is shared. It should be clear that this example falls into the second category.

We assume the bar is in a state of vibration, and at the instant we examine it the displacement is a small angle, θ. In all problems involving the vibration of a bar, assuming a small θ means:

(a) the springs are deformed axially but not sideways;

(b) the spring axes remain perpendicular to the axis of the bar;

(c) the amount of extension or compression of the springs is approximately the arc of the circular sector along which the end of the spring moves.

So we begin by drawing a diagram (a) of the beam with a small deflection θ from the position of static equilibrium, shown by the horizontal chain-line, together with the corresponding free-body diagram, (b). It helps to exaggerate the assumed displacement, to simplify the calculation of the resulting forces.

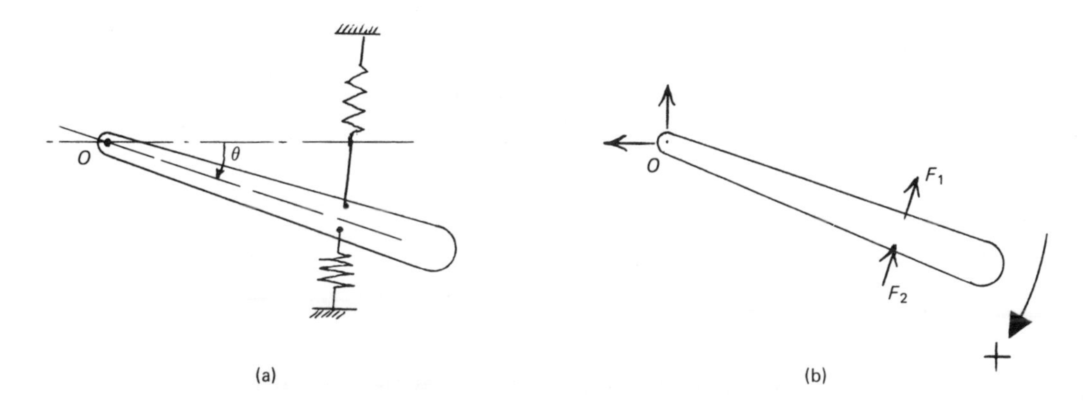

We have taken a liberty here by leaving out the weight of the bar. This is permissible provided we also leave out the initial spring forces which balance the weight in the statically-neutral position. The forces F_1 and F_2 on the free-body diagram, (b), are the *additional* spring-forces arising out of the deflection θ of the bar. This means that the diagram is, strictly speaking, not a true diagram of all the forces acting upon the bar, but it does make the subsequent calculation simpler. Recall from Example 7.1 that the vibration frequency of a simple spring-mass system is independent of whether it vibrates vertically or horizontally. Looking at it another way, our 'free-body diagram' is the same as if we had turned the system round through 90°, allowing the bar to vibrate in a horizontal instead of a vertical plane.

The upper spring will be extended; the effect on the bar will be an additional spring force F_1 upwards, as shown. The lower spring will be compressed by the displacement θ, resulting in an additional force F_2, also upwards. The motion is pure rotation about the pivot; $\Sigma M = I\alpha$ applies.

Taking moments about O,

[a] $$-(F_1 + F_2)\,0.75 = I\alpha. \qquad (1)$$

Force F_1 will be (spring stiffness x extension). The assumption of a small θ means that we may use the simple kinematic formula $x = r\theta$.

$$F_1 = F_2 = k\,(0.75\,\theta).$$

Substituting in equation 1,

$$-2k\,(0.75)^2\,\theta = I\ddot{\theta}.$$

$$\therefore\ \ddot{\theta} = -\left(\frac{2k\,(0.75)^2}{I}\right)\theta, \qquad (2)$$

thus proving s.h.m.

$$\omega = \frac{2\pi}{\tau} = \frac{2\pi}{0.87}.$$

Equating this to the root of the coefficient of θ in equation 2,

$$\frac{2\pi}{0.87} = \sqrt{\frac{2k\,(0.75)^2}{I}}.$$

$$\therefore\ I = \frac{2k\,(0.75)^2\,(0.87)^2}{4\pi^2}$$

$$= \frac{2 \times 650\,(0.75)^2 \times (0.87)^2}{4\pi^2}.$$

$$\therefore\ I = \underline{14.02\ \text{kg m}^2}.$$

It is not necessary to solve the whole problem again completely when both springs are on the same side of the bar. In the first part of the question the springs are in parallel, with an effective resultant stiffness of $(2k)$ (Fact Sheet (e)). With both springs on the one side, they will be in series, with an effective resultant stiffness of $(\frac{1}{2}k)$, i.e. one-quarter the previous value. We may argue as follows:

Referring to equation 2 above, placings placing springs in series will result in a value of $(\omega)^2$ of one-quarter the value.
Hence ω will have one-half the value.
Hence τ will have twice the original value.

$$\therefore\ \tau' = 0.87 \times 2 = \underline{1.74\ \text{s.}}$$

This important example illustrates the general treatment of oscillating bars and beams, the significance of assuming small displacement, and the consequences of this.

The last part of the question is also illuminating. Always beware of solving a second part of a question exactly as the first. Look for the short cut which is usually there to be found.

Example 7.10

A body of mass m hangs from a helical spring attached to the end of a flexible steel bar, which is held in a clamp as shown in the diagram. When the mass is placed on the spring, it is observed that the spring extends by 11.3 mm and that the end of the bar is deflected downwards by 1.72 mm. Neglecting the masses of the spring and the beam, estimate the natural frequency of vertical vibration of the mass-spring system. Also calculate what the natural frequency would be: (a) if the mass were attached directly to the end of the bar; (b) if the spring hung from a fixed support instead of from the end of the bar.

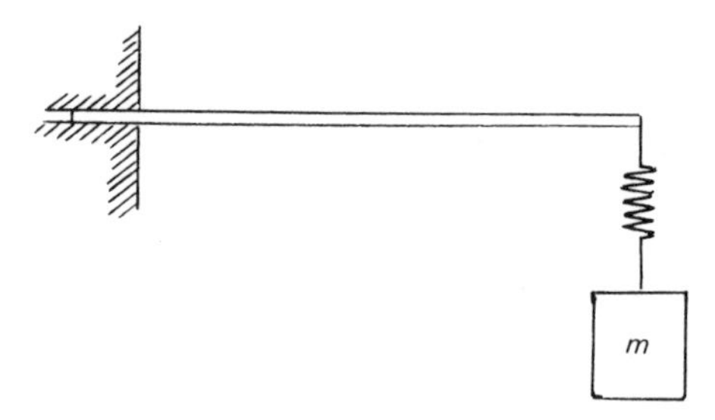

Solution 7.10

The problem should be identified as a simple mass–spring system as analysed in Example 7.1. The complication is that we have two springs – the helical spring and the steel bar – and they are seen to be 'in series', i.e. one attached to the end of the other. The stiffness may be calculated from the deflections of each due to the weight of the mass. The helical spring transmits the weight directly to the end of the bar, so that the same force acts on both springs; this confirms that the springs are in series, according to the definition given in Example 7.9.

Calling the stiffnesses of the bar and of the spring k_1 and k_2 respectively,

[e]
$$k_1 = \frac{F}{x} = \frac{mg}{(1.72 \times 10^{-3})}\,.$$

$$k_2 = \frac{F}{x} = \frac{mg}{(11.3 \times 10^{-3})}\,.$$

For a spring in series,

[f] $$\frac{1}{k} = \frac{1}{k_1} + \frac{1}{k_2}$$

$$= \frac{(1.72 \times 10^{-3})}{mg} + \frac{(11.3 \times 10^{-3})}{mg}$$

$$= \frac{(13.02 \times 10^{-3})}{mg}.$$

$$\therefore\ k = \frac{mg}{(13.02 \times 10^{-3})} \text{ N m}^{-1}.$$

From Example 7.1, $$n = \frac{1}{2\pi}\sqrt{\frac{k}{m}}$$

$$= \frac{1}{2\pi}\sqrt{\frac{mg}{(13.02 \times 10^{-3})\, m}}.$$

$$\therefore\ n = \underline{4.369 \text{ Hz.}}$$

For calculating n with the mass on each spring separately, we substitute k_1 and k_2 in turn in the same formula.

Part (a)

$$n_a = \frac{1}{2\pi}\sqrt{\frac{mg}{(1.72 \times 10^{-3})m}} = \underline{12.020 \text{ Hz.}}$$

Part (b)

$$n_b = \frac{1}{2\pi}\sqrt{\frac{mg}{(11.3 \times 10^{-3})m}} = \underline{4.689 \text{ Hz.}}$$

Example 7.11

A compound pendulum is pivoted at a point O (see diagram). The pendulum mass, m, is 12 kg; its moment of inertia with respect to a transverse axis through O is $I_o = 9.6$ kg m^2, and the mass centre, G, is at a distance $h = 0.85$ m from O. The pendulum is restrained by a spring of stiffness $k = 650$ N m^{-1} which is attached at a point distant $a = 0.7$ m from O.

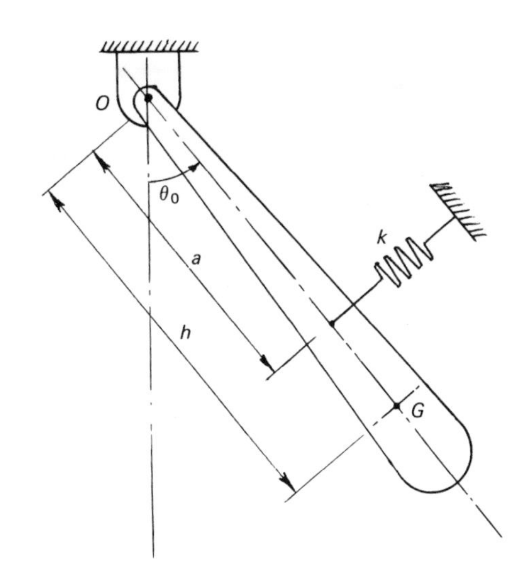

When the system is in a state of static equilibrium, the spring holds the axis OG of the pendulum at an angle of $\theta_0 = 40°$ to the vertical, and the spring axis is perpendicular to OG. Calculate the frequency of small oscillations of the pendulum resulting from giving it a small initial angular displacement from the position of neutral equilibrium.

Solution 7.11

With this problem, we may *not* leave out the pendulum weight and the initial spring force, as we did with Example 7.9, as the subsequent analysis will show: all forces must be correctly evaluated. As with most problems it will be found easier to work algebraically, substituting arithmetical values at the end.

The spring will have an initial tensile force F_0 in the position of static equilibrium. We shall calculate this force first. To save time, the pendulum may henceforward be represented by a thick line.

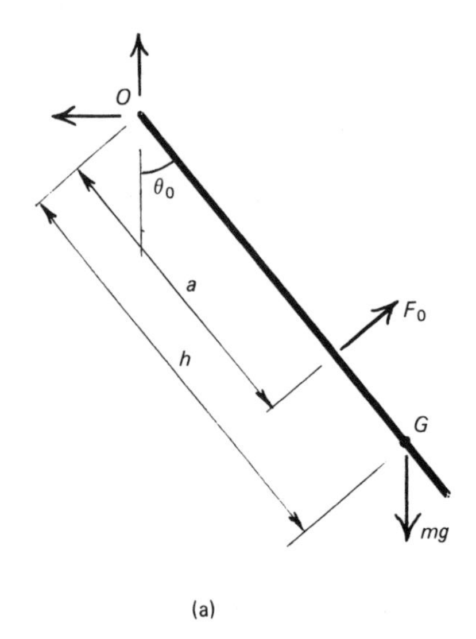

Fig. 7.12

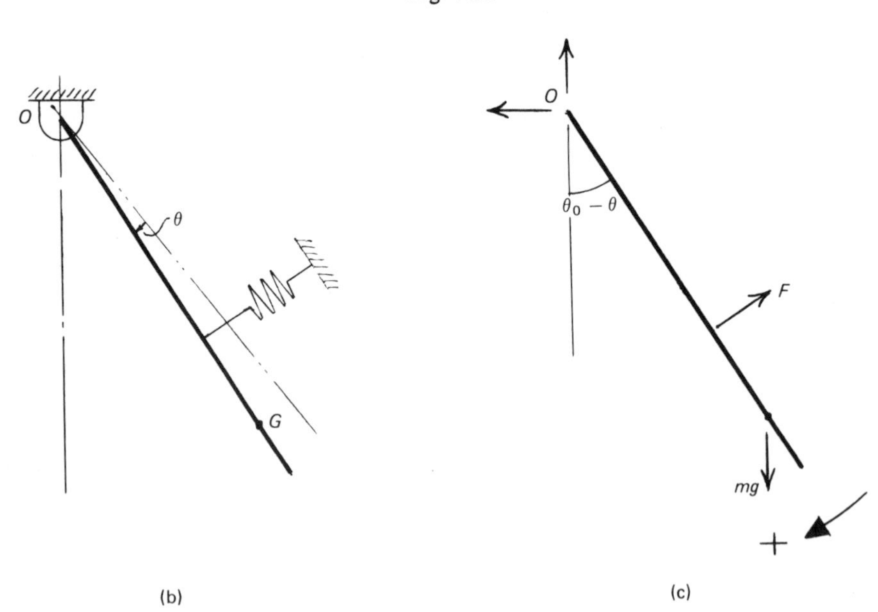

Fig. 7.13

Diagram (a) shows the pendulum in the position of static equilibrium. Taking moments of forces about O,

$$F_0 a = (mg \sin \theta_0) h.$$

$$\therefore \; F_0 = \frac{(mg \sin \theta_0) h}{a}.$$

Assume a *small* displacement θ, arbitrarily clockwise. The system diagram and the free-body diagram are shown as diagrams (b) and (c) respectively.

Rotation about O. $\Sigma M = I\alpha$ applies. Taking moments of forces about O,

$$+ mg \sin (\theta_0 - \theta) h - Fa = I_o \alpha.$$

Note that assuming small displacement permits us to write (Fa) for the spring force moment.

Spring tension will increase by an amount ($k\theta a$), again provided we assume that θ is small. Also replacing α by $\ddot{\theta}$,

$$mg \sin (\theta_0 - \theta) h - (F_0 + k\theta a) a = I_o \ddot{\theta}.$$

To 'bend' this equation round to the desired form $\ddot{\theta} = -k\theta$, we begin by expanding $\sin (\theta_0 - \theta)$, using the identity

$$\sin (A - B) = \sin A \cos B - \cos A \sin B:$$

$$mg[\sin \theta_0 \cos \theta - \cos \theta_0 \sin \theta] h - \left[\left(\frac{mgh}{a}\right) \sin \theta_0 + k\theta a\right] a = I_o \ddot{\theta}.$$

In the first bracket we may use the now familiar approximation $\sin \theta \approx \theta$ and also the not-so-familiar $\cos \theta \approx 1$. Substituting, and expanding brackets,

$$\cancel{mg \sin \theta_0 h} - mg \cos \theta_0 \theta h - \cancel{mgh \sin \theta_0} - ka^2 \theta = I_o \ddot{\theta},$$

giving, on rearranging:

$$\ddot{\theta} = -\theta \left(\frac{ka^2 + mgh \cos \theta_0}{I_o}\right),$$

which proves s.h.m.

The coefficient of $(-\theta)$ is ω^2.

[c]

$$n = \frac{\omega}{2\pi}$$

$$= \frac{1}{2\pi} \sqrt{\frac{ka^2 + mgh \cos \theta_0}{I_o}}. \qquad (1)$$

Substituting values,

$$n = \frac{1}{2\pi} \sqrt{\frac{650 \times (0.7)^2 + 12 \times 9.81 \times 0.85 \cos 40}{9.6}}$$

$$\therefore \; n = \underline{1.021 \text{ Hz.}}$$

Notice in the calculation how the two terms ($mg \sin \theta_0 h$) cancel out. This is the initial spring-force cancelling out the weight component. *But* the frequency of the system is *not* independent of the weight, as the final formula shows, and leaving the weight and the initial spring force off the free-body diagram would lead to an incorrect answer. If we look at the final formula (equation 1 above) we see that

substituting a value for θ_0 of 90° gives us in effect a solution for Example 7.9. If we go to the other extreme and put θ_0 equal to 0 then the frequency becomes

$$n = \frac{1}{2\pi}\sqrt{\frac{ka^2 + mgh}{I_o}},$$

and if we now put $k = 0$, i.e. remove the spring, then we have the compound pendulum formula of Example 7.6.

Example 7.12

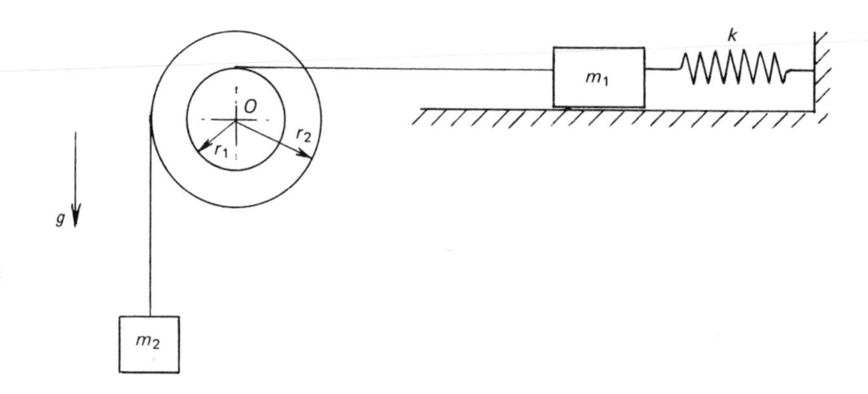

The system shown in the diagram consists of a double pulley, which rotates about a fixed horizontal axis O, a body of mass m_1, which slides on a horizontal surface and is connected to a fixed point by a spring of linear stiffness k and to a cord that is wound around and secured to the periphery of the smaller pulley, and a body of mass m_2, which is attached to a cord that is wound around and secured to the periphery of the larger pulley.

Given that a positive tension always acts in the cords and that friction is negligible, determine an expression for the oscillatory frequency of the system in terms of the given quantities and I, the moment of inertia of the pulley about O.

The body of mass m_2 is pulled down a distance A and released from rest. Determine an expression for the displacement of the body of mass m_1.

U. Lond. K.C.

Solution 7.12

In the free-body diagrams (p. 299) the pulley is assumed to have an instantaneous angular displacement θ anticlockwise; m_1, an instantaneous linear displacement x to the left; m_2, an instantaneous displacement y downwards.

The assumption of anticlockwise displacement for the pulley is arbitrary; we could as simply have chosen clockwise. But having so chosen, the directions of the displacements of the masses must be compatible, and this is seen to be the case. If the pulley is turned anticlockwise, then m_1 moves left, and m_2 moves downwards.

The system consists of connected bodies, the general dynamics of which are analysed in Chapter 4. As always, we need first to relate not only the directions of the corresponding displacements, but also their magnitudes: these are the 'compatibility' relationships.

From the simple equations of motion in a circular path we can say

$$x = \theta r_1$$

and

$$y = \theta r_2,$$

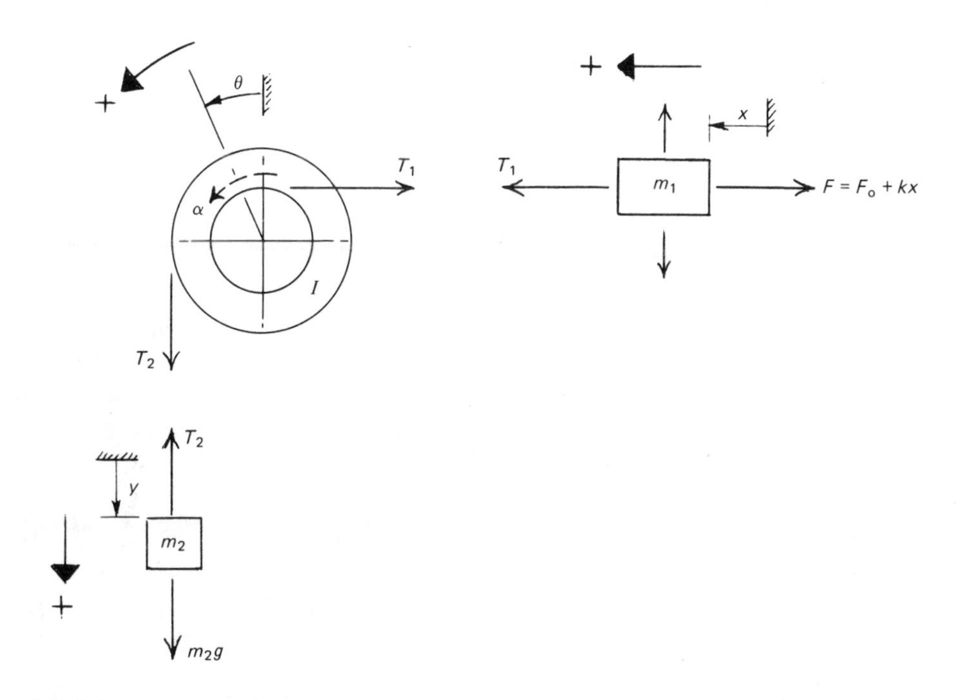

and since we shall also require to relate the corresponding accelerations, differentiating twice gives:

$$\ddot{x} = \ddot{\theta} r_1$$

and
$$\ddot{y} = \ddot{\theta} r_2 .$$

Tension forces T_1 and T_2 have been assumed in the two cords.

In determining the force F in the spring, we have to take care, as with all vibration problems involving gravity forces, to remember that the spring is under load in the position of static equilibrium of the system. The displacement x of body m_1 is reckoned from this statically neutral position. But in this position, the weight of body m_2 causes the spring to stretch, owing to this initial static equilibrium load which we may call F_0. So, for a displacement x from this position, the spring force F will be given by:

$$F = F_0 + kx.$$

F_0 is simply calculated, taking moments of forces about the pulley centre in the statically neutral position (a force diagram is not necessary),

$$m_2 g r_2 = F_0 r_1 ,$$

giving

$$F_0 = m_2 g \left(\frac{r_2}{r_1}\right) .$$

Hence
$$F = m_2 g \left(\frac{r_2}{r_1}\right) + kx$$

$$= m_2 g \left(\frac{r_2}{r_1}\right) + k\theta r_1 .$$

We are now in the position of being able to write the three equations of motion.

For the pulley ($\Sigma M = I\alpha$), taking anticlockwise as positive, and taking moments of forces about 0:

$$T_2 r_2 - T_1 r_1 = I\alpha = I\ddot{\theta}. \tag{1}$$

For body m_1 ($\Sigma F = ma$), taking left as positive,

$$T_1 - F = m_1 a_1.$$

$$\therefore\ T_1 - \left(m_2 g \frac{r_2}{r_1} + k\theta r_1\right) = m_1 \ddot{x} = m_1 (\ddot{\theta} r_1). \tag{2}$$

For body m_2 ($\Sigma F = ma$), taking downwards as positive,

$$m_2 g - T_2 = m_2 a_2 = m_2 (\ddot{\theta} r_2). \tag{3}$$

To obtain the general system equation of motion, we reduce these three equations to a single equation by eliminating the internal forces, T_1 and T_2. Use equations 2 and 3 to substitute for T_1 and T_2 in equation 1:

$$r_2 (m_2 g - m_2 \ddot{\theta} r_2) - r_1 \left[m_1 \ddot{\theta} r_1 + \left(m_2 g \frac{r_2}{r_1} + k\theta r_1 \right) \right] = I\ddot{\theta}.$$

Expanding,

$$\cancel{m_2 g r_2} - m_2 r_2^2 \ddot{\theta} - m_1 r_1^2 \ddot{\theta} - \cancel{m_2 g r_2} - k\theta r_1^2 = I\ddot{\theta},$$

and it is seen that the two terms which do not include either θ or $\ddot{\theta}$ cancel out. This is a familiar result in vibration systems involving gravity which we have come across before (see Example 7.1) and it forms a useful check in the correctness of one's work.

Collecting terms,

$$\ddot{\theta}\,(I + m_1 r_1^2 + m_2 r_2^2) + \theta\,(k r_1^2) = 0.$$

$$\therefore\ \ddot{\theta} + \theta \left(\frac{k r_1^2}{I + m_1 r_1^2 + m_2 r_2^2} \right) = 0 \tag{4}$$

and s.h.m. is therefore proved (Fact Sheet (c)). The coefficient of θ is the value of (ω^2).

(This is always a good point to stop and check that the dimensions of (ω^2) are indeed T^{-2}, as they should be.)

Frequency is given by

[c]
$$n = \frac{\omega}{2\pi} = \frac{1}{2\pi} \sqrt{\frac{k r_1^2}{I + m_1 r_1^2 + m_2 r_2^2}}.$$

You should also note that the denominator of the square root terms comprises the equivalent inertia of the system referred to the pulley centre. See Chapter 4.

For the second part of the question, now that s.h.m. has been proved for the system, we know that each element must move with s.h.m. We can therefore say that for body m_1,

[c]
$$x = C_1 \sin(\omega t) + C_2 \cos(\omega t). \tag{5}$$

(We use C_1 and C_2 for the constants because A has already been used in the question as the initial displacement of body m_2.)

We have two 'initial conditions' of motion:

(a) When $t = 0, y = A$.

$$\therefore\ x = A\left(\frac{r_1}{r_2}\right).$$

(b) When $t = 0, \dot{y} = 0$ (the body is released from rest).

$$\therefore\ \dot{x} = 0.$$

For condition (i),

$$A\left(\frac{r_1}{r_2}\right) = C_1 \times 0 + C_2 \times 1.$$

$$\therefore\ C_2 = A\left(\frac{r_1}{r_2}\right).$$

For condition (ii) we must differentiate equation 5 with respect to *t*.

$$\dot{x} = C_1 \omega \cos(\omega t) - C_2 \omega \sin(\omega t).$$

$$\therefore\ 0 = C_1 \times 1 - C_2 \times 0.$$

$$\therefore\ C_1 = 0.$$

Substituting for C_1 and C_2 in equation 5,

$$x = A\left(\frac{r_1}{r_2}\right) \cos(\omega t)$$

$$= A\left(\frac{r_1}{r_2}\right) \cos\left(\sqrt{\frac{kr_1^2}{I + m_1 r_1^2 + m_2 r_2^2}}\right)t.$$

Example 7.13

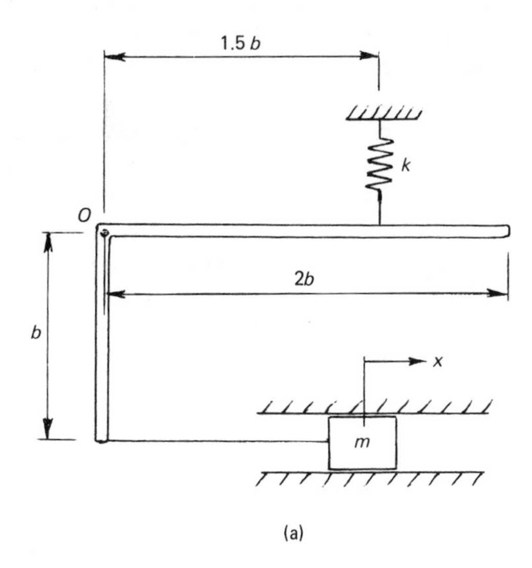

(a)

The system shown in diagram (a) lies in a *horizontal* plane. Two rods, one of length b and mass m, the other of length $2b$ and mass $2m$, are rigidly connected together and pivoted to a fixed point O. A spring of linear stiffness k is attached to the longer rod at a distance $1.5b$

from O. A body of mass m is connected to the end of the shorter rod by a light rigid link and constrained to move in guides. Friction is everywhere negligible.

Determine an expression for the natural frequency, f, of the system for small oscillations about the static equilibrium position shown in the figure.

At time $t = 0$, the body has a displacement $x = 0$ and a velocity $\dot{x} = v$. Determine an expression for the subsequent tension T in the light link.

Given that $m = 0.5$ kg, $b = 100$ mm and $v = 0.25$ m s^{-1}, determine (a) the stiffness k so that the frequency $f = 2$ Hz, (b) the maximum displacement of the body and (c) the maximum tension in the light link.

Comment on whether the assumption of *small* oscillations is reasonable.

U. Lond. K.C.

Solution 7.13

This is another example of a connected system, in many respects similar to Example 7.12. The statement that the system lies in a horizontal plane is intended to inform you that weight forces are not to be taken into account, as they act in a direction perpendicular to the figure.

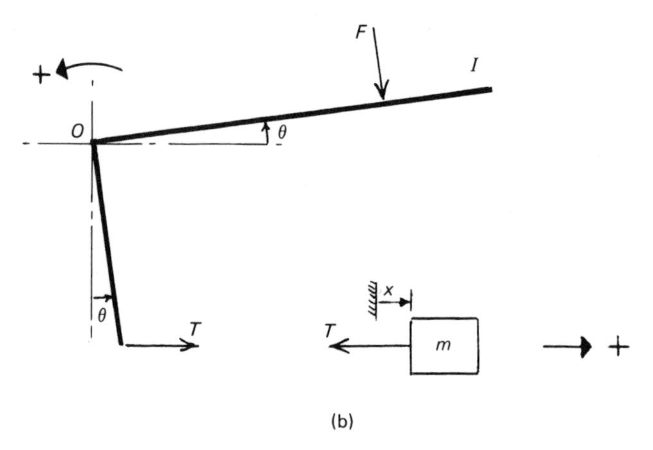

(b)

We show the free-body diagrams (b). A small anti-clockwise angular displacement of the rod assembly has been assumed, and the corresponding displacement x of the sliding body is therefore assumed to be to the right (as suggested in diagram (a)).

The force in the connecting link has been assumed to be a tensile force, labelled T.

(Either tensile or compressive may be assumed, so long as the assumption is consistent on both sliding body and rod assembly.)

First, we establish compatibility. For a *small* rotation θ of the rod assembly, we can say that the corresponding compatible motion x of the body is given by

$$x = \theta b$$

(since x must be the same, approximately, as the movement of the free end of the rod of length b). Differentiating twice,

$$\ddot{x} = \ddot{\theta} b.$$

By the same argument, the spring must compress a length ($\theta \times 1.5b$). Hence, spring force F is

$$F = k\,(1.5\theta b).$$

We may now write the equations of motion. For the rod assembly ($\Sigma M = I\alpha$) reckoning anticlockwise positive, and for the present, calling the moment of inertia of the rod assembly I,

$$Tb - F \times 1.5b = I\alpha.$$

$$\therefore\ Tb - k(1.5\theta b)\,1.5b = I\ddot{\theta}. \qquad (1)$$

For the body of mass m, ($\Sigma F = ma$) reckoning right as positive.

$$-T = ma$$

$$= m\ddot{x}.$$

$$\therefore T = -m\ddot{\theta}b. \qquad (2)$$

Use equation 2 to substitute for T in equation 1:

$$-b(m\ddot{\theta}b) - 2.25kb^2\theta = I\ddot{\theta}.$$

Collecting terms,

$$\ddot{\theta}(I + mb^2) + \theta(2.25kb^2) = 0.$$

$$\therefore\ \ddot{\theta} + \theta\left(\frac{2.25kb^2}{I + mb^2}\right) = 0. \qquad (3)$$

[c] Hence s.h.m. is proved.
We may evaluate I in terms of m and b using the formula $I = \frac{1}{3}mL^2$ for a uniform thin rod rotating about one end.

$$I = \tfrac{1}{3}mb^2 + \tfrac{1}{3}(2m)(2b)^2$$

$$= 3mb^2.$$

Substitute this in equation 3:

$$\ddot{\theta} + \theta\left(\frac{2.25kb^2}{3mb^2 + mb^2}\right) = 0.$$

Cancelling (b^2) gives

$$\ddot{\theta} + \theta\left(\frac{2.25}{4}\frac{k}{m}\right) = 0.$$

$$\therefore\ \ddot{\theta} + \theta\left(\frac{9}{16}\frac{k}{m}\right) = 0,$$

giving

$$\omega = \frac{3}{4}\sqrt{\frac{k}{m}}.$$

[c]

$$f = \frac{\omega}{2\pi}$$

$$= \frac{1}{2\pi} \times \frac{3}{4}\sqrt{\frac{k}{m}}$$

$$= \frac{3}{8\pi}\sqrt{\frac{k}{m}}. \qquad (4)$$

The motion of the sliding body is also s.h.m. of the same frequency. Its displacement x at any instant is therefore given by

[c]

$$x = A\sin(\omega t) + B\cos(\omega t).$$

$$\dot{x} = A\omega\cos(\omega t) - B\omega\sin(\omega t).$$

The two 'initial conditions' are given:

(i) When $t = 0, x = 0$.

(ii) When $t = 0, \dot{x} = v$.

Substituting condition (i),

$$0 = A \sin 0 + B \cos 0$$

giving $B = 0$. Substituting condition (ii),

$$v = A\omega \cos 0,$$

giving $A = v/\omega$.

$$\therefore\ x = \frac{v}{\omega} \sin (\omega t)$$

$$= v\ \frac{4}{3} \sqrt{\frac{m}{k}} \sin\left(\frac{3}{4} \sqrt{\frac{k}{m}}\ t\right). \qquad (5)$$

Differentiating twice gives

$$\ddot{x} = -\ v\omega \sin (\omega t)$$

and writing the equation of motion ($\Sigma F = ma$),

$$-\ T = m\ddot{x}$$

$$= -\ mv\omega \sin (\omega t).$$

$$\therefore\ \underline{T = mv\ \frac{3}{4} \sqrt{\frac{k}{m}} \sin\left(\frac{3}{4} \sqrt{\frac{k}{m}}\ t\right)}. \qquad (6)$$

Substituting the given values in equation 4,

$$2 = \frac{3}{8\pi} \sqrt{\frac{4}{0.5}},$$

giving $\underline{k = 140.4 \text{ N m}^{-1}}$.

$\hat{x}$, the maximum value of x, is obtained by assuming the sine of the expression for x to have its maximum value of 1.

$$\therefore\ \hat{x} = \frac{4v}{3} \sqrt{\frac{m}{k}}$$

$$= \frac{4 \times 0.25}{3} \sqrt{\frac{0.5}{140.4}}$$

$$= 0.0199 \text{ m}$$

$$= \underline{19.9 \text{ mm}}.$$

The same reasoning applies to the maximum value for T, which we obtain from equation 6:

$$\hat{T} = \frac{3}{4}\ mv \sqrt{\frac{k}{m}}$$

$$= \frac{3}{4} \times 0.5 \times 0.25 \sqrt{\frac{140.4}{0.5}}.$$

$$\therefore\ \hat{T} = \underline{1.571 \text{ N}}.$$

The implications of small oscillations are:

(a) that the link connecting the sliding body to the rod assembly must be assumed to remain parallel to the slide (as, otherwise, the compatibility equation would not be valid);

(b) that the axis of the spring remains perpendicular to the longer rod (as, otherwise, the calculation of spring force would not be accurate).

The maximum displacement of the sliding body is 19.9 mm. For b = 100 mm, the corresponding maximum angular displacement of the rod assembly, θ, would be

$$\theta = \frac{19.9}{100} = 0.199 \text{ rad} = 11.4°,$$

and although this angle would not normally be considered a small angle, you can see from an examination of diagram (a) that the accuracy of the two conditions above must be dependent on (i) the length of the light connecting-link, and (ii) the length of the spring. For a very long spring and a very long link, the rod assembly could turn through quite large angles without invalidating the assumption of simple harmonic motion.

Example 7.14

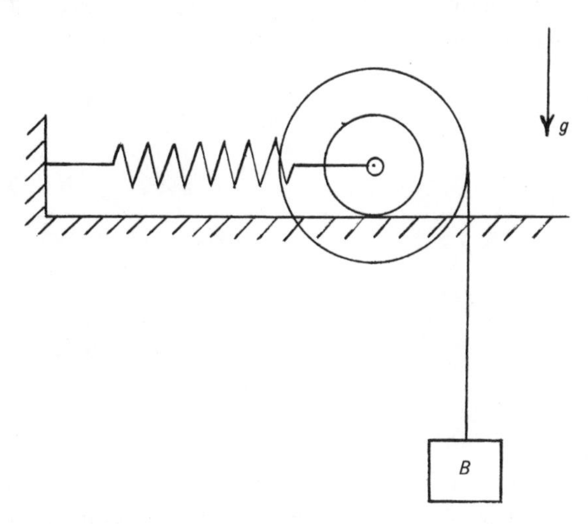

A rotor consists of two uniform discs, one of mass m and radius r and the other of mass $2m$ and radius $2r$, rigidly fixed together. The rotor rests on a horizontal rail as shown in the diagram. Attached to the centre of the rotor is a spring of stiffness k, and to the periphery of the larger disc a cord which is connected to a body B of mass m.

The system is at rest in its static equilibrium position. From this position the body B is given a downward displacement δ and released from rest. For the conditions that the rotor rolls on the rail *without* slipping and the cord always remains in tension, determine expressions, in terms of the given quantities, for the frequency of oscillations and the displacement of the body B as a function of time.

Show that the variation with time t of the tension in the cord is:

$$mg + \frac{2k}{23} \cos \sqrt{\frac{2k}{23m}}\, t.$$

U. Lond. K.C.

Solution 7.14

We have another example (see Example 7.12) of a system of connected elements. An additional complication is that the rotor does not turn about a fixed axis but

rolls along a rail. When a rigid body both moves and turns, two equations of motion are needed – one for translation and one for rotation. (See Chapter 2.) Also, the presence of the hanging weight alerts us to the fact that the spring will be in tension in the static equilibrium position.

We first need to establish compatibility equations relating rotation of the rotor, and corresponding movements of rotor and hanging weight. If the rotor turns clockwise through an angle θ the body B moves downwards a distance of $(2r\theta)$. And it is not difficult to see that this vertical displacement is not affected by the fact that the rotor rolls instead of turning about a fixed axis. Hence:

$$y = 2r\theta$$

and, differentiating twice,

$$\ddot{y} = 2r\ddot{\theta}.$$

Since the rotor rolls without slip, the corresponding horizontal movement of the rotor, x, is

$$x = r\theta.$$

$$\therefore\ \ddot{x} = r\ddot{\theta}.$$

This allows us to calculate the spring force. Calling the initial force (in the static equilibrium position) F_0, the spring force F is

$$F = F_0 + kx.$$

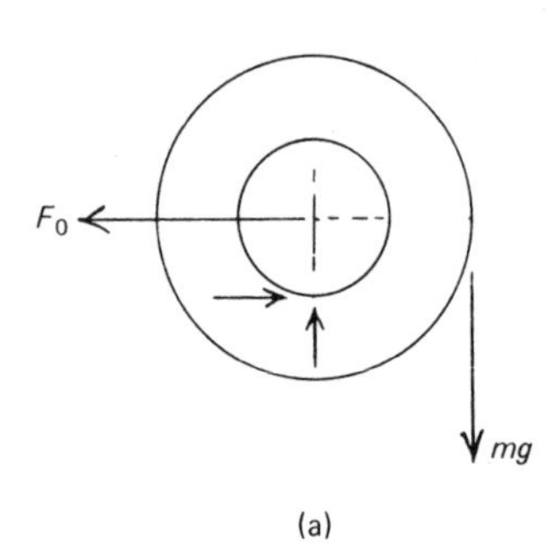

(a)

Shown here is the free-body diagram (a) of the rotor in the static equilibrium position. Taking moments of forces about the contact point with the rail:

$$F_0 \times r = mg \times 2r.$$

$$\therefore\ F_0 = 2mg.$$

We can now draw the free-body diagrams for the assumed displacements. These are shown in diagram (b).

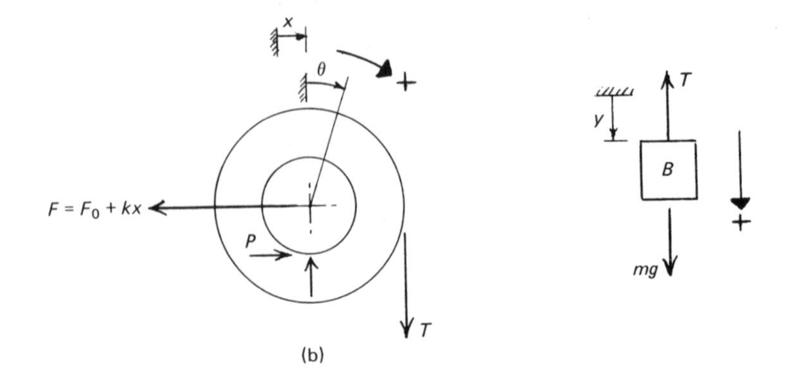

(b)

On the free-body diagram for the rotor, the horizontal force *P* is particularly to be noted. Because the rotor is assumed to be rolling clockwise and moving to the right, and because it does not slip, the smaller disc must thrust backwards on the rail with a friction force. The rail must thus thrust forwards on the rotor; this, it can be seen, is the actual force which moves the rotor to the right.

The equations of motion are:

For the rotor in rotation ($\Sigma M + I\alpha$), taking clockwise as positive and taking moments about the rotor centre,

$$T \times 2r - Pr = I\alpha = I\ddot{\theta}. \qquad (1)$$

For the rotor in translation ($\Sigma F = ma$), taking right as positive,

$$P - (F_0 + kx) = (3m)\,a = (3m)\,\ddot{x}.$$

$$P - (2mg + k\theta r) = 3mr\ddot{\theta}. \qquad (2)$$

For body B ($\Sigma F = ma$) taking down as positive:

$$mg - T = m\ddot{y}$$

$$= m \times 2r\ddot{\theta}. \qquad (3)$$

Substituting for T and P in equation 1,

$$2r\,(mg - 2mr\ddot{\theta}) - r\,(3mr\ddot{\theta} + 2mg + kr\theta) = I\ddot{\theta}.$$

Expanding,

$$\cancel{2mgr} - \ddot{\theta}4mr^2 - \ddot{\theta}3mr^2 - \cancel{2mgr} - \theta kr^2 = I\ddot{\theta},$$

and, as we expect, the terms not containing either θ or $\ddot{\theta}$ cancel.

Rearranging,

$$\ddot{\theta}\,(I + 3mr^2 + 4mr^2) + \theta\,(kr^2) = 0.$$

$$\ddot{\theta} + \theta\left(\frac{kr^2}{I + 7mr^2}\right) = 0.$$

s.h.m. is thereby proved (Fact Sheet (c)). The natural circular frequency ω is given by

$$\omega^2 = \left(\frac{kr^2}{I + 7mr^2}\right).$$

Using the formula for the moment of inertia of a uniform disc $I = \frac{1}{2}mr^2$,

$$I = \tfrac{1}{2}mr^2 + \tfrac{1}{2}(2m)\,(2r)^2$$

$$= 4\tfrac{1}{2}mr^2.$$

$$\omega^2 = \frac{kr^2}{4\frac{1}{2}mr^2 + 7mr^2}$$

$$= \frac{k}{11\frac{1}{2}m}$$

$$= \frac{2k}{23m}.$$

[c]

$$n = \frac{1}{2\pi}\sqrt{\frac{2k}{23m}}.$$

So the hanging weight B moves with s.h.m. with this frequency. Hence the displacement y of B is:

[c]
$$y = A \sin(\omega t) + B \cos(\omega t).$$
$$\therefore \dot{y} = A\omega \cos(\omega t) - B\omega \sin(\omega t).$$

The two 'initial conditions' are:

(i) when $t = 0, y = \delta$, and

(ii) when $t = 0, \dot{y} = 0$ (since B is released from rest).

For the second condition,

$$0 = A(1) - B(0).$$
$$A = 0.$$

For the first condition

$$\delta = B \cos(0).$$
$$\therefore B = \delta$$
$$\therefore y = \delta \cos(\omega t)$$
$$= \delta \cos\left(\sqrt{\frac{2k}{23m}}\, t\right).$$

Differentiating this twice gives

$$\ddot{y} = -\delta\left(\frac{2k}{23m}\right) \cos\left(\sqrt{\frac{2k}{23m}}\, t\right).$$

Referring back to the free-body diagrams and the equation of motion for B,

$$mg - T = m\ddot{y}.$$
$$T = mg - m\ddot{y}$$
$$= mg - m\left[-\delta\left(\frac{2k}{23m}\right)\cos\left(\sqrt{\frac{2k}{23m}}\, t\right)\right]$$
$$= mg + \frac{2k\delta}{23} \cos\left(\sqrt{\frac{2k}{23m}}\, t\right).$$

There remains the complication that because the body B hangs from a moving rotor, it will also have sideways motion and acceleration. However, the answer expected is clearly not intended to take this into account.

Example 7.15

The diagram shows a rigid body of mass M supported horizontally by two vertical inextensible strings of length h. The centre of mass is distance d_1 from one string, d_2 from the other. Find the natural frequency of small amplitude torsional oscillations about a vertical axis through the centre of mass, if the moment of inertia of the body about this axis is I.

Suggest possible engineering applications of this result.

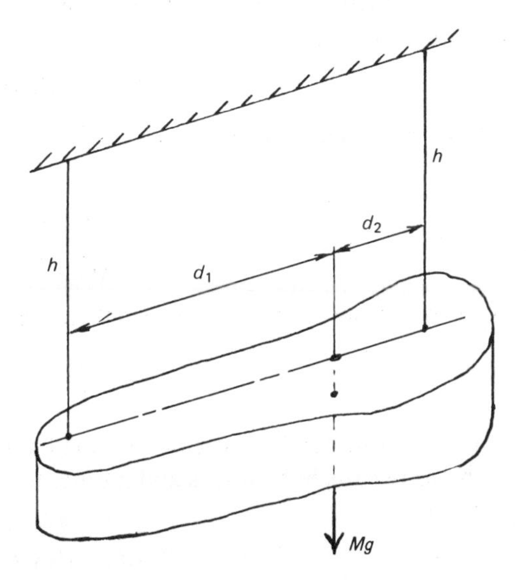

U. Manchester

Solution 7.15

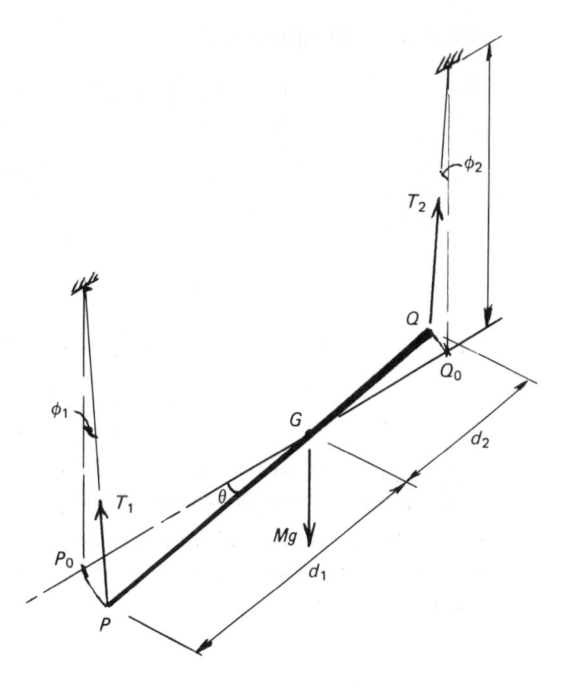

The system is shown as a free-body diagram. Such a system is called a bi-filar suspension. ('Bi-filar' means having two wires.) G is the mass centre and the body is assumed to have an instantaneous small angular displacement θ anticlockwise as viewed from above. The resulting inclinations of the two wires means that the tension forces T_1 and T_2 will each have a horizontal component tending to turn the body in the direction opposite to the displacement. The angles of inclination of the wires are ϕ_1 and ϕ_2.

The equation of motion ($\Sigma M = I\alpha$), taking moments about G and reckoning clockwise positive, is

$$-T_1 \sin \phi_1 (d_1) - T_2 \sin \phi_2 (d_2) = I\alpha = I\ddot{\theta}. \tag{1}$$

The displacement arcs P_0P and Q_0Q can be expressed in terms of both θ and ϕ, thus:

$$P_0P = h\phi_1 = d_1\theta,$$

$$Q_0Q = h\phi_2 = d_2\theta,$$

for small angles. Replacing the sines by angles and substituting in equation 1,

$$-T_1 d_1^2 \frac{\theta}{h} - T_2 d_2^2 \frac{\theta}{h} = I\ddot{\theta}. \tag{2}$$

Considering vertical forces, for small angles the body will be in vertical equilibrium, as regards both forces and moments. Taking a moment equilibrium equation about Q,

$$T_1 (d_1 + d_2) = mgd_2.$$

$$\therefore T_1 = \frac{mgd_2}{d_1 + d_2},$$

and, similarly,

$$T_2 = \frac{mgd_1}{d_1 + d_1}.$$

Substituting in equation 2,

$$-\left(\frac{mgd_2}{d_1 + d_2}\right) d_1^2 \left(\frac{\theta}{h}\right) - \left(\frac{mgd_1}{d_1 + d_2}\right) d_2^2 \left(\frac{\theta}{h}\right) = I\ddot{\theta}.$$

$$\therefore -\frac{mgd_1 d_2}{d_1 + d_2} \frac{\theta}{h} (d_1 + d_2) = I\ddot{\theta}$$

$$\therefore \ddot{\theta} + \theta\left(\frac{mgd_1 d_2}{I h}\right) = 0.$$

[c] This proves s.h.m. The frequency of the oscillation is given by

$$n = \frac{\omega}{2\pi} = \frac{1}{2\pi} \sqrt{\frac{mgd_1 d_2}{I h}}.$$

The significance of the result is that this method of suspension and vibration allows an accurate determination of the moment of inertia of an irregularly shaped body. Determination of moment of inertia by a vibration method is accurate because of the high accuracy with which the frequency of oscillation may be measured.

Example 7.16

A circular cylinder of radius r rolls on a curved surface of radius R as shown in the diagram. As point A moves to point B, show that the angles θ and ϕ are connected by $r\phi = R\theta$. By considering the conservation of energy, or otherwise, prove that for $\theta \ll 1$ the natural frequency of oscillation about the lowest point of the motion is

$$\sqrt{\frac{2g}{3(R - r)}}.$$

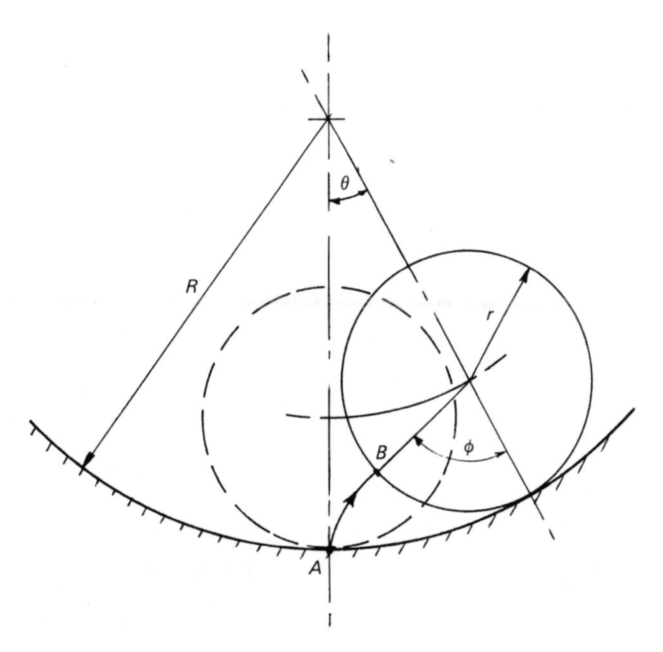

U. Manchester

Solution 7.16

The principle of energy may be used in a problem of this kind because once such a system is set vibrating, no energy is subsequently added, or removed (if, as is normally assumed, friction effects are neglected). Such a system is called a conservative system. Since the total energy remains constant, the differential of energy with respect to time will be zero. This is a useful alternative method of arriving at the system equation of motion.

The energy comprises:

(a) translational k.e. of the cylinder mass centre;
(b) rotational k.e. about the mass centre;
(c) potential energy.

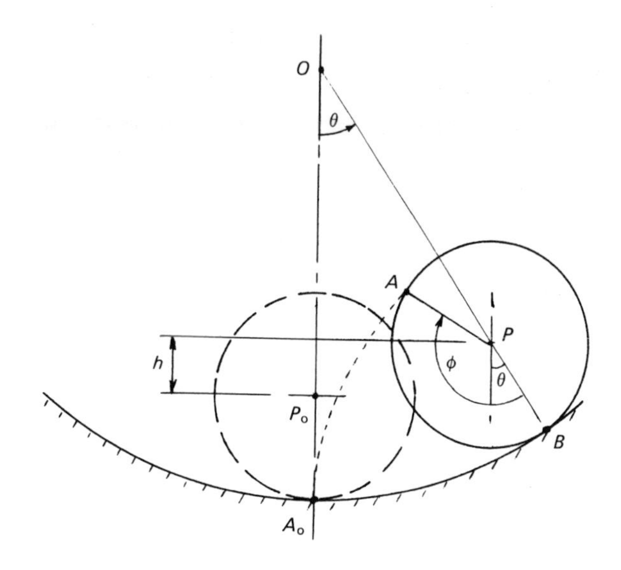

The system kinematics require very careful analysis.

In the diagram the cylinder centre P_0 has moved to P, and the point A_0 has moved to A, resulting in the radius OP_0 turning through an angle θ. It is necessary to assume that the cylinder rolls without slip. Thus, the arc AB round the cylinder must be the same length as the arc A_0B round the curved surface. Hence

$$\phi r = \theta R. \tag{1}$$

The cylinder has *not* turned through an angle ϕ as at first might be supposed.

We see that the radius PA has turned from its original vertical position P_0A_0 through an angle $(\phi - \theta)$. Thus the angular speed of the cylinder, Ω, is

$$\Omega = \dot{\phi} - \dot{\theta}.$$

The translational speed v of the cylinder centre is given by

$$v = \dot{\theta}\,(R - r).$$

The vertical height h of the cylinder mass centre is

$$h = (R - r) - (R - r)\cos\theta$$

$$= (R - r)(1 - \cos\theta).$$

The total energy E is

$$E = \tfrac{1}{2}I\Omega^2 + \tfrac{1}{2}mv^2 + mgh$$

$$= \tfrac{1}{2}(\tfrac{1}{2}mr^2)(\dot{\phi} - \dot{\theta})^2 + \tfrac{1}{2}m(\dot{\theta})^2(R - r)^2 + mg(R - r)(1 - \cos\theta)$$

(using the familiar expression for the moment of inertia of a solid uniform cylinder).

By rearranging and differentiating equation 1,

$$\dot{\phi} = \dot{\theta}\left(\frac{R}{r}\right).$$

$$\therefore\ E = \tfrac{1}{4}mr^2(\dot{\theta})^2\left(\frac{R}{r} - 1\right)^2 + \tfrac{1}{2}m(\dot{\theta})^2(R - r)^2 + mg(R - r)(1 - \cos\theta)$$

$$= \tfrac{1}{4}m(\dot{\theta})^2(R - r)^2 + \tfrac{1}{2}m(\dot{\theta})^2(R - r)^2 + mg(R - r)(1 - \cos\theta)$$

$$= \tfrac{3}{4}m(\dot{\theta})^2(R - r)^2 + mg(R - r)(1 - \cos\theta).$$

As stated, $\mathrm{d}(E)/\mathrm{d}t = 0$.

$$\therefore\ 0 = \tfrac{3}{4}m(R - r)^2 \times 2\dot{\theta} \times \ddot{\theta} + mg(R - r)\sin\theta \times \dot{\theta}.$$

We make the customary approximation that for small angles $\theta \approx \sin\theta$, giving

$$0 = \tfrac{3}{2}(R - r)\ddot{\theta} + g\theta.$$

Hence

$$\ddot{\theta} + \theta\left(\frac{2g}{3(R - r)}\right) = 0,$$

proving s.h.m. The coefficient of θ is ω^2, the square of the natural circular frequency of the motion. Hence frequency n is

$$n = \frac{1}{2\pi}\sqrt{\frac{2g}{3(R - r)}}.$$

Although in the interests of strict accuracy, ω should be called the natural *circular* frequency, to distinguish it from the natural frequency (in hertz), the word

'circular' is often omitted, particularly in more advanced treatment of vibration. It is clear that the examiner here requires the expression for ω.

7.3 Problems

Problem 7.1

A machine has a mass of 3450 kg. It is to be mounted on six identical springs arranged in parallel. Calculate the required maximum stiffness of each spring if the natural frequency of vertical vibration of the system is not to be greater than 3 Hz. It may be assumed that the machine is constrained to move vertically only. When the machine vibrates at this frequency, calculate the maximum velocity and acceleration of the motion when the amplitude of the vibration is 4 mm. Determine the maximum and minimum forces in each spring under these conditions. It may be assumed that the weight of the machine is shared equally by the six springs. (Refer to Example 7.1.)

Problem 7.2

A uniform disc of mass 12.73 kg has a radius of 0.2 m. It is attached to one end of a rigid rod of length L and of negligible mass, the upper end of the rod being attached to a frictionless pivot, as shown in the diagram. Calculate the frequency of small oscillations about the vertical when L is: (a) 0.4 m; (b) 2.0 m. In each case, estimate the percentage error resulting from assuming the system to be a simple pendulum with all the mass concentrated at the disc centre. (Refer to Examples 7.6 and 7.4.)

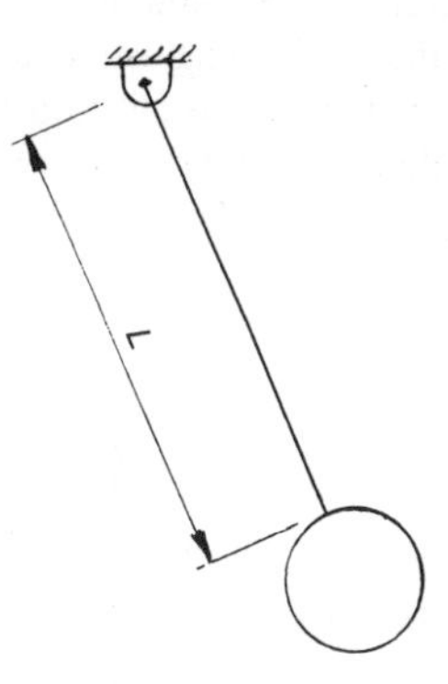

Problem 7.3

A mass of 6.0 kg hangs by a light cord from a spring of negligible mass and stiffness 1275 N m^{-1} as shown in the diagram. Calculate the frequency of small vertical oscillations if the mass is pulled down a small distance and released from rest. Show also that if the initial displacement of the mass exceeds 46.16 mm then the subsequent motion will not be s.h.m. and state why.

Hint: When the mass moves downwards, the only downward force acting on it is its weight; the cord cannot push downwards. So maximum possible acceleration is due to weight only and hence cannot exceed 9.81 m s^{-2}. Greater initial downward displacement would result in the string going slack and the mass 'bouncing'. Use Fact Sheet (c) to determine amplitude A.

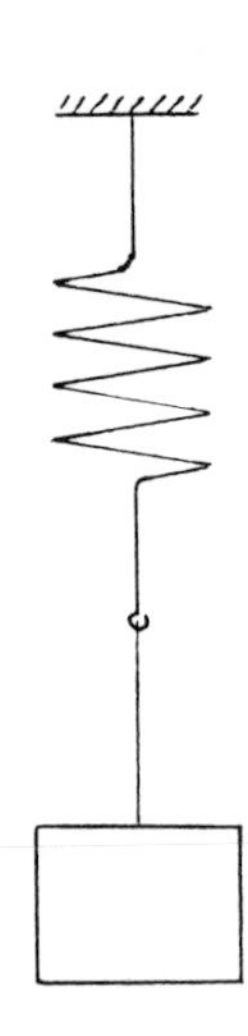

Problem 7.4

An aircraft arrester gear may be assumed to be a linear spring of stiffness 112 kN m^{-1}. The aircraft has a mass of 32 000 kg and it is travelling at 75 m s^{-1} when it first contacts the arrester, the spring then being unstrained in the statically neutral position. Calculate (a) the time for the aircraft to come to rest; (b) the distance in which it comes to rest; (c) the maximum acceleration; (d) the displacement and velocity of the aircraft 0.5 seconds after first contacting the arrester. (Refer to Example 7.5.)

Problem 7.5

A flywheel of mass 17 kg is suspended from a knife edge as shown in the diagram, and oscillated with small amplitude. It is found that ten oscillations are completed in 13.42 seconds. The wheel is to be mounted on the end of a solid steel shaft of 24 mm diameter, the other end of the shaft being assumed fixed. Calculate the minimum length of the shaft if the frequency of torsional oscillation is not to exceed 10 Hz. Assume a value of 80 GN m^{-2} for G, the modulus of rigidity of the shaft material. (Refer to Examples 7.6 and 7.2. Use Fact Sheet (b), (c) and (d).)

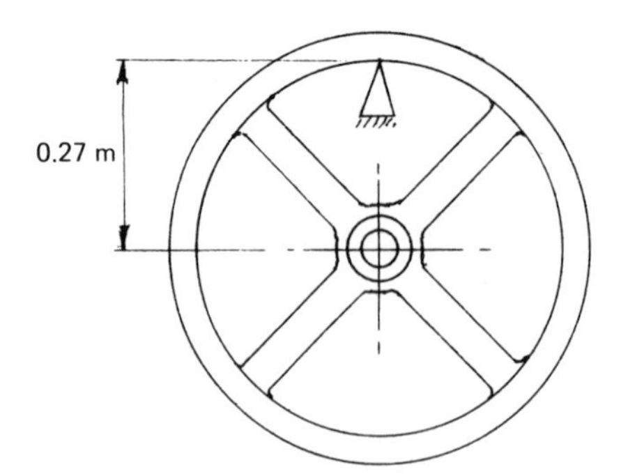

Problem 7.6

A bar is pivoted at a point O. The bar has a mass of 12.8 kg and the mass centre G is distant h = 0.56 m from O. A spring of stiffness k = 1200 N m^{-1} is attached at a point distant a from O, as shown in the diagram, the other end of the spring being attached to a fixed point, and the spring axis being perpendicular to the axis OG of the bar. In equilibrium, OG is hori-

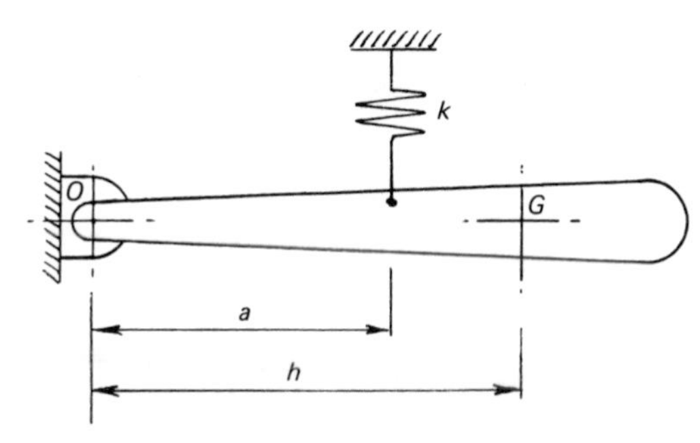

zontal. When the distance a is 0.48 m the system is caused to oscillate freely, and the frequency of the motion is observed to be 2.2 Hz.

(a) Assuming the spring can be moved, what value of a would result in a frequency of oscillation of 1.0 Hz?

(b) If the spring is removed and the bar allowed to hang vertically, what will be the frequency of small oscillations about the vertical?

(Refer to Examples 7.9 and 7.6.)

Problem 7.7

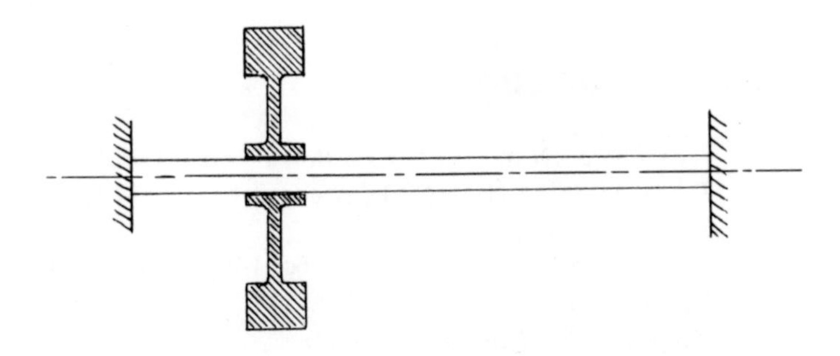

The diagram shows a flywheel having a moment of inertia about its axis of 6.2 kg m^2, keyed to a horizontal steel shaft of diameter 36 mm and length 2.6 m. The position of the wheel is adjustable. The modulus of rigidity, G, for the shaft is 80 GN m^{-2}.

(a) Calculate the frequency of torsional oscillations when the wheel is at the centre of the shaft.

(b) Calculate the required position of the wheel if the frequency of torsional oscillation is to be twice the above value.

Hints: The shaft forms in effect two torsional springs, one either side of the flywheel; these springs are in *parallel*. Use Fact Sheet (d), (e) and (f) to determine k_t. Treat then as Example 7.2. For part (b), assume the wheel to be a distance x from one end, and use the same formulae to evaluate k_t in terms of x. You should obtain a quadratic equation in x: $x^2 - 2.6x + 0.4225 = 0$.

Problem 7.8

A uniform solid disc of mass 6.5 kg and radius 0.1 m is attached to the end of a uniform thin rod of mass 2.4 kg and length 0.7 m (see diagram). The rod is pivoted at point 0 and is constrained by a spring of stiffness $k = 800$ N m^{-1} as shown; the spring axis is perpendicular to the rod axis. When the system is in static equilibrium, the rod axis makes an angle of $\theta_0 = 35°$ to the vertical. Calculate the frequency of small angular oscillations of the system about the neutral position.

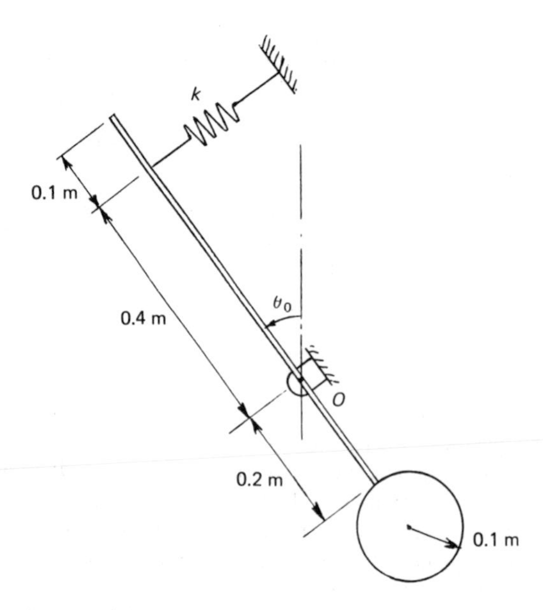

Hints: Use Fact Sheet (b) to evaluate I_O, the moment of inertia of the composite system about the transverse axis through O, calculating rod and disc separately and adding. Answer: 0.7695 kg m^2. Calculate the position of the mass centre G. Answer: 0.1787 m from O. Assume a small displacement θ anticlockwise. Note that the initial spring force F_0 will be *compressive* in this case.

7.4 Answers to Problems

7.1 204.3 kN m^{-1}; v = 75.4 mm s^{-1}; a = 1.421 m s^{-2}; 6458 and 4824 N.
7.2 (a) 0.6264 Hz; +2.74 per cent. (b) 0.3354 Hz; +0.206 per cent.
7.3 2.320 Hz.
7.4 (a) 0.8396 s; (b) 40.09 m; (c) 140.3 m s^{-2}; (d) 32.27 m; 44.51 m s^{-1}.
7.5 0.81 m.
7.6 (a) 0.2182 m; (b) 1.109 Hz.
7.7 (a) 9.106 Hz; (b) 1.126 m from shaft centre.
7.8 2.153 Hz.

Index